Nature Loves To Hide Itself

Heraclitus 500 BC

NE
NAT

Today the human impact on our planet can hardly be underestimated. Climate change, population explosion, genetic manipulation, digital networks, hurricane control and engineered microbes. Untouched old nature is almost nowhere to be found. "We were here" echoes all over. We are living in a time in which the 'made' and the 'born' are fusing. This does not mean however, that we have become gods and have control over our own destiny. Rather, our relationship with nature is changing.

Edited and designed by
Koert van Mensvoort
and Hendrik-Jan Grievink

Where technology and nature are traditionally seen as opposed to each other, they now appear to merge or even trade places. While old nature, in the sense of trees, plants, animals, atoms, or climate, is increasingly controlled and governed by man – it is turned into a cultural category – our technological environment becomes so complex and uncontrollable, that we start to relate to it as a nature of its own. Will we be able to improve our human condition, or will we outsource ourselves for good? People are catalysts for evolution, yet we are only have a small inkling of our new job description.

This book explores our changing notion of nature. How nature has become one of the most successful products of our time, yet much of what we perceive as nature is merely a simulation: a romanticized idea of a balanced, harmonic, inherently good and threatened entity. How evolution continues nonetheless. How technology – traditionally created to protect us from the forces of nature – gives rise to a next nature, which is just as wild, cruel, unpredictable and threatening as ever. How we are playing with fire again and again. How we should tread carefully; yet how this is simultaneously what makes us human.

Where to Find Untouched Nature?

CANCER CELLS?

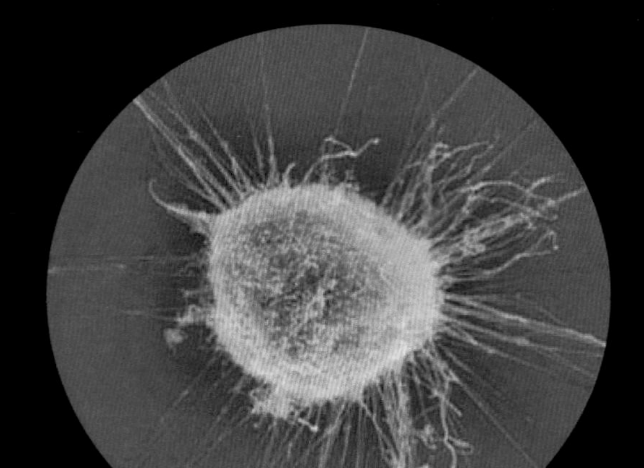

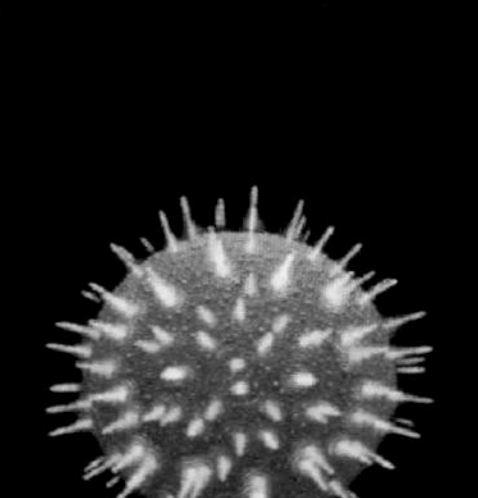

VIRUSES?

WE CRAVE FOR NATURE UNSPOILED BY PEOPLE. PARADOXICALLY, WE EVOLVED OUT OF THAT VERY SAME NATURE.

THE OUTBACKS OF THE UNIVERSE?
THE HIMALAYAS?

You are Born in a World that is Designed Already

LOOK AROUND YOU AND TRY
TO FIND THE MOST
NATURAL THING IN THE
ROOM YOU ARE IN NOW. IT
IS YOU.

We Co-evolve with Technology, Like Bees with Flowers

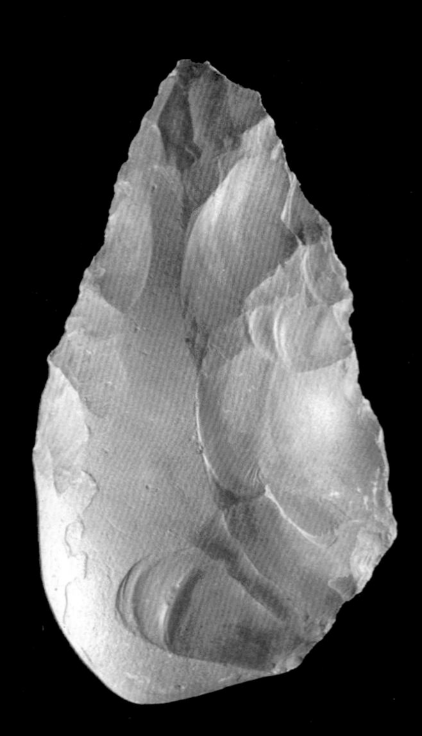

STONE TOOLS BECAME PRIMITIVE EXTENSIONS OF OUR BODIES AND THE BASIS OF MAN'S DEVELOPMENT.

THE FIRST FORCE OF NATURE WE LEARNED TO USE FOR OUR OWN BENEFIT. BUT HE WHO PLAYS WITH FIRE...

AGRICULTURE MEANT THAT A NOMADIC LIFESTYLE WAS NO LONGER NECESSARY. WE LEARNED TO GROW ON DEMAND.

- 2 600 000 YEARS AGO: STONE TOOLS
- 375.000 YEARS AGO CONTROLLED FIRE
- 100.000 YEARS AGO: BURIAL
- 31.000 YEARS AGO: DOMESTIC DOG
- 9500 BC: BEER
- 7000 BC: AGRICULTURE
- 3500 BC: WHEEL
- 3500 BC: WRITING
- 3000 BC: TOOTHBRUSH
- 700 BC: ARTIFICIAL TEETH
- 600 BC: PLASTIC SURGERY
- 600 BC: COINS
- 274 BC: COMPASS
- 150 BC: CLOCKWORK
- 490 AD: PSYCHIATRIC HOSPITAL
- 700 AD: BANKNOTES
- 826 AD: UNDER-ARM DEODORANT
- 895 AD: CHEMOTHERAPY
- 1150 AD: HOMING PIGEONS
- 1200 AD: CENTRAL HEATING
- 1277 AD: CONDOM
- 1450 AD: PRINTING PRESS
- 1642 AD: CALCULATOR
- 1761 AD: PENCIL
- 1831 AD: ELECTRICAL GEN
- 1837 AD: MAGNET
- 1860 AD

RITING BECAME A WAY TO CAPTURE OUR
HOUGHTS. THE BEGINNING OF POWER
HROUGH INFORMATION.

MONEY CREATED A VIRTUAL SYSTEM
TO EXPRESS VALUE FOR EVERYTHING
AROUND US.

WE DISCOVERED MORE AND MORE
ABOUT HOW LIFE WORKS AND
BIO-TECHNOLOGY ALLOWS US TO
SHAPE BIOLOGICAL SYSTEMS.

EXPLORING MATTER ON AN ATOMIC LEVEL
BRINGS US ENORMOUS POSSIBILITIES. BOTH
GOOD AND BAD...

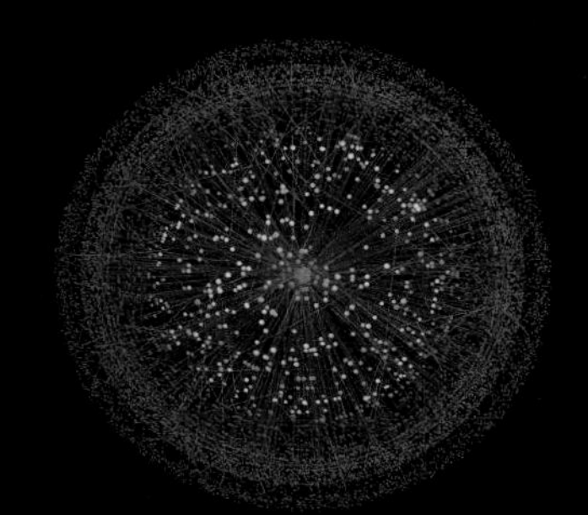

INFORMATION BECOMES EASY TO SHARE
WITH A WIDE AUDIENCE. EVERYONE IS
CONNECTED TO EVERYONE.

- ED WIRE
- AD: TELEPHONE
- 1886 AD: DISHWASHER
- 1887 AD: CONTACT LENS
- 1892 AD: MILKING MACHINE
- 1895 AD: BREAST IMPLANT
- 1902 AD: AIR CONDITIONER
- 1928 AD: ELECTRIC DRY SHAVER
- 1928 AD: ANTIBIOTICS
- 1945 AD: NUCLEAR WEAPONS
- 1949 AD: ARTIFICIAL HEART
- 1950 AD: CREDIT CARD
- 1955 AD: TV REMOTE CONTROL
- 1958 AD: MICROWAVE OVEN
- 1958 AD: PACEMAKER IMPLANT
- 1960 AD: INSTANT NOODLES
- 1960 AD: ORAL CONTRACEPTIVE PILL
- 1961 AD: SPACE FOOD
- 1965 AD: E-MAIL
- 1969 AD: INTERNET
- 1973 AD: MOBILE PHONE
- 1973 AD: GPS
- 1978 AD: IN-VITRO FERTILIZATION
- 1984 AD: DNA FINGERPRINTING
- 1987 AD: PROZAC
- 1998 AD: GENETIC SEQUENCING
- 2008 AD: RETAIL DNA TEST

Evolution Goes On

GEOSPHERE
4,5 BILLION YEARS OLD

BIOSPHERE
3 BILLION YEARS OLD

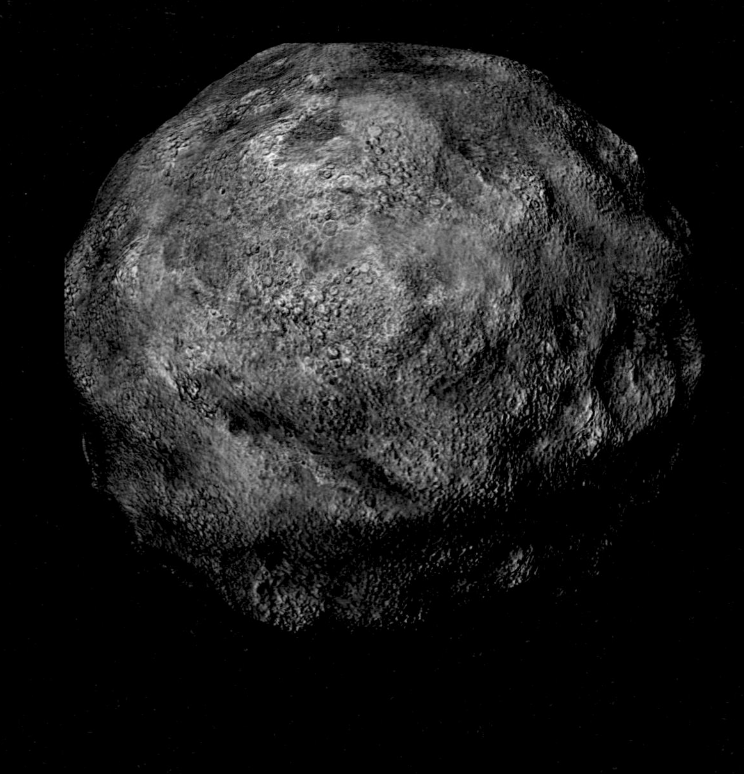

OLD NATURE

Since inception our planet took about a billion years to form a biosphere of life around its geosphere. Some three and a half billion years later humankind emerged. We in turn, initiated a new sphere: of human thought and activity. Vladimir Vernadsky and Teilhard de Chardin proposed the term noosphere – derived from the Greek (*nous* "mind") and (*sphaira* "sphere") – while others described it as the mediasphere, datasphere or technosphere. According to Vernadsky, the noosphere is the third in a succession of phases of development of the earth, after the geosphere and the biosphere. Just as the emergence of life fundamentally transformed the geosphere, the emergence of human cognition fundamentally transforms the biosphere.

NOOSPHERE
10,000 YEARS OLD

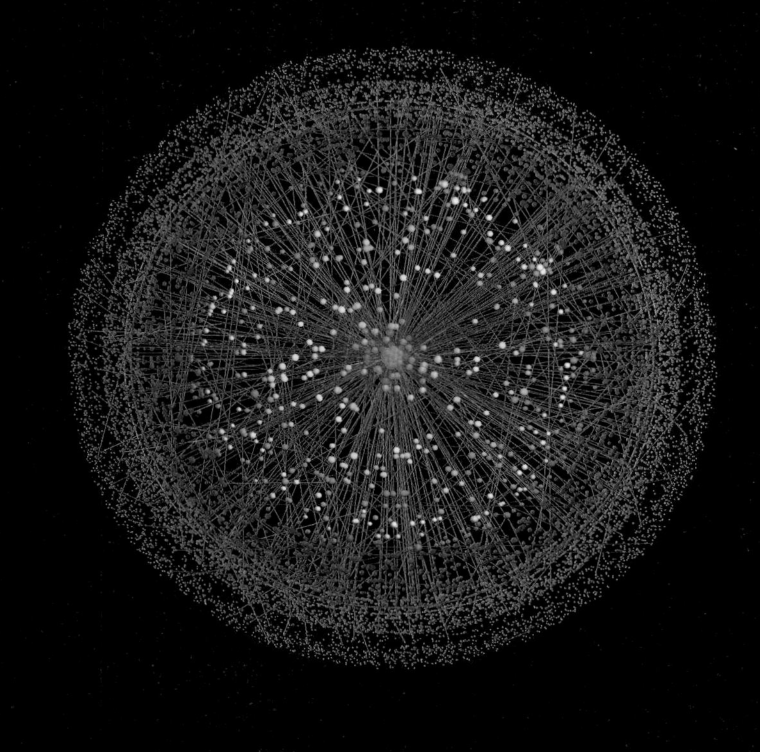

NEXT NATURE

We Should Reconsider the Divide Between Nature and Culture

If we consider our notions of nature throughout history, it is immediately apparent that they have altered over time. The word nature stems from the Latin term, *natura*, which in turn was derived from the Greek word *physis*. For the Romans *natura* was associated with ‘everything born’, while the Greek associated *physis* with ‘growth’. For the past few centuries, our notion of nature has been in line with the Roman interpretation: nature is born, while culture is everything made by man.

Yet, as we are living in a time of genetic modificatio and climate control this distinction between ‘made and ‘born’ becomes less meaningful. At the sam time, man-made systems become so complex an autonomous that we start to perceive them as nature Hence, our notions of nature and culture seem to b shifting from a distinction between born and made t a distinction between controlled and beyond control - aligning with the original Greek interpretation of *physis*

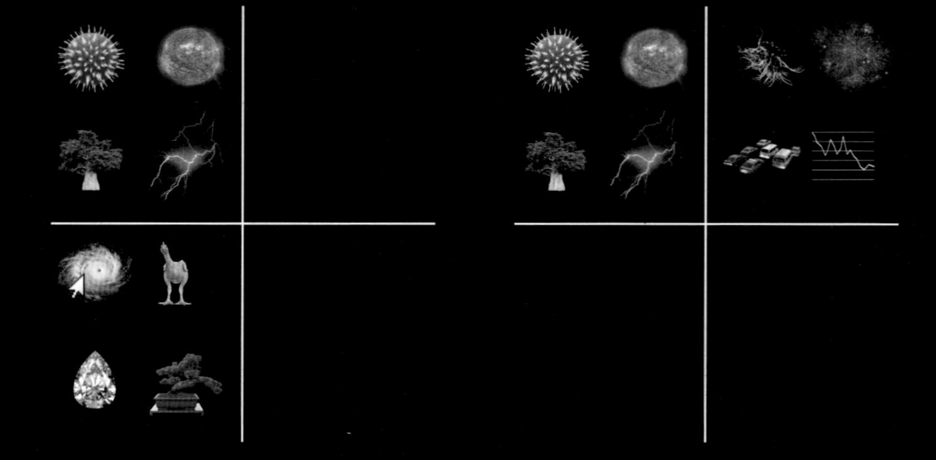

THE OLD VIEW OF NATURE
EVERYTHING BORN

OUR NEW VIEW OF NATURE
EVERYTHING BEYOND OUR CONTROL

Born and
beyond our control
Old nature

Beyond Control

Made and
beyond our control
Next nature

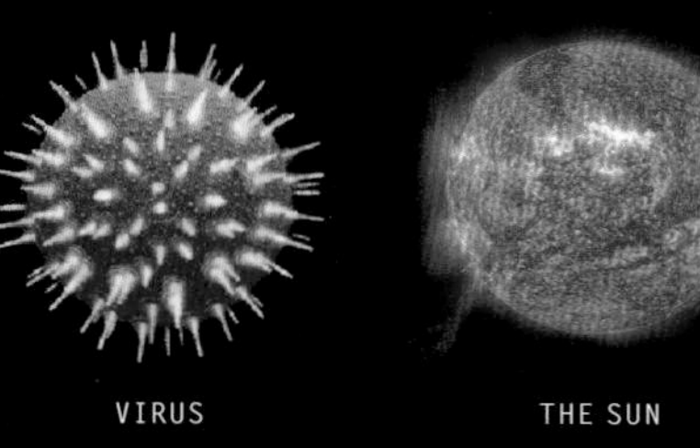

VIRUS THE SUN

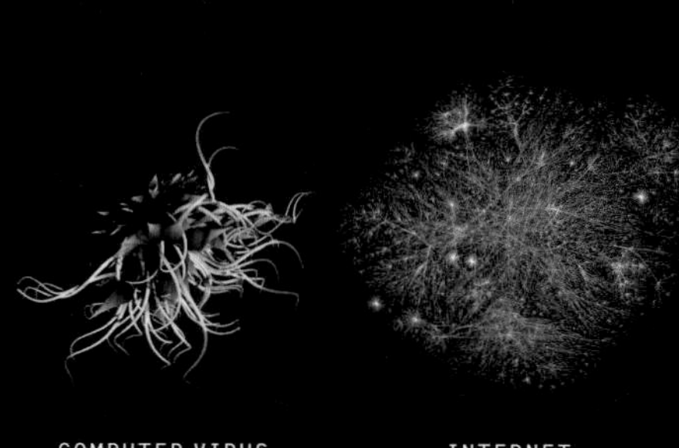

COMPUTER VIRUS INTERNET

BAOBAB TREE LIGHTNING

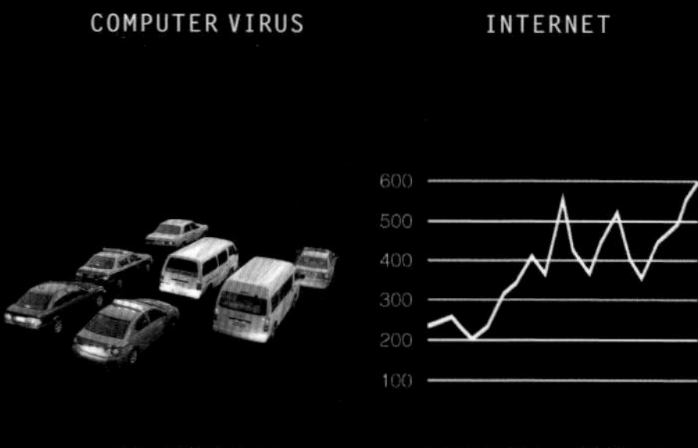

TRAFFIC JAM FINANCIAL SYSTEM

Born **Made**

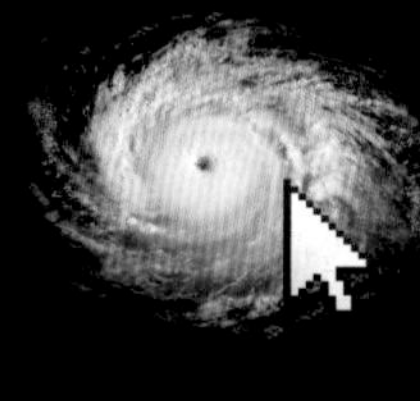

HURRICANE CONTROL

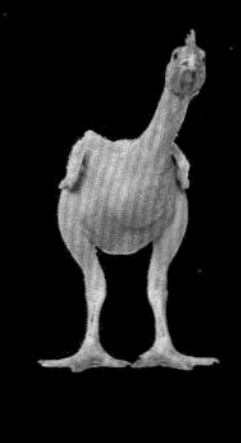

FEATHERLESS CHICKEN

AIBO ROBOTIC DOG

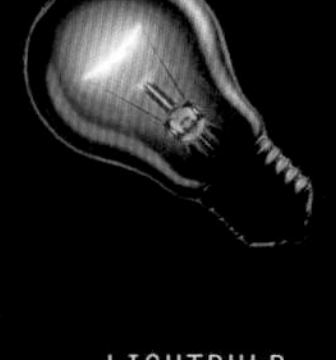

LIGHTBULB

DIAMOND

BONSAI TREE

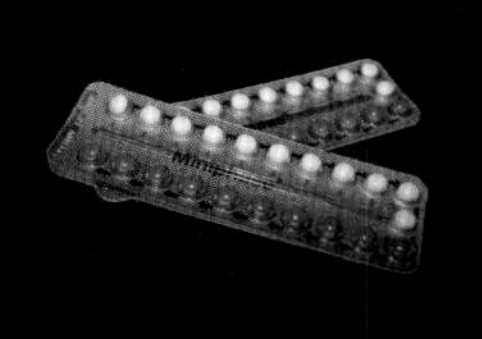

BIRTH CONTROL

TELEPHONE

Controlled

Born and
in our control
Cultivated nature

Made and
in our control
Culture

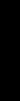

PREFACE

By Bruce Sterling

This book is about nature's brand image. One might surmise that "nature," being 100% all natural, can't have any brand image. The facts suggest otherwise. Try it for yourself: tell a friend that something seemingly 100% natural is actually "96% natural". Not a great difference, apparently, yet a profound unease arises.

This book is a study in why we feel uneasiness when the nature brand is violated. It's also about the exciting new-and-improved varieties of unnatural unease that have come to exist quite recently. It explains why this sensibility is spreading, and what that implies for who we are, and how we live with nature. Now, when nature is slightly artificialized – say, by installing a park bench under a tree – we rarely get any dark suspicious frisson about that. The uncanny can only strike us when our ideological constructs about nature are dented. We're especially guarded about our most pious, sentimentalized notions of nature.

Nature as a nurturing entity that is harmonious, calm, peaceful, inherently rightful and all-around "good for you." This vaguely politicized attitude about nature never came from nature. It was culturally generated. Nature didn't get her all-natural identity branding until the Industrial Revolution broke out. Then poets and philosophers were allowed to live in dense, well-supplied cities, where they could recast nature from some intellectual distance. Before that huge effusion of organized artifice, people lived much closer to the soil.

These farmers rarely spoke of "nature" in the abstract. They were too deeply involved in a lifelong subsistence struggle with natural events, such as inclement weather, bad harvests, weeds, pests, and blights. They certainly never mistook their existing state of affairs for the biblical Eden: their theological utopia in which nature was always harmonious, calm, peaceful and good-for-you.

However, that was back then, and this is now. Under the emergent regime of next nature, the potential for nature to behave in a sweet-tempered mother nature-ly fashion has been stripped away. The dame is running an ever-mounting fever from climate change, and there are no humanly untouched landscapes anywhere on the surface of the planet. We've entered the Anthropocene epoch, in which humanity and its instrumentalities are the most potent and influential geological force. Most available sunlight and soil goes for crops. The ever-increasing tonnage of human flesh outweighs all other wild mammals. Nature becomes a subset of culture, rather than vice versa.

We also have an exciting suite of new technical interventions – bio-chemical, genetic, roboticized, nano-technological – which are poorly understood. They can all interfere radically in what we construe as the "natural order." They change nature faster than our ideas about nature can change. The result is Tofflerian future shock with a leafy green tinge. It's unclear whether there is any tenable way, or even any further need, to separate "nature" from "culture" – on the surface of this planet, anyway. That commingled, hybridized, chimeric future is already here, and awaiting distribution – with operators standing by.

As evolved beings produced by a biosphere, we're not capable of perceiving reality unassisted.

Some level of "normal failure" in technological systems is as "natural" as the sun rising. Next Nature is an investigative enterprise by a set of mostly Dutch researchers. Next nature is haunted by previous nature, or rather, by the ghostly gothic absences of a vanished natural world. Next nature also bears many premonitions about the seething, favela-like, feverish state of our planet tomorrow. Next nature offers us few reassurances. It refuses to view nature as a given, solid, static entity to be discovered, dissected and destroyed by human agency. Instead, next nature is a dynamic entity that is fated to change right along with us.

There is an ontological crisis involved in our ignorance of what the Earth was like before we humans altered it. It's hard for us to establish a comfortable sense of our place in the world when the world itself is so outworn and bedraggled by so many previous human efforts. It's degrading to work creatively on hand-me-downs: the writer whose page is a scraped-down palimpsest, the artist whose canvas is torn and worn, the architect engaged in endless renovations, the actress in thrift-shop clothes. That's what it's like for a civilization existing in a natural milieu that has been irretrievably damaged. And yes, that is our future. Worse yet is to gaze with a fatuous satisfaction on a seemingly untouched sylvan scene, without realizing that the whole thing is a put-up job. At its best, it can be a superb put-up job, such as Holland: a nation of artifice that still clings to a pretty myth of tulips, clogs and contented cows while, in some anxious corner of the Dutch psyche, the dykes leak endlessly and the laboring windmills creak in a fitful breeze. Next nature is about the planet becoming Dutch: Nature made the world, but mankind made Holland.

We rarely allow ourselves any tender, reverential, nurturing attitude toward technology.

At its worst, though, our ignorance of the human effect on nature has Lovecraftian aspects. We become our own unnatural monsters, an eerie half-glimpsed force of archaic destruction. How many of the "primeval jungles" of Central and South America were cultivated places, once? How many alien species have been shipped around the planet by humanity, disrupting ecologies in ways we fail to see and don't suspect? How many seemingly pristine landscapes have been transformed by fire and overgrazing? What have antibiotics done to the unseen bacterial world, and dissolved plastics done to the seas?

Could it be true that our scattered ancestors, equipped with nothing more than fire and pointed sticks, briskly wiped out all the Pleistocene megafauna? Did we cause an abject collapse of the natural order before we were even literate? We are clearly culpable in the massive wave of extinction today – but could it be that human beings actually evolved in a mass extinction? Has that been our role in the planet since our species took shape?

Our tainted atmosphere proves that we'll never see a pristine world again, but, in the meantime, we will also have to come to terms with the ever-lengthening human legacy. Our previous attitudes are no longer tenable; they are actively harmful to us and to nature. We no longer have any way to leave a "natural reserve" alone, to be "reserved" and stay "natural." These relict biomes have been chopped up into unsustainable island fragments, and are severely stressed by rising temperatures, water shortages, invasive weeds and admiring tourists.

Abandoned areas of the planet can no longer "revert to nature" as they once supposedly did. Instead, they must revert to next nature, becoming weird "involuntary parks" such as the Cypriot Green Line, a long, human-free strip of flammable weeds and weed-trees, junkyards and landmines. Nor can we trust our means of technical control – our systematic, bureaucratic, commercial and analytical artifices. These artificial systems are not natural. Yet they can all manifest organic forms of behavior within a technological matrix. Our technology commonly manifests feral, eruptive, untamable qualities.

In what sense is an abject surrender to mysterious "market forces" any different than an abject surrender to the Mayan rain gods? The market is often seen and described as stormy, witchy, inaccessible, overwhelming in power – in short, as primal, wild and fearsome, a force of nature. Not because markets necessarily must have those natural attributes, but because we've trained ourselves to propitiate a feral market after freeing it from government control. It's common for greens to boast that "nature bats last," and she does – but technology bats as well.

When society is disrupted by a Chernobyl-scale event (leading to the world's largest involuntary park), we have the moral luxury of searching for a human scapegoat – in engineering, in design, in a political system, or in "human error." But technology is not merely about us: it's also about laws of Nature. Entropy requires no maintenance. All technological systems must age and decay. Extreme "black swan" events cannot possibly be outguessed even in principle. Some level of "normal failure" in technological systems is as "natural" as the sun rising.

"Next nature" cannot be fathomed without a similar study of next artifice, which this book carries out

We blind ourselves to the nature of technology by segregating certain classes of systemic behavior as "natural." We also stigmatize technology by denying its "natural" aspects: its mortality, fragility, complex interactivity, and its utter dependence on sometimes fitful flows of energy and material sustenance. We rarely allow ourselves any tender, reverential, nurturing attitude toward technology. The mass extinctions of entire classes of objects and services go almost unnoticed. We can surely do better than that. In conclusion, we may ask ourselves: in a world of next nature, what has become of "real" nature? Where is the objective reality behind this clever study of natural imagery and social attitudes toward nature? Next Nature maven Koert van Mensvoort likes to quote Heraclitus – "nature loves to hide." How hidden is nature? Is it possible that we have never seen nature, but only our notions of nature?

Here, I think, we can take some cold comfort in lifting our gaze to the stars. Despite what we've done to the surface of this planet, we're still a speck on a rock. The planet has been repeatedly wrecked by asteroids – sudden mass extinctions that dwarf anything humanity has yet triggered. We have most of the lesser beings in our biosphere at our mercy (to the point where we know them better as corporate logos than as living entities), but we don't run nature or even as yet grasp its, well, real nature.

Modern astrophysics suggests – more than "suggests," it asserts, with much painstaking accumulated evidence – that the Cosmos is mostly "dark energy" and "dark matter." These two rather ineffable substances are, presumably, the realest things in the universe. We human cannot manipulate, control, pollute or industrialize "dark energy" or "dark matter." They are natural, and yet it seems that they will remain forever closed to any form of human intervention. We can more or less get to terms with the edgy sensibility of "next nature." It's not beyond mankind to conceptualize ideas like the ones in this book, and even, eventually, domesticate them and even find them charming. Many of the ideas and images in next nature are experimental probes, which may seem far-out right now, but which may some day seem as endearingly corny as the Sputnik. However, nature has never existed for our convenience. Our society is mentally light-years away from metabolizing the bizarre assertions of dark matter theory. A universe in which nature is mostly darkness is a Copernican-scale dethroning of everything we once thought was natural. And that may be the objective truth. If so, then our notion of "natural" is a foam-like four percent of a severely alien universe. In other words, real true cosmic objective nature is 96% otherness, while we are, and always have been, the four percent adulterated whatever. Quite an odd sensation, thinking that. We seem to lack an unease than can get any more profound.

So nature clearly has her surprises in store. We have artificialized most everything we can grip, but there are still innumerable worlds well beyond our opposable thumbs. We can view some worlds other than the Earth, and we can measure them. Obedient to the Laws of Nature, they still remain serenely detached from us.

What we know of those worlds, we know by severely unnatural means. And only by unnatural means. There never was, and never could be, any entirely "natural" way to understand all of "real" nature. There is no direct, intuitive, unmediated, "real and genuine" experience of the actually existing universe. As evolved beings produced by a biosphere, we're not capable of perceiving reality unassisted. There can only be our technical instrumentalities. Our weak, decaying, flawed, falsifiable, even pitiable instrumentalities. But that's how we learn what's natural and real – through the unnatural.

There is a mental world in which these seeming oxymorons make good sound brisk common sense. Adjusting to demonstrable reality, no matter how mind stretching, is generally a praiseworthy effort. It would mean a lot of change in our ideas of the natural. It would mean a more fully humane mental world which was less notional, less delusional, less self-indulgent, and more attentive to the genuine otherness of nature. We're not there – we may never get there, for we may lack the time, and the will. But we ought to go there, to the extent that we can.

This book will help.

WE ALL LOVE NATURE, BUT WHAT IS IT? AND HOW IS OUR NOTION OF NATURE CHANGING?

RESEARCHERS AT THE UNIVERSITY OF BUFFALO MANAGED TO CREATE THE WORLD'S FIRST GENETICALLY MODIFIED BUTTERFLY WITH MAN-MADE MARKINGS ON THE WINGS. PERHAPS ONE DAY, BUTTERFLIES WILL BE STATE-OF-THE-ART ADVERTISING BILLBOARDS. JUST DO IT?

PEOPLE KNOW MORE BRANDS AND LOGOS THAN BIRDS OR TREE SPECIES.

ONCE YOU DEVELOP AN EYE FOR IT, IT IS QUITE ASTONISHING TO SEE HOW MANY PRODUCTS AND BRANDS REFER TO NATURE THROUGH THEIR NAME OR LOGO. WE CALL THE USE OF IMAGES OF NATURE TO MARKET A PRODUCT: 'BIOMIC MARKETING'. READ MORE ABOUT THIS PHENOMENON ON PAGE 292.

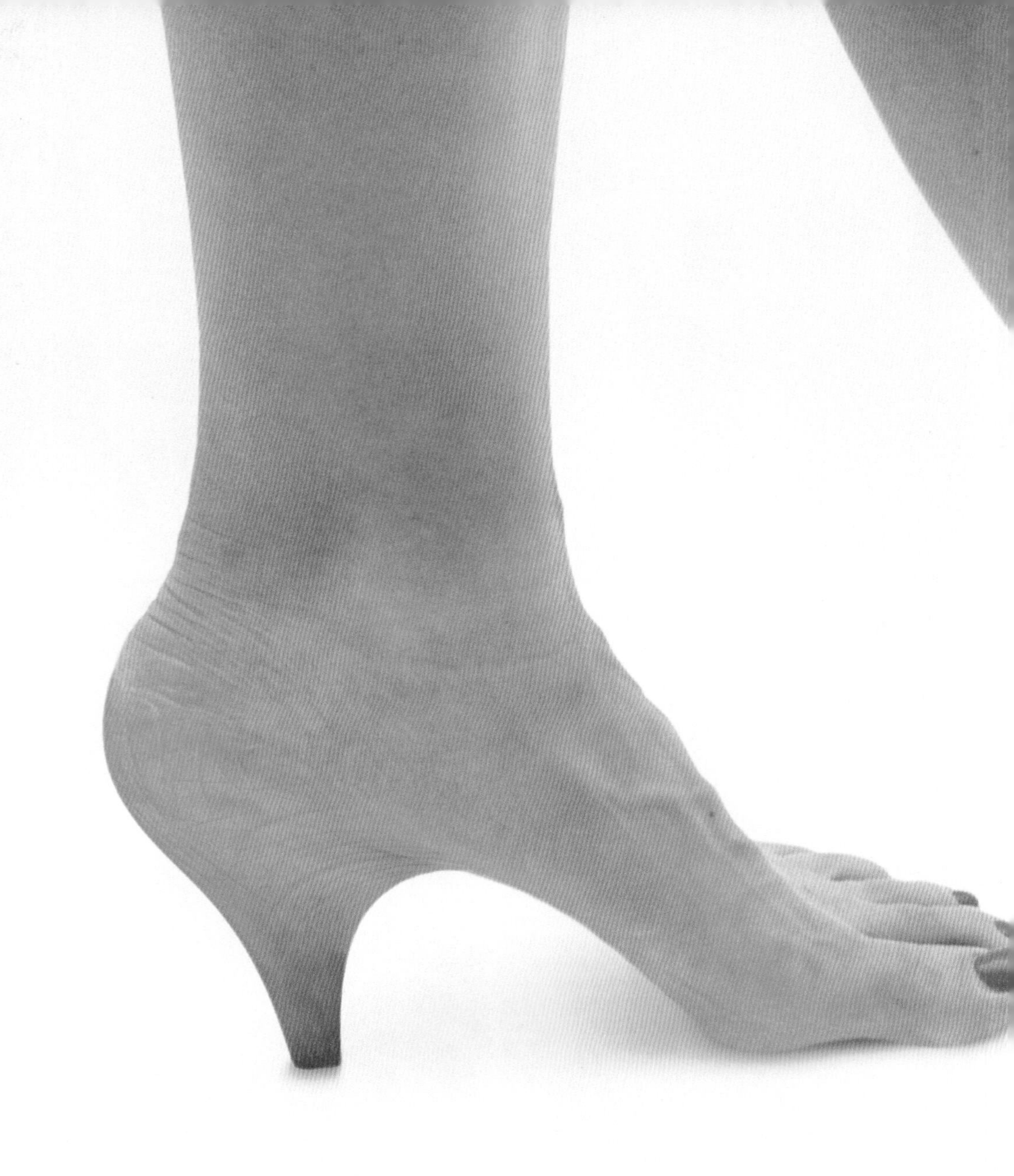

THE MADE AND THE BORN ARE FUSING; TRADITIONAL NOTIONS OF NATURE HAVE TO BE RECONSIDERED.

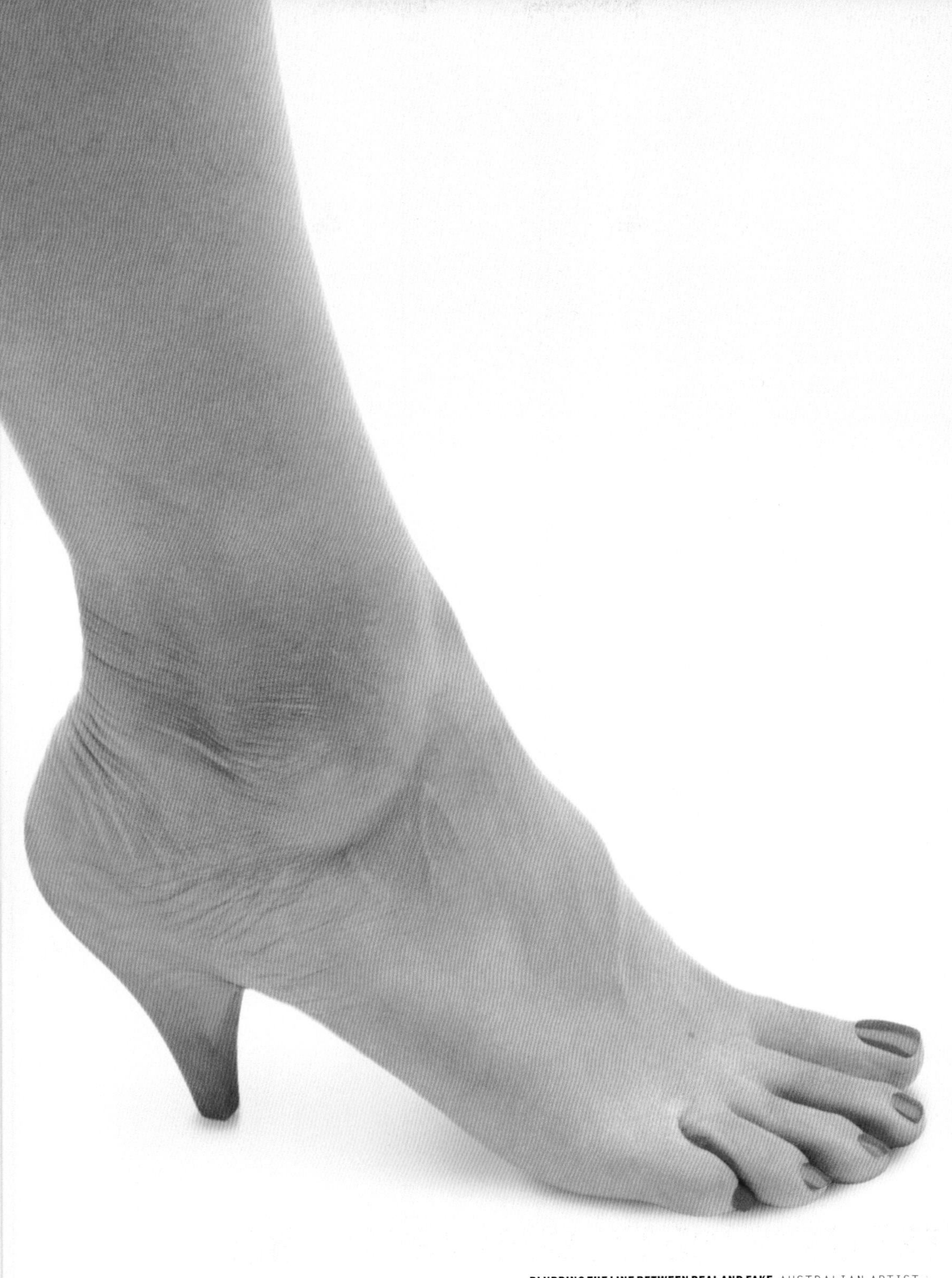

BLURRING THE LINE BETWEEN REAL AND FAKE, AUSTRALIAN ARTIST JULIE RRAP (61) CREATED THIS DIGITAL PRINT TITLED *OVERSTEPPING* IN 2001. EXPLORING FETISH AND MANIPULATION, THE IMAGE REPRESENTS EITHER THE NEXT SOLUTION TO SHOE-WEAR OR THE ULTIMATE FEMALE AGONY.

OUR NATURAL ENVIRONMENT IS REPLACED BY A WORLD OF DESIGN.

CONCRETE SPIDER WEB OR A NEW VARIETY OF CLOVERLEAF? ALTHOUGH WE RADICALLY TRANSFORM OUR NATURAL ENVIRONMENT, IT DOESN'T MAKE US OMNIPOTENT. EXEMPLIFYING HOW WE CONSTANTLY LOSE CULTURAL CONTROL AS MUCH AS WE GAIN IT; TRAFFIC JAMS RENDER US ENTIRELY IMPOTENT.

OUR TECHNOLOGICAL WORLD BECOMES A NATURE OF ITS OWN.

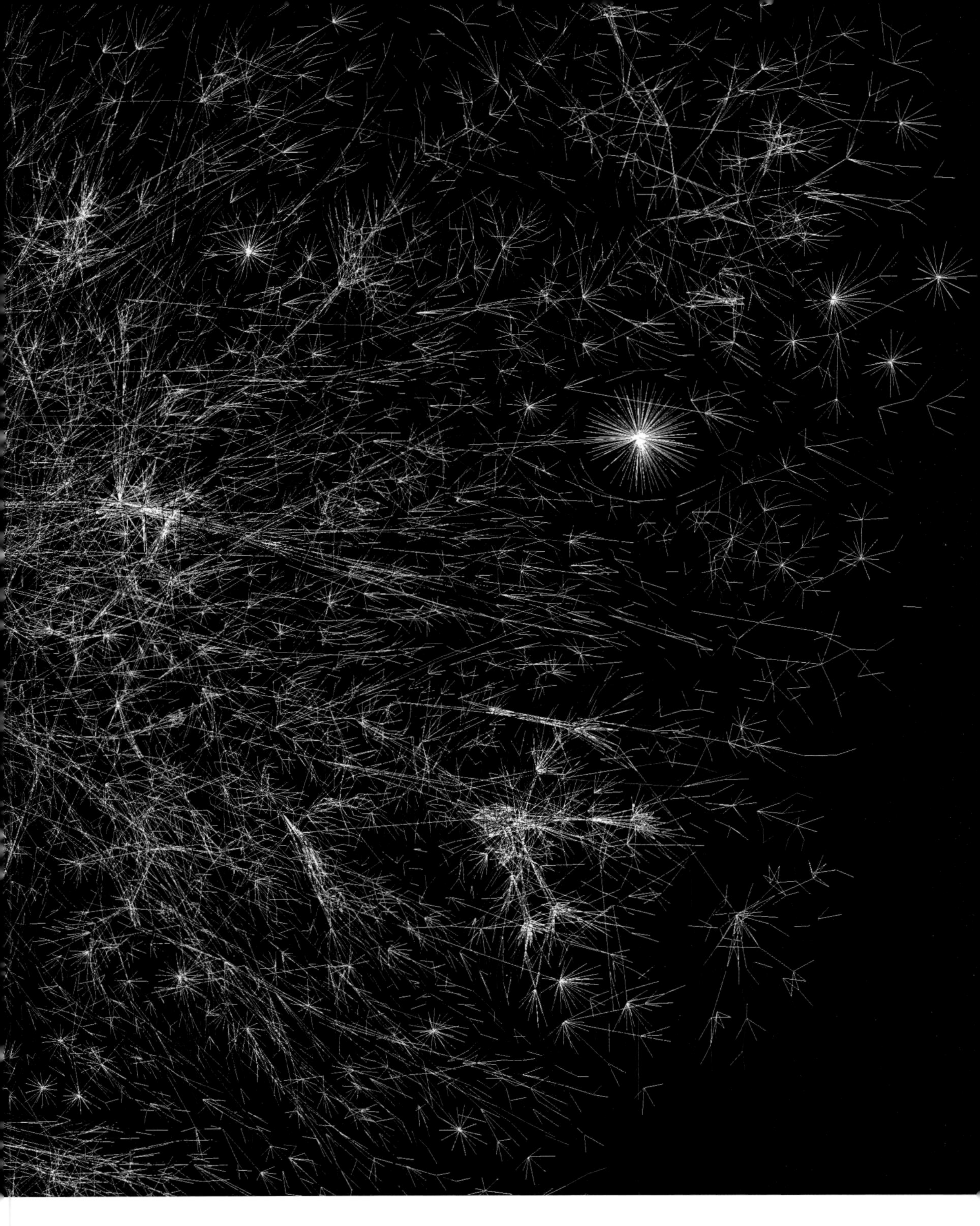

NO, THIS IS NOT SOME FAR-FLUNG GALAXY. CLOSER THAN YOU THINK, THIS IS A MAP OF THE INTERNET, CREATED IN 2003 BY BARRET LYON (33) IN AN ATTEMPT TO CAPTURE THE INTERNET IN ONE IMAGE. OBVIOUSLY THE MAP IS ALREADY OUTDATED, AS THE INTERNET IS CONTINUOUSLY SPAWNING NEW ARMS AND LEGS.

NATURE CHANGES ALONG WITH US.

FIFTEEN YEARS AGO, YOU WONDERED IF YOU REALLY NEEDED A MOBILE PHONE. FIVE YEARS AGO EVEN YOUR GRANDMOTHER HAD ONE. NOWADAYS, IF YOU LEAVE YOUR HOUSE WITHOUT IT, IT FEELS LIKE YOU'RE MISSING A LIMB.

REAL NATURE IS NOT GREEN

By Koert van Mensvoort

Many agree that nature is vanishing. Untouched nature is increasingly rare. But while the wilderness retreats, the hard boundaries between nature and culture are disappearing too. The stock market is controlled by an ecosystem of autonomous computer programs. A ‘natural’ tomato is a marvel of selective breeding and genetic engineering. Could it be that we’re just not looking for nature in the right places?

At the edge of the woods along the motorway near the Dutch town of Bloemendaal is a mobile telephone mast disguised as a pine tree. This mast is not nature: at best, it is a picture of nature. It is an illustration, like a landscape painting hanging above the sofa. Do we have genuine experiences of nature anymore? Or are we living in a picture of it?

In the Netherlands, every square meter of ground is a man-made landscape; original nature is nowhere to be found. The Oostvaardersplassen – which makes up one of the Netherlands' most important nature reserves – was, after the land was reclaimed, originally an industrial site; it was only turned into a nature reserve later. Even the 'Green Heart' at the centre of the most densely populated part of Netherlands is a medieval industrial area, which was originally reclaimed for turf-cutting. Our 'nature reserves' are thus in fact 'culture reserves' shaped by human activity. "God created the world, with the exception of the Netherlands. That the Dutch created themselves", as Voltaire put it in the eighteenth century.

In our culture, nature is continually presented as a lost world.

And ever since, we have been doing everything we can to live up to his pronouncement. Today, we even actively design and build nature in the Netherlands. Prehistoric forests are being planted in locations designated by bureaucrats: our image of nature is being carefully constructed in a recreational simulation (a 'regeneration of our lost heritage', as the nature-builders call it themselves).[1] Traditional cattle breeds are even being placed in this so-called 'new nature'.[2]

The original wild ox unfortunately became extinct in 1627, but the Scottish Highlander is an acceptable alternative. These cattle know what they're supposed to do: graze, under orders of the forestry service. Thanks to them, the landscape stays clear instead of becoming overgrown (we find this attractive, as it reminds us of famous seventeenth-century landscape paintings). In theory, the animals are supposed to look after themselves, but in winter the forestry service is willing to give them a bit of extra food. It also removes dead animals, lest walkers be offended by a carcass rotting on the footpath. In our culture, nature is continually presented as a lost world. It is associated with originality, yet appears only once it has disappeared. Our experience of nature is a retro effect.[3]

It is a widespread misconception that nature is always calm, peaceful and harmonious: genuine nature can be wild, cruel and unpredictable. Our contemporary experience of nature is chiefly a recreational one:[4] Sunday afternoon scenery; Disneyland for grown-ups. Indeed, lots of money is required to maintain the illusion. But nature is also a terrific marketing tool: there are Alligator garden tools, Jaguar convertibles, Puma trainers. Natural metaphors give us a familiar feeling of recognition. In commercials cars always drive through beautiful untouched landscapes. Strange that in this make-believe countryside there is not a billboard in sight, while logos and brands are so omnipresent in our environment, we can probably tell them apart better than we can bird or tree species. In my neighborhood, four-wheel drives have become an integral part of the street scene. These SUVs (sport utility vehicles, previously known as Jeeps or all-terrain-vehicles) have formidable names like Skyline, Explorer, Conquerer and Landwind. Luckily, you can buy spray-on mud for spattering your wheel rims, since SUVs rarely go off road. There are no hills around here, nor snow or other weather conditions that could justify a four-wheel drive. It's merely cool to join the urban safari.[5]

Nature Becomes Culture

The dividing line between nature and culture is difficult to draw. When a bird builds a nest, we call it nature, but when a human puts up an apartment building, suddenly it's culture. Some try to sidestep the problem by claiming that everything is nature, while others claim that nature is only a cultural construction. It's tempting just to lump the two together and give up thinking about it.

The word 'nature' is derived from the Latin word *natura*. This was a translation of the Greek *physis*. Natura is related to Latin terms meaning 'born' (and the Greek *physis* to Greek words for 'growth'). By the time of the ancient Greeks, the distinction between nature and culture was already considered important. Various things have changed since then; nature in the sense of physical matter unaltered by humans hardly exists anymore. We live in a world of petrochemical cosmetics, microprocessors and synthetic clothing (all things whose conditions of existence I know nothing of). New shower-gel scents are put on the market faster than I can use the stuff up. Shopping centers, websites and airports dominate our environment. There's precious little nature left that has remained untouched by humans: perhaps a bit here and there on the ocean floor, the South Pole, or the moon.

Logos and brands are so omnipresent in our environment, we can probably tell them apart better than we can bird or tree species.

Old concepts like nature and culture, human and animal, and body and mind seem inadequate for understanding ourselves and the technological society we live in.[6,7] Cloned babies, rainbow tulips, transgenic mice afflicted with chronic cancer to serve medical science: are they natural or cultural? In an evolutionary sense, every distinction between culture and nature has something arbitrary about it; both have been part of the same evolutionary machine since Darwin's day. When we speak about nature, we are always in fact talking about our relationship with nature, never about nature itself. nature is always 'so-called nature'.[8] The terms 'natural' and 'cultural' are usually deployed to justify one position or another. In the thirteenth century, Thomas Aquinas (the Catholic priest) believed art imitated nature, because human intellect was based on all things natural. Oscar Wilde (the homosexual writer), on the other hand, claimed that nature imitated art.[9] From this thought, it is only a small step to the idea that nature exists only between our ears and is in fact a cultural construction. Jacques Lacan (the postmodernist) claims that we cannot see nature.[10] A moderate constructivism is currently widely accepted among philosophers and scientists. Our image of nature has changed greatly over the centuries. It is likely that in the future we will adapt it further. This does not release us from our need to keep looking for nature. The manner in which we distinguish between nature and culture remains relevant, because it says something about the human perspective: what is our place in nature? An alternative approach is to distinguish between natural and artificial processes. Some processes can take place as a result of human action, others cannot. For example, a room can be lit through the flick of a switch or a sunrise. Sunrise is a natural process; flipping a light switch is an artificial one. In this view, cultural processes are the clear consequences of purposeful human action, and culture is whatever human beings invent and control. Nature is everything else. But much of the 'so-called nature' in our lives has taken on an artificial authenticity. Genetically manipulated tomatoes are redder, rounder, larger, and maybe even healthier than the ones from our gardens. There are hypoallergenic cats, and nature reserves laid out with beautiful variety.

In spite of all our attempts and experiments, it is still hardly practicable to mold life

You can buy specially engineered living beings in the supermarket. Human design has made nature more natural than natural: it is now hypernatural.[11] It is a simulation of a nature that never existed. It's better than the real thing; hypernatural nature is always just a little bit prettier, slicker and safer than the old kind. Let's be honest: it's actually culture. The more we learn to control trees, animals, atoms and the climate, the more they lose their natural character and enter into the realms of culture.

Culture Becomes Nature

Thus far I have said nothing new. Everyone knows that old nature is being more and more radically cultivated. However, the question is: is the opposite also possible? I think it is. In contrast to optimistic progress thinkers who believe human beings' control of nature will steadily increase until we are ultimately able to live without it, I argue that the idea that we can completely dominate nature is an illusion. Nature is changing along with us.[12]

It is said Microsoft founder Bill Gates lives in a house without light switches. His house of the future is packed with sensors and software that regulate the lighting. Nature or culture? The average Dutch person worries more about mortgage interest deductions than about hurricanes or floods. Do you control the spyware and viruses on your computer? In their struggle against nature, human beings have become increasingly independent of physical conditions, it is true, but at the same time they are becoming more dependent on technological devices, other people, and themselves. Think of the dependence that comes with driving a car. We need motorways, for which we pay road tax. A supply of petrol must be arranged. Once you're on the road, you have to concentrate so you won't crash into the guardrail. You must take account of other drivers. Youneed a driving license. All this is necessary in order to get your body from point A to point B more quickly. Along with physical deconditioning comes social and psychological conditioning.

Hypernatural nature is always just a little bit prettier, slicker and safer than the old kind. Let's be honest: it's actually culture.

I believe the way we draw the boundary between nature and culture will change. The domain of origin, of 'birth', previously belonged to nature, while culture encompassed the domain of the 'made'. Thanks to developments in science and technology, this distinction is blurring.[13] Origin is playing a smaller and smaller role in human experience, because everything is a copy of a copy. Insofar as we still wish to make a distinction between nature and culture, we will draw the line between 'controllable' and 'autonomous'. Culture is that which we control. Nature is all those things that have an autonomous quality and fall outside the scope of human power. In this new classification, greenhouse tomatoes belong to the cultural category, whereas computer viruses and the traffic jams on our roads can be considered as natural phenomena. Why should we call them nature? Isn't that confusing? We categorize them as nature because they function as nature, even though they're not green.

Human actions are not nature, but it can cause it – real nature in all its functioning, dangers and possibilities. In spite of all our attempts and experiments, it is still hardly practicable to mold life. Every time nature seems to have been conquered, it rears its head again on some other battlefield. Perhaps we should not see nature as a static given, but as a dynamic process. It is not only humans that are developing; nature, too, is changing in the process. Thus, I am proposing a new approach to distinguish nature and culture. At first – as is usual with paradigm shifts – it takes some getting used to, but after a while things become clear again. Real nature is not green.

REFERENCES

1 WWW.NIEUWENATUUR.NL, STICHTING DUINBEHOUD LEIDEN'S WEBSITE.

2 METZ, TRACY (1998). NEW NATURE: REPORTAGES OVER VERANDEREND LANDSCAPE. AMSTERDAM: AMBO, 1998, ISBN 90-263-1515-5.

3 WARK, MCKENZIE (2005). 'N IS FOR NATURE', IN VAN MENSVOORT, GERRITZEN, SCHWARZ (EDS.) (2005), NEXT NATURE, BIS PUBLISHERS, ISBN 90-636-9093-2.

4 METZ, TRACY (2002) PRET! LEISURE EN LANDSCHAP. ROTTERDAM: NAI, 2002, ISBN 90-5662-244-7.

5 CATLETT WILKERSON, RICHARD (2006). POSTMODERN DREAMING: INHABITING THE IMPROVERSE (WWW.DREAMGATE.COM/POMO).

6 BACON, FRANCIS (1620). 'NOVUM ORGANUM', TRANSLATED BY JAMES SPEDDING, ROBERT LESLIE ELLIS AND DOUGLAS DENON HEATH, IN THE WORKS (VOL. VIII), PUBLISHED IN BOSTON BY TAGGARD AND THOMPSON IN 1863 (WWW.CONSTITUTION.ORG/BACON/NOV_ORG.HTM).

7 HARAWAY, DONNA (1994). 'EEN CYBORG MANIFEST', TRANSLATED BY KARIN SPAINK ('A MANIFESTO FOR CYBORGS, 1991), AMSTERDAM: DE BALIE, 1994.

8 SCHWARZ, MICHIEL (2005). 'NATURE SO CALLED', IN VAN MENSVOORT, GERRITZEN, SCHWARZ (EDS.) (2005), NEXT NATURE, BIS PUBLISHERS, ISBN 90-636-9093-2.

9 WILDE, OSCAR (1889). THE DECAY OF LYING: AN OBSERVATION. NEW YORK: BRENTANO, 1905 [1889].

10 LACAN, JACQUES (2001). ECRITS, TRANSLATED BY ALAN SHERIDAN, LONDON: ROUTLEDGE, 2001.

11 OOSTERLING, HENK (2005). 'UNTOUCHED NATURE', IN VAN MENSVOORT, GERRITZEN, SCHWARZ (EDS.) (2005), NEXT NATURE, BIS PUBLISHERS, ISBN 90-636-9093-2, PP 81-87.

12 VAN MENSVOORT, KOERT (2005). 'EXPLORING NEXT NATURE', IN VAN MENSVOORT, GERRITZEN, SCHWARZ (EDS.) (2005), NEXT NATURE, BIS PUBLISHERS, ISBN 90-636-9093-2, PP. 4-43.

13 KELLY, KEVIN (1994). OUT OF CONTROL: THE RISE OF NEO-BIOLOGICAL CIVILIZATION, READING, MASSACHUSETTS: ADDISON-WESLEY, 1994, ISBN 0-201-57793-3.

NATURE BECOMES CULTURE

Who Designed the Landscape?

The Dutch Polder

While humans have been altering the landscape from the day we first took down a large herbivore, the Dutch polder is one of the most impressive examples of successful terraforming. Made of land reclaimed from the sea, polders are maintained by the country's iconic windmills and dikes, although now electric pumps have largely replaced the mills. The distinctive aesthetic of the polder is no accident. In the 16th and 17th century, the reclaimed lands were explicitly designed to express cohesion, both visual and social, through a rectilinear grid. The rich farmland of the polder typifies the aesthetic of the surveyors who literally dredged the Netherlands from the sea and the marshes. Flat as a pancake, the polder is dominated by rectangular fields, long, straight roads, and orderly lines of trees that visually defined boundaries. The nation's engineers extended the principles of the planned city into a planned environment. Not all polders have remained so rigid. The Naardermeer polder, for instance, was deliberately flooded to protect the city of Amsterdam from Spanish troops in 1629, almost immediately after it had been pumped dry. In the early 20th century, the city of Amsterdam planned to turn the Naardermeer into a garbage dump, however after a heated debate it was decided to give it 'back to nature'. Today it is one of the most important nature resorts in the Netherlands, featuring unique wetlands and breeding grounds for birds that exist thanks to the work of humans 400 years ago. | AG

Dubai Islands

The World, launched in 2003, recreates the world's nations as private island retreats for the wealthy. It presented an idealized image of the earth on permanent vacation, without any pesky poor folk or inclement weather. The project is now an emblem of the global economy: based on a reckless faith in appearances that slumped in the absence of any real substance. Without vegetation or a solid basis in the local ecosystem, the engineering project has fallen into disrepair. It is prey to both a lack of investors and rising water levels. In the end, global warming may finally sink The World.

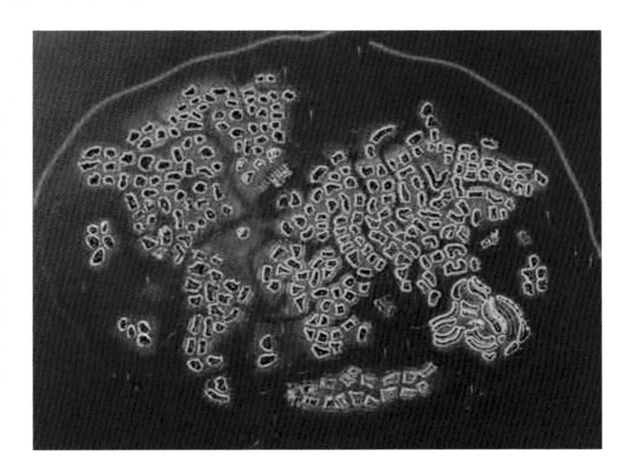

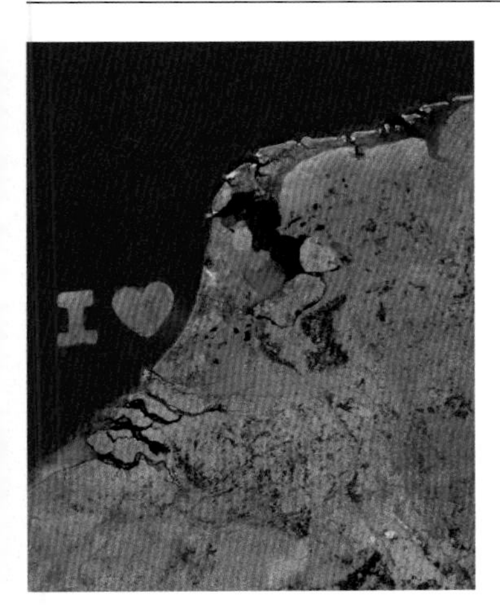

Nation Branding

Already masters of the polder, the Dutch may consider a more light-hearted way of wresting land from the sea. In 2007, the Christian-Democratic political party proposed building a tulip-shaped island off the Dutch coast. Not just a way to protect the coast and provide 600,000 m^2 of living space, the tulip island would use the land itself to brand the nation. More tongue-in-cheek responses came in the form of a joint, a miniature replica of the country or, as seen here, a logo derived from the famous 'I Love New York' campaign. Instead of hiding artifice in pseudo-nature, the constructed island calls attention to the country's man-made landscape. The Netherlands might be the first country to distill its national character for orbiting satellites.

Welcome Home

Unox uses the land itself to greet travelers landing in Amsterdam's Schiphol Airport. Their 'Welkom Thuis' (Welcome Home) creates a brandscape, co-opting a farmer's fields for an advertisement on a massive scale. A Unilever product, Unox is a ubiquitous Dutch brand known for its soups and smoked sausages. This advert links the traveler's warm feelings of returning to familiar pastures with the warm feelings she might experience, for instance, eating one of Unox's instant soups. Here, the company is not just a participant in the land. The brand asserts that it is the land. To the consumer, Unox is as Dutch as the distinctive flat countryside or the fields of spring tulips. A company that manufactures artificial pre-packaged foods is integrated into the natural, wholesome goodness of the farm, and by extension the national identity.

Who Designed the Banana?

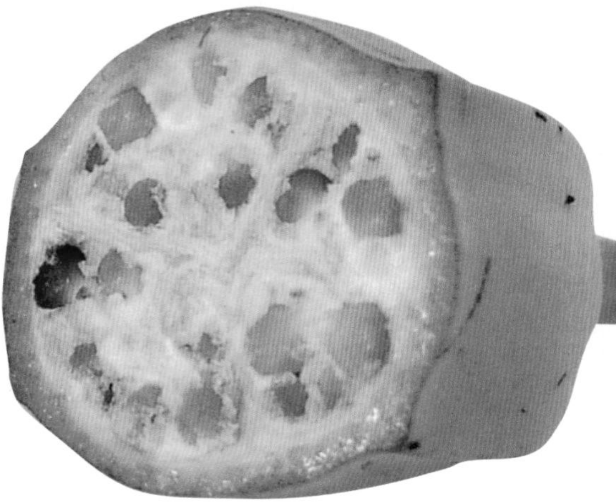

A WILD BANANA: NOT VERY ERGONOMICAL OR TASTY

Looking at a banana from a design perspective, one immediately notices the fruit is highly ergonomic and sophisticated: bananas fit perfectly in the human hand, they come with a non-slip surface, a bio-degradable packaging that is easy to open, and they have an advanced informative skin that turns yellow when the product is ready for consumption – green means not yet, brown means too late.

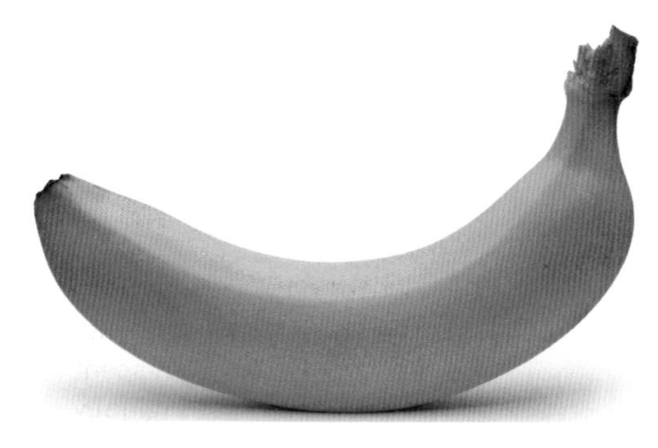

A 'REGULAR' BANANA: VERY ERGONOMICAL AND TASTY

The design of the banana is so good, some evangelists present it as evidence that an 'intelligent designer' must have created the fruit. These evangelists however, make a quintessential mistake on the static origins of 'nature', as they ignore that the bananas we eat today are hardly products of old nature. Rather, they are the result of thousands of years of domestication by people.

Archaeological and palaeo-environmental evidence suggests that banana cultivation goes back to at least 5000 BCE. The designed bananas we eat today cannot even reproduce without the hand of man, as they have no seeds – they are all clones, which makes the species highly vulnerable to diseases. Wild bananas are still around, yet they are much less ergonomically adjusted for human consumption as they have numerous large, hard seeds. Perhaps in the far future evangelists will present Coke bottles as evidence for their 'intelligent designer' argument? | KVM

FLUORESCENT FISH BY GLOFISH

And who made our fish transgenic?

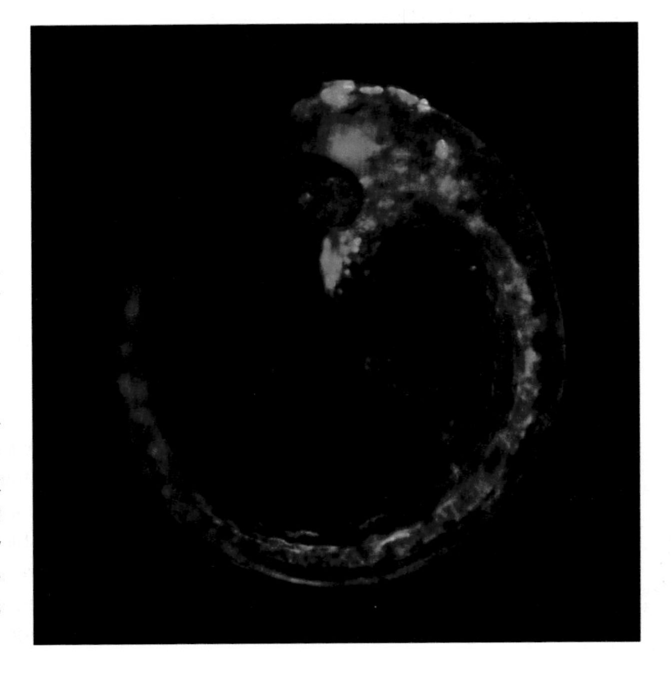

The artist Eduardo Kac commissioned the first transgenic pet, Alba, a rabbit that supposedly glowed green under blue light. The same technology is now available in *GloFish*, descendants of zebrafish implanted with Green Fluorescent Protein (GFP). The GFP gene was originally isolated from jellyfish, although certain colors are also derived from coral.

GloFish are billed as fun accessories for classrooms and blacklight enthusiasts, but they are not without controversy. Due to concerns about genetically modified animals, GloFish are banned in the European Union. While the fish are happy to reproduce, the GloFish license prohibits the copyrighted critters from doing what comes naturally. | AG

Who Arranges the Atoms?

YOUR LOGO HERE

ON APRIL 1990 SCIENTISTS AT IBM'S ALMADEN RESEARCH CENTER IN SAN JOSE, CALIFORNIA DEMONSTRATED HOW TO MOVE AND POSITION INDIVIDUAL ATOMS ON A METAL SURFACE USING A SCANNING TUNNELING MICROSCOPE. THE RESEARCHERS CREATED THE WORLD'S FIRST ATOMIC SCALE STRUCTURE: THE

MAN-MADE SUN

THE PHOTO OF THIS 'MAN-MADE SUN' WAS TAKEN ON JULY, 8, 1956 DURING AN APACHE H-BOMB TEST ON ENIWETOK ATOLL. IN 1963, HEALTH CONCERNS ABOUT RADIOACTIVE FALLOUT LED TO A BAN ON ATMOSPHERIC TESTING OF ATOMIC BOMBS. SINCE THEN, WE HAVEN'T HAD A CHANCE TO ENJOY THE VIBRANT RADIATION OF ATOMIC

Bloodcells 2.0

Author and futurist Ray Kurzweil hypothesizes that human bodies will soon be augmented by advanced medical nano-technology. Using the work of American nanotechnologist Robert Freitas he proposes a theoretical nano-robot the same size as a human blood cell, but far more effective. These 'respirocytes' would mimic the action of natural red blood cells. Only one micrometer in diameter, the spherical robots consist of 18 billion atoms arranged into a tiny pressurized tank. The respirocytes tank up on oxygen in the lungs, transfer it to the body's tissues, and gather up carbon dioxide for the lungs to exhale.

BLOODCELLS ENHANCED WITH BLOODBOTS

Respirocytes are able to store about one-and-a-half billion oxygen molecules, compared to the one billion that a blood cell can manage. While only 25% of the oxygen molecules are available from natural red blood cells, the nano-bots are designed to make 100% of their stored oxygen accessible to the body. These hyper-efficient respirocyctes could act as a universal blood supply and reduce deaths from asphyxia. Bot-enhanced divers could remain underwater for hours on a single breath of air, while athletes could break records in endurance events. Kurzweil is optimistic that it will be possible to manufacture respirocytes within 10 years. If so, there may be many new doping scandals to watch for in the 2020 Olympics.

Organ Printer

Researchers may soon be able to 'print' three-dimensional copies of damaged organs using stem cells and hydrogels as ink. Organovo, a San Diego-based company, is already shipping out production models of its bio-printers. Similar to rapid prototyping machines, these machines use inkjet technology to build up an organ around a sugar-based scaffold. The printers take advantage of a useful biological trait: cell clusters spontaneously organize themselves when placed together, like the polymer droplets that fuse when coming out of conventional 3D printers. Researchers have only been able to produce simple tissues such blood vessels, but more complex organs are not too far down the road.

ORGAN PRINTER BY VLADIMIR MIRONOV MD, PHD

Eventually, hospitals around the world may be able to download instructions for any number of body parts and print them on demand. Using human growth factors, a patient's stem cells can be turned into new teeth, lungs, or even designer meat for the truly curious or morbid. Printed organs do not even need to look like the real thing to function just as well. Some day, our bodies, like our homes, may have interior designers to modify flow and functionality. | AG

Bacteria that Eat Waste and Poop Petrol

WILL WE SOON BE DRIVING ON BACTERIA EXCRETE?

If peak oil hasn't arrived already, no doubt it's coming fast. Entrepreneurs are racing to find cheap, renewable forms of fuel to replace the fossil crude that runs the industrialized world. The answer may come from bacteria that normally have us running to the bathroom, not running to the pump. LS9, a biotech company in Silicon Valley, engineers non-pathogenic *E. coli* and yeast strains that excrete a one-to-one replacement for petroleum. Instead of fussing with other renewable energy forms like hydrogen cells or electricity, the genetically altered microbes produce a substance that is compatible with conventional combustion engines. Since the bacteria and yeast excrete fatty acids that are chemically similar to crude oil, it is not a complicated process to modify the bugs to produce the desired result. While ethanol fermentation requires energy-intensive refining, LS9's technique needs only minimal processing before it goes in the tank. The technology is something of a holy grail for a warming world: the company claims that production of the bio-fuel will use up more carbon than it emits.

However, problems arise when scaling up the operations from the lab to the consumer. Even with efficient plants, a nation the size of America would still need 530 square kilometers (329 square miles) of factories to cover its weekly oil consumption. More unsettling is the possibility of the microbes escaping from their tanks to the wild, though they would need specific conditions to survive. If the *E.coli* crossbred with their infectious relatives, we might discover that food poisoning gives us enough fuel to drive to the hospital. Technology turns nature into culture, but even culture can't be controlled. |KVM|AG

Fake DNA Evidence

DNA evidence used to be unassailable. A stray hair or lipstick-smudged cup can be planted on the scene, but DNA itself can't lie about where, or who, it comes from. A lab in Tel-Aviv has upended this assumption, creating artificial DNA that passes forensic muster. The lab, Nucleix, offers a paid test that can distinguish between in-vivo DNA and the artificial kind. The trick lies in looking for a lack of methyl groups, which occur naturally in the genome, but not the lab's pseudo-genes. Luckily, the existence of fabricated DNA will not result in a sudden upswing in doctored crime scenes. While the process is fairly simple for a trained biologist, fabricating custom DNA is still beyond the abilities of the standard murderer or molester. Human-made DNA, however, is more proof that we live in a designed environment, down to a microscopic level. For forensic science, biology is no longer the final arbiter of truth. | AG

TAKING FAKE DNA SAMPLES

Bioluminescent Plants

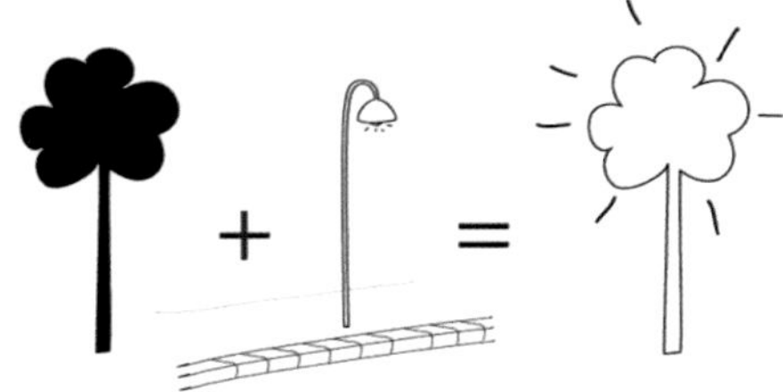

We have all seen the cellphone masts disguised as trees, created in an attempt to blend technology with the 'natural' landscape. Now Taiwanese scientists have created plants that could eventually pave the way for luminescent roadside trees used to replace streetlights. Scientists infused the leaves of *Bacopa caroliniana* with gold nano-particles that cause the chlorophyll to produce a reddish luminescence. This phenomenon is awkwardly named bio-LED by the scientists. According to Yen Hsun Su of the Research Center for Applied Sciences at the Academia Sinica in Taiwan: 'The bio-LED could be used to make roadside trees luminescent at night. This will save energy and absorb CO_2 as the bio-LED luminescence will cause the chloroplast to conduct photosynthesis'. This means that while the tree is 'lit' more CO_2 is consumed from the atmosphere, therefore the glowing trees could reduce carbon emissions and cut electricity costs while still lighting streets. | MVB

CitySounds™

Relax Anywhere

Losing yourself in silence? Miss the sound of honking cars, airplanes passing over, the sirens in the background? Don't miss them any longer. Get away to the city you love. Do it. With CitySounds, you have the world's greatest sounds in the palm of your hand. Select from a range of 12 themes, including Amsterdam, Baghdad, New York and Mumbai. Available now, on your iPhone or Android mobile.

CULTURE BECOMES NATURE

The New Forces of nature

Economy = Ecology

Westerners worry more about stock prices and mortgage rates than hurricanes or floods. The global economy seems autonomous, with its own dynamics and its own natural disasters. Like old nature, the economy operates in an amoral space, cruel and kind in equal measures. Should we begin to view the economy as an ecology, as something vast and unpredictable, rather than as a human institution governed by rational thought? Treating the economy as a natural system does not imply giving up control. Instead, economists will need to become more like farmers, creating patches of order within a system that's distinctly averse to control. Farmers and economists have to be prepared for the unpredictable, be it boll weevils or housing bubbles. In economies, like ecosystems, even the slightest adjustment to the laws of nature can produce widespread, unforeseen consequences. | KVM

Westerners Worry More About Stock Prices and Mortgage Rates than Hurricanes or Floods

Viruses

MALWAREZ BY ALEX DRAGULESCU

Contemporary computers are complex devices even for specialists, and entirely mysterious to the average user. We only care when they stop working. Yet, even a seemingly healthy computer can be infected with discreet parasites. Viruses and worms can create chaos through spam for thousands of innocent users. When next nature evolves beyond our control it forgets our interests and starts serving its own. *Malwarez* is an art project that uses computer viruses, spyware code and worms as the DNA for a virtual 'organism.' The creature grows according to an algorithm that takes its inputs from the grouping and frequency of malicious code. According to the artist Alex Dragulescu, patterns in the data dictate the development of the 3D organism. The seediest, creepiest side of the Internet, barely controlled by law and software, is given visual representation as an independent lifeform. | AG

When your Second Skin Becomes your First Skin

In Your Face

It's hardly a coincidence that the rounded Volkswagen Beetle looks so cute you want to hug it, or that those BMW headlights in the rear view mirror seem to growl at you. Car fronts or "faces" show their personalities and significant work goes into this design. As Chris Bangle, former Chief of Design at BMW said, 'A person can have lots of different faces, but with a car, you can only have one. When you put on that face, it's there forever. It becomes the car's expression'.

A 2008 study, published in the journal *Human Nature,* confirmed what many people have already felt – that cars seem to have personality traits associated with them, and that this is similar to the way people perceive facial expressions.

According to the study, "Throughout evolution, humans have developed an ability to collect information on people's sex, age, emotions, and intentions by looking at their faces." When it came to cars, "30% of the subjects associated a human or animal face with at least 90% of the cars. All subjects marked eyes (headlights), a mouth (air intake/grille), and a nose in more than 50% of the cars." The authors found that the better the participants liked a car, the more it bore shape characteristics corresponding to high values of what the authors termed "power", indicating that both men and women like mature, dominant, masculine, arrogant, angry-looking cars.

In the (near) future could car customization take extreme forms? Will we send a picture of ourselves to a vehicle manufacturer to tailor-make the lights, logo and license plates in our own image? Why not? We can already choose most of the car's features, so customizing it's face only seems natural... | ND | MB

"THAT CAR IS ALIVE! YOU GOTTA BELIEVE ME. IT'S MOCKING ME! I'M TELLING YOU! THAT CAR IS ALIVE!"
FROM: *HERBIE FULLY RELOADED* FILM, 2005.

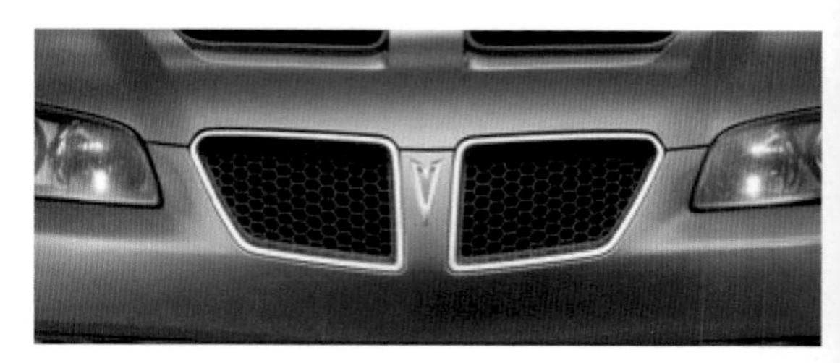

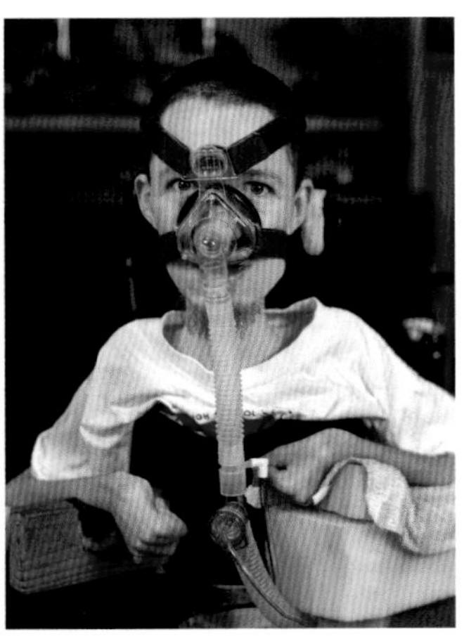

Alter Ego

During a photoshoot with the CEO of a company in 2003, photojournalist Robbie Cooper heard the divorcé's interesting story. He played in a virtual reality game in order to spend time with his children who he had limited real world access to. Cooper's subsequent fascination with virtual reality took him on a three-year journey photographing real-life players and picturing them next to their online avatars for his book *Alter Ego* published in 2007. Featured here is Jason Rowe (35) who has Duchenne Muscular Dystrophy. Rowe can move his thumbs, which enables him to play the multiplayer game *Star Wars Galaxies* an average of 80 hours a week. His avatar is a large, faceless, fearsome robot character. "In virtual worlds, everyone is on common ground," said Rowe. | KVM

Your Fifth Limb?

Nearly all 18–29-year-old Americans sleep with theirs. Barack Obama once said somebody would have to "pry it from my hands." Mobile phones have become a pervasive, invasive and essential part of millions of people's daily communication practice.

Ironically early predictions in the 1980s by consultancies like McKinsey put the global mobile phone market at close to one million by the year 2000. However, at the end of 2010, there were an estimated 5.3 billion cellular subscriptions worldwide. Despite its many celebrated benefits, like creating social cohesion, the technology is not without its drawbacks. In 2006 the *Independent* newspaper ran a story about Blackberry users and their display of "textbook addiction" symptoms – denial, withdrawal and anti-social behavior – and that family time was being eroded by users attending to the red flashing light of the Blackberry.

Some employees have gone so far as to sue employers, claiming health and standard issues with regards to the Blackberry. One employee claimed her marriage fell apart thanks to her Blackberry addiction. The same article quoted professor Kakabadse of Northampton Business School who suggested employers should go as far as developing policies for Blackberry usage to avoid addiction and lawsuits. According to reports, some companies have settled out of court with employees to avoid negative publicity. What seemed like a dream symbiosis between man and mobile phone, usage has now produced some interesting side effects.

ANYBODY ELSE HEAR THAT RINGING?

IN A STRANGE SYMBIOTIC TURN, MOBILES HAUNT US EVEN WHEN THEY'RE NOT AROUND. THAT IMAGINED PHONE RING OR VIBRATION NOW HAS A NAME: PHANTOM RINGING OR PHANTOM VIBRATION. THANKS TO A STATE OF CONSTANT PHONE ALERTNESS, EVEN EVERYDAY TASKS LIKE TAKING A SHOWER, USING A HAIR DRYER OR WATCHING COMMERCIALS MAKE US REACH INTO OUR POCKETS.

SEXANDTHECELLPHONE?

IN 2009, SAMSUNG MOBILE FOUND THAT NEARLY ONE THIRD OF MEN AND WOMEN IN DENVER (USA) WOULD RATHER FORGO SEX FOR AN ENTIRE YEAR RATHER THAN GIVE UP USING THEIR CELL PHONES FOR THE SAME AMOUNT OF TIME. THE AVERAGE DENVER MOBILE PHONE USER SPENDS NEARLY TWENTY PERCENT OF THEIR WAKING TIME ON THEIR PHONES.

The more humorous amongst them are conditions like "Blackberry Thumb" resulting from overuse of the keyboard. More serious ones as revealed in a report by MIT's Sloan Business Management Lab is that mobile phones and specifically Blackberries represent a comfort blanket for owners and according to Melissa Mazmanian (one of the authors of the report): "Spouses find it frustrating and aggravating and to avoid problems couples have to negotiate rules and boundaries over use." In a similar scenario, there are claims that the line between personal and workspace has been eroded by mobile technology.

The "always on" nature of the medium allows for work calls and messages out of office hours and adds towards normalizing the notion that people should be available and accountable to others, visibly and transparently, at any time and place. The Blackberry handset uses a red flashing light (normally reserved for cases of emergency rather than communication) to indicate the presence of new messages and creates a sense of urgency amongst users to respond to the waiting email or text message. Studies on the impact of mobile devices on work and family life have implied that the employer literally becomes embodied in the device.

The British Association for Counseling and Psychotherapy claims that the human brain is not properly adapted to the multi-tasking that modern life demands. They call for eight hours of work, eight hours of rest and eight hours of play a day. | ND | KVM

A new environment, a new vocabulary

In response to a "change in the environment" in 2008, the Oxford Junior Dictionary opted to drop some terms pertaining to old nature and include others pertaining to technology.

The more than ten thousand words and phrases in the Oxford Junior Dictionary were selected by using several criteria', including how often words would be used on a daily basis by young children.

Some of the words taken out where: 'Adder', 'Ass', 'Beaver', 'Boar', 'Budgerigar', 'Bullock', 'Cheetah', 'Colt', 'Corgi', 'Cygnet', 'Doe', 'Drake', 'Ferret', 'Gerbil', 'Goldfish', 'Guinea Pig', 'Hamster', 'Heron', 'Herring', 'Kingfisher', 'Lark', 'Leopard', 'Lobster', 'Magpie', 'Minnow', 'Mussel', 'Newt', 'Otter', 'Ox', 'Oyster', 'Panther', 'Pelican', 'Piglet', 'Plaice', 'Poodle', 'Porcupine', 'Porpoise', 'Raven', 'Spaniel', 'Starling', 'Stoat', 'Stork', 'Terrapin', 'Thrush', 'Weasel' and Wren.

Some of the words added where: Blog', 'Broadband', 'Mp3 Player', 'Voicemail', 'Attachment', 'Database', 'Export', 'Chatroom', 'Bullet Point', 'Cut And Paste' and 'Analogue'.

Intimate Technology

Our lives are increasingly regulated by ever-shrinking electronics. The next logical step could see electronic circuits on the surface of or even inside the body. | KVM

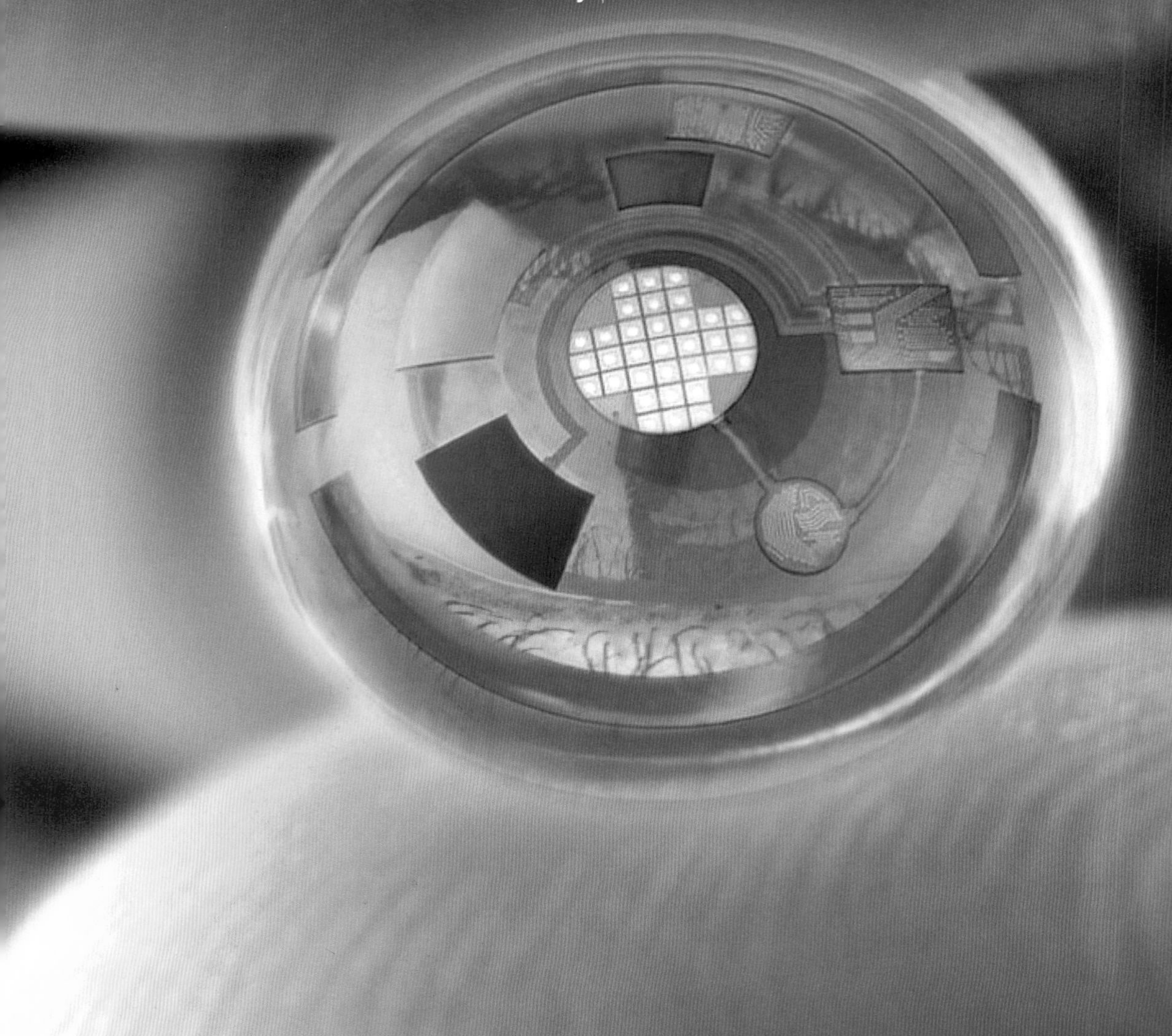

Silky Membranes

Scientists at the University of Pennsylvania have created implantable electronics on silk substrates that dissolve inside the body. The thin silk membrane eventually 'melts,' leaving behind stretchy silicon circuits that match the efficiency of traditional, inflexible circuits. Unlike other medical electronics, the silicon-based circuits do not need to be shielded from the body. Since silk, silicon, and the metals used in the transistors are bio-compatible, the body does not need to be protected either.

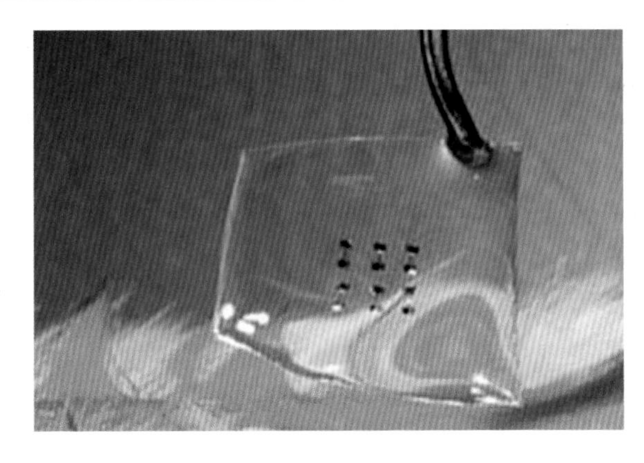

So far, the electronics have only been tested in mice, but they show promise for medical applications. The silicon-silk may be used for implantable LED monitors that display blood sugar level for diabetics, and for electrodes that interact with the nervous system. The devices are so tiny that they can be wrapped around individual nerve fibers to help control prosthetic limbs. For body-art enthusiasts, LED implants may give tattoos that futuristic edge.

Ring My Tooth

The 'tooth phone' is a theoretical device, consisting of a tiny vibrator, receiver and generator implanted in a tooth during routine dental surgery. Sound comes into the mouth as a digital radio signal. Bone resonance - vibration at a molecular level - transfers sound to the inner ear. The user can privately receive phone calls and stream music. The designers James Auger and Jimmy Loizeau collaborated on the Audio Tooth Implant, first shown in the 'Future Products' exhibition in 2002. Describing it as causing "techno-schizophrenia," they intended to stimulate debate about bio-technology. The media enthusiastically picked up on the idea; many treated the Audio Tooth Implant as a working device.

No More Rejection

For damaged or malformed heart valves, medical replacements don't perform as well as the real deal. They cannot respond to changes in blood flow like a natural valve. They are made of plastic, metal or static human cells that don't grow along with the body. Especially for children, this requires multiple high-risk operations. British researchers are now creating living valves by placing stem cells on collagen scaffolds. Scientists may eventually be able to culture replacement heart valves from a patient's own cells,

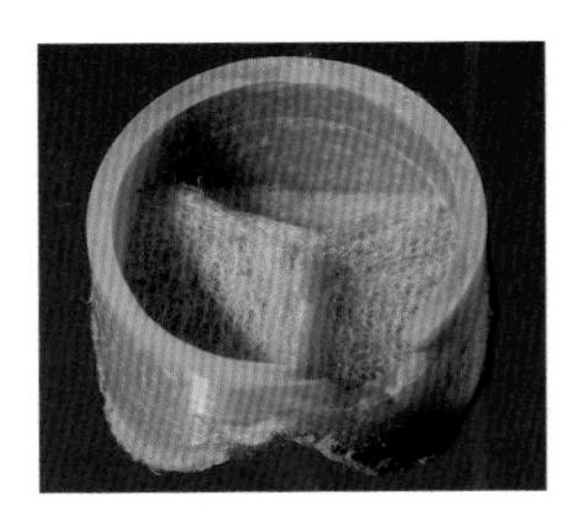

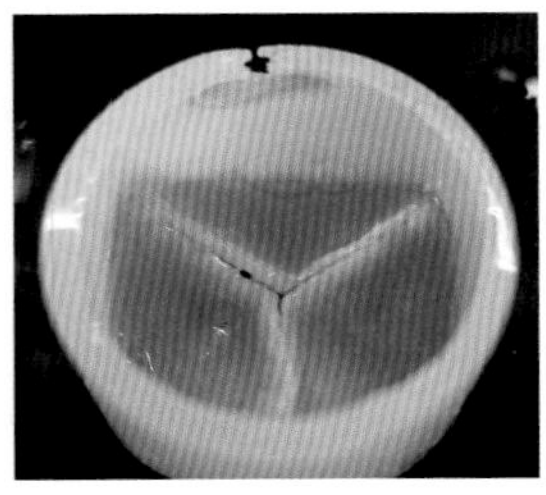

eliminating the need for anti-rejection drugs and a long wait for a donated organ. In the future, stem cell technology may be able to recreate whole hearts from scratch. | AG

Next Nature avant la Lettre

Your Parents' Next Nature

Mobile phones are one of the most ubiquitous consumer technologies on the planet. Leaving home without your mobile often renders us "lost" or even suffering from phantom limb syndrome. Contemplating a time before their existence seems like an alien notion and yet the first mobile phone was only developed 38 years ago by Motorola engineer Martin Cooper (picture right) in 1973.

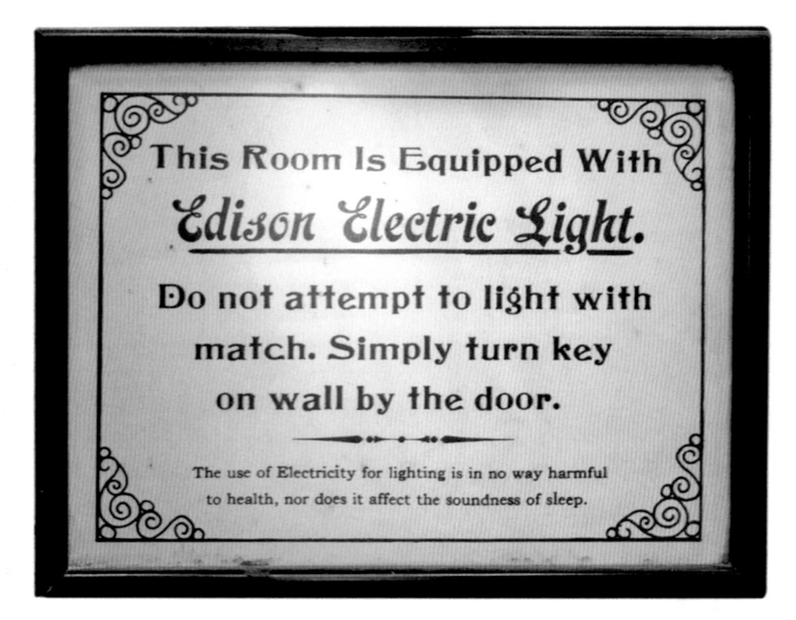

Your Grandparents' Next Nature

Reminiscent of warnings against smoking or excessive cellphone use, it seems our forefathers needed instructions for handling electrical lights. The small print on this 1878 disclaimer is the most poignant: "The use of Electricity for lighting is in no way harmful to health, nor does it affect the soundness of sleep". For all we know 19th century philosopher Friedrich Nietzsche was inspired by the developments in lighting technology when he wrote: "... in the end, every victorious second nature becomes the first".

NEXT NATURE 10,000 YEARS AGO >
SOMEWHERE AROUND 9500 BC PROTOFARMERS BEGAN TO SELECT AND CULTIVATE FOOD PLANTS WITH DESIRED CHARACTERISTICS . WHILE IN THEIR OWN DAYS THEY WERE 'REVOLUTIONARY TECHNOLOGISTS' WHO RADICALLY CHANGED THEIR RELATION WITH THE ENVIRONMENT, TODAY WE WOULD DESCRIBE THESE PEOPLE AS 'ORGANIC FARMERS'. HOWEVER, WITH EVERY NEXT NATURE THAT EMERGES, AN OLDER NATURE DIES OUT.

WE CAN ONLY BEGIN TO IMAGINE WHAT WAS LOST WITH THE RISE OF AGRICULTURE. SOME HAVE SUGGESTE THE END OF HUNTER-GATHERER LIFE BROUGHT A DECLINE IN SIZE, STATURE, SKELETAL ROBUSTNESS AN INTRODUCED TOOTH DECAY, NUTRITIONAL DEFICIENCIES, AND MOST INFECTIOUS DISEASES. IN THE MIND OF A HUNTER-GATHERERS WE - CONTEMPORARY HUMANS - WOULD PROBABLY BE CONSIDERED 'POST-HUMAN' ALREADY. THINK ABOUT IT NEXT TIME YOU ARE GATHERING FOOD IN YOUR LOCAL SUPERMARKET.

WHY HUNT?

WHY GATHER?

JOIN THE NEOLITHIC REVOLUTION!

How goes the **hunt?**
Not so great. How's **gathering?**
So-so.

Look! A **village!** **I wonder** what they **do** over there ...?

Excuse me. I couldn't help but **overhear.** Let me tell you about living the **Neolithic Way!**

First off — we don't just **look around** for our food ... we actually **grow** some of it ourselves, **where we live!**
Gasp!

Plant and animal **domestication** is **the key.** We grow **edible plants** ourselves, right out of the **ground,** time after time!

Yum!

Animals, too! We **control** their reproduction to select **desirable characteristics** and eliminate bad ones.
Wow! How can we live the **Neolithic** way?

You can start by **joining us** in the village! **Leave your troubles behind!***

*Some hunting and gathering may be necessary to maintain dietary variety and avoid famine.

Build permanent structures!

Enjoy regular meals!

Be civil!

Settle down!

Reshape your environment!

Be sociable!

Form complex societies!

Special offer! Free booklets!

The Pleasures of Porridge

Earn Your Animals' Respect

How to Tell a Weed

Your KEYS to a BETTER LIFE!

Harness Plant Power!

- Learn how the seeds you drop can become next fall's crop!
- Use seed selection to make future plants more productive and easier to harvest!
- Preserve and store surpluses for hard times!
- Invent new ways of preparing and cooking plant foods!

Put Animals To Work For You!

- Learn which species are slow and submissive!
- Use food and fences to keep them around!
- Influence their choice of mates!
- Breed the best and eat the rest!

Disclaimer: Plant and animal domestication can lead to overpopulation, deforestation, erosion, flooding, desertification, materialism, diminished nutrition, cavities, and television. Caution advised.
YOUR RESULTS MAY VARY

Tomorrow's Fossils

Ode to the car in 40 years' time? A future Museum of Obsolete Objects? Inspired by Stonehenge while living in England, Jim Reinders, an experimental American artist, originally built Carhenge in Western Nebraska as a memorial to his father. Created in 1987 with the help of his family, it is now a free tourist attraction. It uses 38 vehicles, including a 1962 Cadillac, to mirror the position of the rocks that comprise Stonehenge. England's 'natural' past, an idealized place of agrarian idyll and legendary deeds, is transported to contemporary America. Reinders argues for the mythological resonance of the automobile, both as a continuation of past traditions, and as a progenitor of myth itself. As much as we live exclusively in next nature, we look to old nature, and old culture, for context. Will vintage gas-guzzlers prove as enduring as Stonehenge's boulders? |RC|AG

Past Perfect

Certain technologies, already obsolete in our time, may be as inscrutable in the distant future as long-extinct species are to us. When presented as a natural part of the geological record, a cellphone or a Playstation controller becomes a rare oddity. The skeletons of videogame and cartoon characters are just as disorientating, conjuring a life (and death) for the patently fictional. Yet these imagined artifacts recognize the same premise: the fossil record of our species will not be distinguished by our bones, but by our technologies. |AG

CARHENGE (1987), JIM REINDERS

PECULIAR FOSSIL SPECIMEN, CREATED BY CHRISTOPHER LOCKE

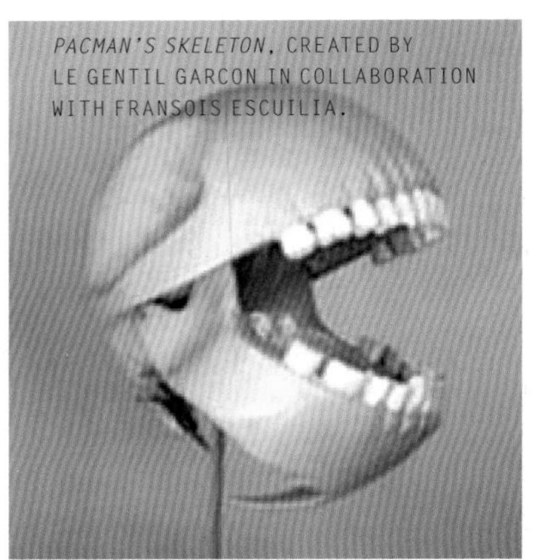

PACMAN'S SKELETON, CREATED BY LE GENTIL GARCON IN COLLABORATION WITH FRANSOIS ESCUILIA.

vorsprung durch technik.
standardized shape
precise color management
genetic protection system
neutral flavor guarantee
no-smell technology
designtomato.com

PEOPLE ARE CATALYSTS OF EVOLUTION

RAZORIUS GILLETUS: ON THE ORIGIN OF NEXT SPECIES

By Koert van Mensvoort

Is the evolution of the single bladed razor into an exorbitant five–bladed vibrating gizmo the outcome of human needs, or is there another force in play? Say hello to *Razorius Gillettus*, one of the new species emerging from our techno-economic ecology. Proof that evolution should be understood as a universal principle rather than a DNA-specific process. Yet if this is the case, how can we become responsible stewards of these new, non-genetic forms of life?

I received my first razor when I was 15. It had two blades on a simple metal stick, and, as I recall, it gave me a really close, comfortable shave. In the 20 years since my first shave, I've used seven different models. This morning I shaved with the Gillette Fusion Power Phantom, a rather heavy, yet ergonomically designed, battery-powered razor that looks a bit like a vacuum cleaner and has five vibrating blades with an aloe-moisturizing strip. So what happened? This is a story about design, technology, marketing and evolution.

Welcome to the 21st century. We don't travel in spaceships, but we do have five-bladed razors!

First, a personal disclaimer (in case you were wondering): Yes, I agree that shaving technology was already sufficiently developed when I got my first razor 20 years ago. In fact, the U.S television show *Saturday Night Live* had already parodied the Gillette Trac II for its excessive design (it was the first two-bladed men's razor) shortly after it was first advertized in 1975. The creators of the satirical program played on the idea of a two-bladed razor as a sign of a growing consumption culture and made a fake commercial for a fictitious razor with the ridiculous amount of three (!) blades, emphasizing the fact that the consumer is gullible enough to believe, and then buy, everything they see on TV. Of course, the *Saturday Night Live* comedians couldn't have known that a three-bladed razor would actually enter the consumer market in the late 1990s. Nor could they have anticipated that I would have shaved with a five-bladed razor this very morning. Welcome to the twenty-first century, folks! No, we don't travel in spaceships... but we do have five-bladed razors!

Fortunately, it is still possible to buy new blades for my very first razor model. These older blades are not only cheaper (they're sold in a box of 10 that costs less than a box of four-blade cassettes that fit the latest model), they're also more durable. And yet, in the years since my first shave, I have bought over a dozen different razors. I must confess I've bought some models from the competing brand as well. So why did I buy this whole collection of razors over the years? Maybe I'm the type of person who is keen on new things: I am a sucker for innovation.

Copy paste mix breed delete evolve

Before we analyze my own behavior as a buyer, let's first study the razors. If we look at the development of razor technology over time, we can distinguish some similarities with evolutionary development as we know it from the biological world.

1. Every new model builds upon the properties of the previous model.

2. Successful adaptations are preserved in future generations, whereas unsuccessful ones will be phased out.

3. The shift in emphasis from functional elements like the pivoting head, to the seemingly functionless aesthetics of the newer models, whose color changes seem to have no purpose other than to make them stand out amidst competing models, reminds us of the exuberant tail of the male peacock.

4. The unique click-on systems for replacement blades on different models resemble biological immune systems, preventing intruders from entering and feeding on the body.

5. Different survival strategies are being tested, and may even result in separate species over time – think of the parallel branches in more recent models that come with and without batteries. The marketers don't seem sure whether the future is in electric or non-electric shaving and have therefore decided to gamble on the two strategies. Once again, I confess, I bought them both.

Now it may seem quirky, corny even, to consider the development of razors from an evolutionary perspective. After all, these are industrial products assembled in factories. Yet I propose to look at them as the result of an evolutionary process. I already hear your rebuttals: "These razors didn't evolve, people designed them! How can that be an evolutionary process?" Well, allow me to elaborate – and this is where we can learn something about our symbiotic relationship with technology. It is indeed true that all the individual razors were created by engineers and designers. However, if we look at the design of the entire series of shavers as it has developed over my personal shaving career, it is difficult to pinpoint one creator. Where is that one big mind, that 'intelligent designer' responsible for the transformation of the razor from a simple blade on a stick to a five-bladed electric beast?

Obviously many designers and engineers have been involved in the creation of my razors over the years. No doubt these are all decent, friendly people (with good incomes, too). But what more are these creators of individual razors than tiny cogs in the Gillette Machine? Calling them engineers and designers may be giving them too much credit for the work they do. They merely sketch out the next model whose 'innovative' new properties are wholly predictable, that is, slight variations on current models – a nanotech-sharpened blade, an extra-moist strip, an anti-slip grip or perhaps even a customizable color scheme. The razor designers don't have a lot of room for truly creative design work. They aren't in a position to think deeply on the meaning and origins of shaving, to contemplate how this ancient ritual might be reinvented or improved upon. Like bees in a beehive, their work is determined by the logic of the larger structure. The chair belonging to that one great 'intelligent designer' steering the entire development of shavers over time is sitting empty. Rather, a larger design gesture emerges from the closely interrelated forces of the consumer market, technological affordances and, of course, competition. With regard to the latter, just think of the Wilkinson brand, the first to introduce a four-bladed shaving system, and forcing Gillette to answer with a five-bladed system. Together, these contextual influences constitute an ecosystem of a sort, which (again) closely resembles the environmental forces known to play a part in the evolutionary development of biological species.

Evolution, but not as we know it

There are certainly arguments against this evolutionary view on the development of razor technology – so let's consider both sides of the coin here. The most common objection is that "people play a role in the process, so it can't be evolution." While this reasoning is tempting, it also positions people outside of nature – as if we are somehow outside of the game of evolution and its rules don't apply to us. There is no reason to believe this is the case: after all, people have evolved just like all other life forms. The fact that my razors are dependent on people to multiply is also not unprecedented. The same is valid nowadays for many domesticated fruits, such as bananas, as well as a majority of cattle. Moreover, we see similar symbiotic relationships in old nature: just think of the flowers that are dependent on bees to spread their seeds.

There is no reason why evolutionary processes could not transfer themselves to other media

Another objection to the suggestion that my razors are the result of evolutionary development is because they are made of metal and plastic and are not made up of carbon-based biological matter. Underneath this argument lies the assumption that evolution only takes place within a particular medium, that of carbon-based life forms. A variation of this argument states that we are only dealing with evolution if there are genes involved – as with humans, animals and plants. This way of thinking demonstrates a limited understanding of evolution, however; evolution should be understood as a principle and not as a process contrained to a single medium. In fact, the genetic system of DNA that underlies our species is itself a product of evolution – DNA evolved from the simpler RNA system as a successful medium for coding life. There is no reason why evolutionary processes could not transfer themselves to other media: building on the notion of genes, Richard Dawkins sees 'memes' as the building blocks of cultural evolution, and Susan Blackmore suggests that 'temes' are the building blocks of technological evolution.

In the end, the question we should ask ourselves is: are the environmental forces of the economy and technology at least equally or perhaps even more important for the shaping of razor technology than the design decisions made by the 'inventors' of the individual models? I, for one, am quite sure this is the case. I therefore propose we consider the development of razors as a truly evolutionary process – not metaphorically, but in fact. The species this process has brought into being we shall now classify as: *Razorius Gillettus*. It is just one of the many new species emerging within the techno-economic system, a system that is fast evolving.

Technodiversity is increasing

Once we allow that the development of razor technology is an evolutionary process, let's zoom in a bit on our own role in the evolutionary game. How do we relate to *Razorius Gillettus* and its many fellow evolving techno-species? Are we like bees, who feed on nectar from flowers and, in return, spread their pollen, enabling the flowers to reproduce? Are we heading towards a similar symbiotic relationship with the technosphere, one that feeds upon our labor and creativity and gives us *Razorius Gillettus* in return? Should we take pride in our role as catalysts of evolution? As propagators of a technodiversity unlike the world has ever seen? As the one and only animal that transfers the game of evolution to another medium? Indeed, we can. Yet, as in every symbiotic relationship, we should also be alert to whether both parties are actually getting a good deal. And although I did buy all these razors and they have been providing me with an ever-smoother-closer shave throughout my life, I am not entirely sure there is an appropriate balance in this relationship.

Innovation without a cause

Too many so-called 'innovations' are directed primarily towards intensifying the growth and increasing the well-being of the technosphere – bigger economy, bigger corporations, more technological devices – and not towards actually improving people's lives. Indeed, my latest razor shaves just that tiny little bit more smoothly than the previous model. Still, if you asked me whether the device has 'upgraded' my life, I'd have to say no.

Let's face it: Gillette's new razors are primarily created for the sake of Gillette Corporation, creating higher turnover and bigger profit and more shareholder value. Now that's not all bad, as good business does provide people with good jobs and steady incomes, which allows them to live happy lives – and to buy more razors. So far, it's a win-win situation. But the manufacture of all these devices also uses an amazing amount of resources, putting immense pressure on the biosphere. Remember the 'old nature' that used to surround us before the emergence of the technosphere? We should not be naïve about the fact that corporations (I know they'll tell you otherwise) do not intrinsically care about the health of the biosphere. Being able to breathe clean air is simply not important for *Razorius Gillettus* and its unique digestive system. Clean air is only a requirement for carbon–based life forms, like algae, plants, birds, polar bears, and, of course, people.

Catalysts of evolution

So how should we proceed? I would be the first to admit there is a certain appeal in the development of *Razorius Gillettus*. The notion that human activity is causing the rise of such a peculiar new species and that we are now co-evolving towards a shared future is intriguing to say the least. I wonder what Charles Darwin would have thought about this. Perhaps he would have pointed out the serious risks involved in this evolutionary leap. Certainly, our awareness of our own role as 'catalysts of evolution' has yet to mature. It's quite a job description we've got our hands on there, with a great deal of responsibility. If we feel we're not suited to the job, we might be better off growing our beards and returning to our caves. At least that's what some people have proposed we do. However, to turn back the clock of civilization would be to deny what it means to be human, or at least reveal a certain cowardliness towards the unknown. On the other hand, adopting a purely techno-utopian attitude of 'letting grow' would likewise not serve the long-term benefit of humanity or its fellow biosphere-dependent species – we'd run the risk of being outsourced altogether.

Should we take pride in our role as catalysts of evolution?

The mature thing to do as catalysts of evolution would be to assume a stewardship role that focuses on maintaining a balance between the declining biosphere and the emerging technosphere – between old nature and next nature – and aspires towards an environment in which both can live in relative harmony. I am not saying it will be easy. But if we are able to do this, we will have something to be truly proud of.

Gillette
Gillette
Gillette
Gillette
Gillette
Gillette
Gillette
Gillette

#1 NEXT NATURE

wwww.nextnature.net/
themes/nextnature

NEXT NATURE

RECREATION

ISBN 978-84-92861-53-8
9 788492 861538

IN THE 17TH CENTURY, PAINTERS TAUGHT US TO APPRECIATE THE QUALITIES OF NATURE.

THE FIRST TRUE EUROPEAN LANDSCAPE WAS PAINTED IN 1520 BY ALBRECHT ALTDORFER, BUT THE GENRE ONLY ACHIEVED REAL POPULARITY LATER. IT REACHED ITS PEAK WITH THE WORK OF ROMANTIC PAINTERS LIKE CASPAR DAVID FRIEDRICH. AS MAN BECAME SECONDARY TO THE LANDSCAPE, NATURE WENT FROM BACKDROP TO CENTER STAGE. NATURE IS PASTORAL, NOBLE, OR SPIRITUAL, BUT ABOVE ALL, IT IS AESTHETIC. ONLY A SOCIETY THAT IS NO LONGER IMMERSED IN THE LANDSCAPE IS ABLE TO TREAT IT AS ART.

NOWADAYS, WE RECREATE THE LANDSCAPE AFTER OUR IMAGE OF NATURE.

WE ALL KNOW WHAT A NATURAL LANDSCAPE SHOULD LOOK LIKE: GREEN AND LEAFY. A CELLPHONE TOWER RUINS THE VIEW. THE SAME TOWER DRESSED UP AS A REDWOOD TREE, NO MATTER HOW ECOLOGICALLY INACCURATE, SOMEHOW LOOKS LESS ARTIFICIAL. LANDFILLS ARE COVERED OVER WITH MEADOWS, AND CITY PARKS ARE PLANTED TO LOOK LIKE OLD-GROWTH GROVES. REAL NATURE IS NOT GREEN, BUT MANUFACTURED NATURE SURE IS.

WE TRAVEL THOUSANDS OF MILES TO INDULGE IN A NATURAL EXPERIENCE.

WHEN NATURE IS SEPARATE FROM CIVILIZATION, IT ONLY MAKES SENSE THAT WE HAVE TO LEAVE HOME TO FIND IT. THE FARTHER THE PARK IS FROM THE CITY, THE MORE AUTHENTIC IT BECOMES. ONLY WHEN WE ARRIVE DO WE REALIZE THAT 'REAL' NATURE CANNOT BE BOUGHT WITH A PLANE TICKET. ON SAINT MARTIN ISLAND THE AIRPORT IS NOW AS MUCH AN ATTRACTION AS THE BEACH ITSELF.

OR WE STAY AT HOME AND SETTLE FOR A SIMULATION...

IF WE CAN'T GO FIND NATURE, WE WILL BRING NATURE TO US. GERMANY IS HOME TO TROPICAL ISLANDS, A BEACH RESORT ENCASED IN AN AIRCRAFT HANGAR. IT BOASTS THE WORLD'S LARGEST INDOOR RAINFOREST, CAMPING UNDER THE PALM TREES, AND PIPED-IN BIRDCALLS FOR A NATURAL FEEL. YET THE PROJECTED CLOUDS BY THE FAKE SEA DON'T QUITE MATCH THE MAJESTY OF THE HANGAR'S STEEL RIBS.

DO WE STILL HAVE A REAL EXPERIENCE OF NATURE, OR ARE WE LIVING IN A PICTURE OF IT?

NOWADAYS, OUR NOTION OF NATURE IS FILTERED THROUGH PAINTINGS, TELEVISION DOCUMENTARIES AND MAGAZINES. IT IS ALMOST A GUARANTEE THAT WE WILL SEE HUNDREDS OF PHOTOGRAPHS OF A FAMOUS MOUNTAIN OR CANYON BEFORE WE EVER STEP FOOT NEAR THE REAL THING. AND ONCE WE ARE THERE, WILL WE LOOK AT, SMELL, AND TOUCH THE LANDSCAPE, OR WILL WE HOLD UP A CAMERA FOR A PERFECT SHOT?

NATURE IS AN AGREEMENT

By Tracy Metz

Is nature the product of a collective illusion? From all-natural soaps to milk cartons with images of friendly farmyard cows on them, our expectation of nature as unspoiled and pure is being neatly packaged and sold back to us at a premium.

Nature is an agreement. Just like the nude beach. Here you keep your breasts and your crotch covered, there you drop everything and act like it is the most ordinary thing in the world that everyone is suddenly walking around naked. That is also how we deal with nature. We make an agreement with each other that this or that piece of the country is 'nature,' and put a sign next to it and a fence around it. Nature itself must of course stick to this agreement – no thorns, please, no bites and certainly no flooding! – and it must stick to the budget. After all, we have invested a lot of time and money in making nature.

Nature is also a feeling, a symbol for an escape from the rush of everyday life, a time and a place for reflection. We project onto nature our yearning for that which is larger than ourselves, larger than life, something that is not subject to the latest hype, fads and fashions. Nature represents eternal values. There is something in us which longs for nature precisely because it is beyond our control. At the same time, paradoxically, we cannot bear the fact that it is beyond our control. So while we find the naturalness of nature attractive, at the same time we feel the urge to get a grip on it, to control it. Wildness must be tempered.

Nature is also an image. You might even say: without images and representations... no nature. Without Discovery Channel no wild animals in the Serengeti, without amateur video no tsunami, without scented candles no autumn smells, without a CD or a ringtone no birdsong, without screensavers no sunsets! Take the new milk cartons of the Dutch creamery Campina: smart blue cartons with drawings on the side, almost as big as the cartons themselves, of an old-fashioned milk bottle. The suggestion is that this is real milk like it used to be – pure nature. You can superimpose these pictures on any surface, on any object.

Given that we now hold nature as the embodiment of all that is good and pure, even with redemptive power, this image is ultimately best suited as a marketing tool. At the American supermarket chain Whole Foods, the Valhalla of all that is organic, I recently bought a bar of soap. $5.99 plus tax. Not cheap, but with this Sierra Cedar *wildcrafted* soap (whatever that may mean) you do have a piece of real nature in your very hands. And 10% percent of that $5.99 plus tax goes to defending that wilderness – by who, it doesn't say.

The accompanying text on the recyclable brown chipboard box is overwhelming proof of the marketing power of the *buy-good-feel-good* principle. "California's Sierra Nevada, the deep heart of wilderness, has been the fuel of inspiration for generations of wilderness activists. This soap smells just like the Sierra because it is made from the spicy leaves of the cedar which blankets the mountains' western flanks. This is the real smell of a warm summer evening along the Tuolumne River or of stumbling out of your sleeping bag in the middle of the night, with the impossibly cold, wood-smoke Sierra air lighting up your senses." In the meantime I am choking for air in my study, that's how strong the spicy Sierra cedar and smoke smell is that emanates from the soap in the box. But this way I always have a symbolic bit of wilderness at hand, *fresh from the mountains*.

We project onto nature our yearning for that which is larger than ourselves, larger than life, something that is not subject to the latest hype, fad and fashion.

Americans are masters at commoditizing and marketing a concept – and for most urban dwellers, nature is mostly a concept. In her book *Flight Maps: Adventures with Nature in Modern America* Jennifer Price writes with wry humor and surprise about the Nature Company that is found all over the country in (indeed) shopping malls. It was founded in 1972 when the environmental movement was budding by a couple from Berkeley for, "a population that in increasing numbers goes into the wilderness." But that makes you wonder why the people are in the mall buying 'nature' rather than going outside to actually enjoy it. Perhaps real nature is scary? Unpredictable? Far away? A visit to nature in the mall is certainly easier to fit into your Saturday afternoon errands; the experience is predictable but still makes you feel good, especially if you go home afterwards and put on your CD with whale sounds and get your kids started with their new butterfly-raise-and-release kit.

Idealizing a distant wilderness too often means not idealizing the environment in which we actually live, the landscape that for better or worse we call home.

Somehow, Price writes with a feeling of slight discomfort, the Nature Company undermines precisely the authenticity of the experience that it is trying to promote. (Or maybe it's not promoting it at all, but primarily sees a really great market in our yearning for wildness and wilderness and our feeling of guilt about having lost it and/or destroyed it). The people at Nature Company have us all figured out, as we can tell by their mission statement: "Authenticity and knowledge are balanced with sufficient humor to give our customers an experience which makes them feel good about themselves and world in which they live."

Where does our fascination with wilderness come from – a fascination which we occasionally express in real time but usually takes the form of an image, a product. Not that long ago wilderness was a threat, a vague but dangerous place one did well to avoid. The moors of the Dutch province of Drenthe, now domesticated into the Florida of the Netherlands, well deserved their name of *woeste gronden*, wild grounds. It was dangerous there, you would get lost and disappear and other than the sheep that grazed there no one would ever find you again. After all, doesn't Exodus say that Christ fought with the demons in the wilderness? Now, however, wilderness has transformed into a kind of beckoning perspective, untouched by the banalities of our daily life such as suburbs, traffic jams, office parks and commerce – a buzzword for the unsullied. In America, wilderness symbolizes the founding myth of the nation, part and parcel of the epic struggle of the (white) inhabitants of the New World to make the land their own, to prove that they really deserved it. Wilderness then, is a cultural, i.e. human creation, a mirror in which we think to see nature but onto which we in fact project our own unfulfilled longings.

"The irony, of course, was that in the process wilderness came to reflect the very civilization its devotees sought to escape," writes William Cronon in the book *Uncommon Ground: Rethinking the Human Place in Nature*. Blinded by our self-created idealized image of wilderness, we disdain and ignore our daily surroundings. "Idealizing a distant wilderness too often means not idealizing the environment in which we actually live, the landscape that for better or worse we call home. Most of our most serious environmental problems start right here, at home, and if we are to solve those problems, we need an environmental ethic that will tell us as much about using nature as about not using it."

Perhaps real nature is scary? Unpredictable? Far away?

The paradox according to Cronon: in our romanticized view of nature and our image of the wild, the human stands completely separate from the natural. That has horrendous implications, namely that our presence in nature by definition implies the downfall of nature. "The place where we are is the place where nature is not." If there is no place for mankind in the concept of wilderness that we have created, writes Cronon, then the wilderness can never offer a solution for the struggle between us and our man-made surroundings. Our 'love' of nature actually sets man and nature against one another.

The Dutch fascination with wilderness is different. The adulation for a now largely man-made nature has two sources: the awareness that we live in an artificial landscape, and the guilt we feel about the way we have dominated nature in order to create that landscape of suburbs, traffic jams and office parks. We are more than willing to give ourselves over to a *suspension of disbelief*, to a collective mirage in which we, in all seriousness, agree that a climbing wall in a business park is the Matterhorn, that a trash dump covered with snow is a ski resort, that a soggy corner of a pasture is nature, that a painted expedition hall is a tropical beach or a flower field.

We give ourselves over to a suspension of disbelief, to a collective mirage in which we agree that a climbing wall in a business park is the Matterhorn

With an almost childlike enthusiasm we go along with every new illusion, in every new (re)incarnation of that landscape. The true celebration of the Dutch landscape was the pavilion that architecture practice MVRDV designed for the 2000 World Expo in Hannover, where all the various landscape typologies were piled on top of the other – a cross section and a portrait of the Netherlands at the same time. Here 'new nature' is made. Dutch-style wilderness is slotted in between a train track, a housing development, a road and a military shooting range.

Agreement = feeling + image. It's a surefire combination. Let's call it a collective delusion. It's all a matter of convention and mutual agreement. Not just in the arts, but apparently also in nature the replica and the spin-off are now just as good as, if not better than, the original, if only because they are more easily controlled.

Nature is made to order, cut to size, tailored to fit the needs of the hasty mobile consumer. A nature of convenience, a green camping mattress that we can roll out wherever it suits us. Shortly after the tsunami in Asia the Dutch literary critic Arnold Heumakers wrote: "It is hard to deny that western man (and thanks to globalization this no longer applies only to western man) has succeeded better and better at humanizing nature, at adapting it to his own needs. Our most intimate interactions with nature all pass through the filter of culture. Our nature consists of pets, milk from a bottle or a carton, meat that is pre-cut to a convenient size, the mowing of the lawn, a stroll in a park or a reserve, a vacation in a tropical paradise."

So what is the next nature? It is no longer exclusively a physical place that you have to actually go to in order to experience it. The next nature comes to you, and you can take it with you anywhere, on your mobile phone with multi-thousand pixel color screen, on your laptop or your DVD-player, under the shower. *Nature is going virtual*. It can be summoned at any moment and it has a handy on-off button. That image and that feeling – those are the next nature, and they are just as mobile as you are.

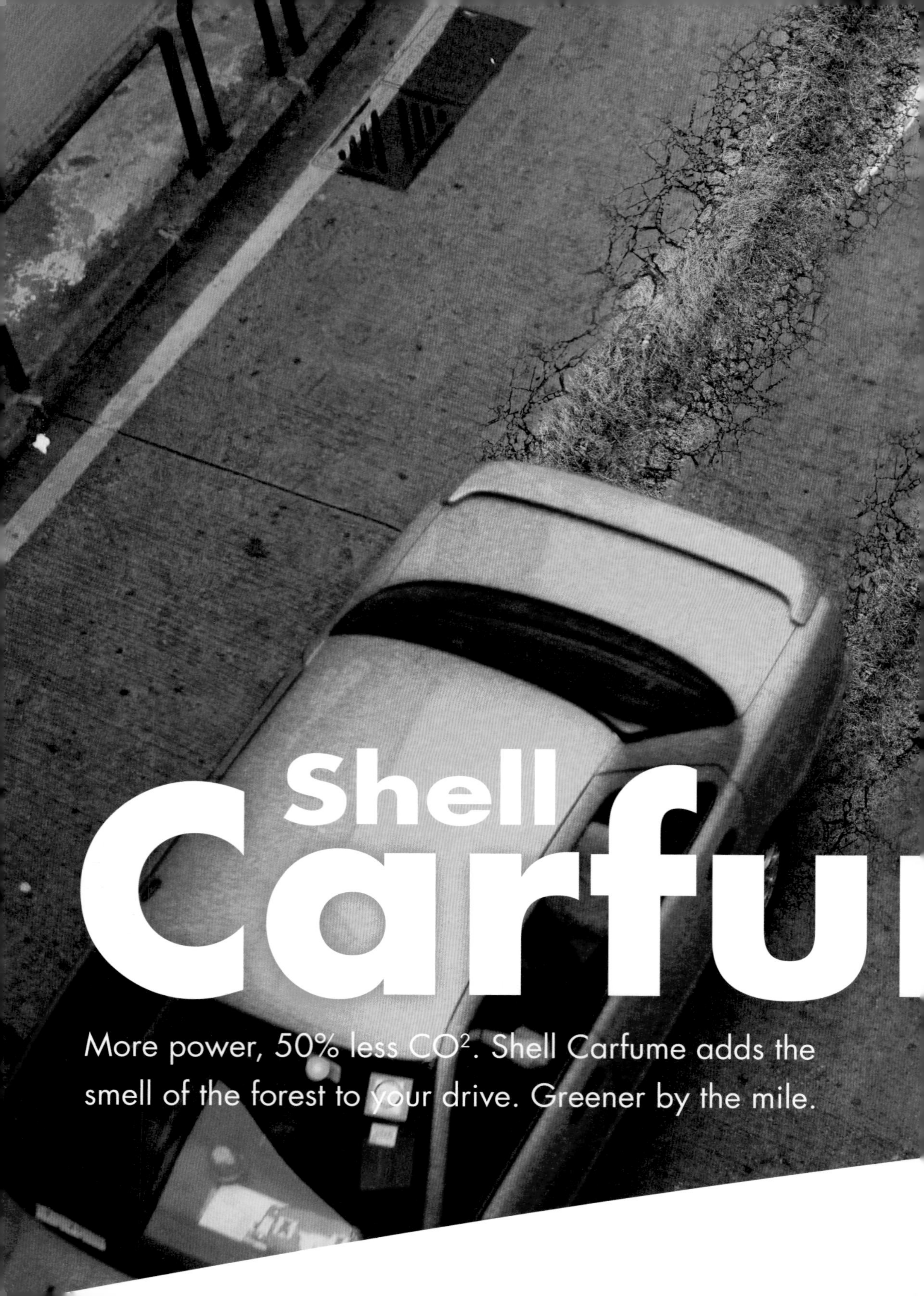
Shell
Carfu
More power, 50% less CO². Shell Carfume adds the
smell of the forest to your drive. Greener by the mile.

me

Who determines our Image of Nature?

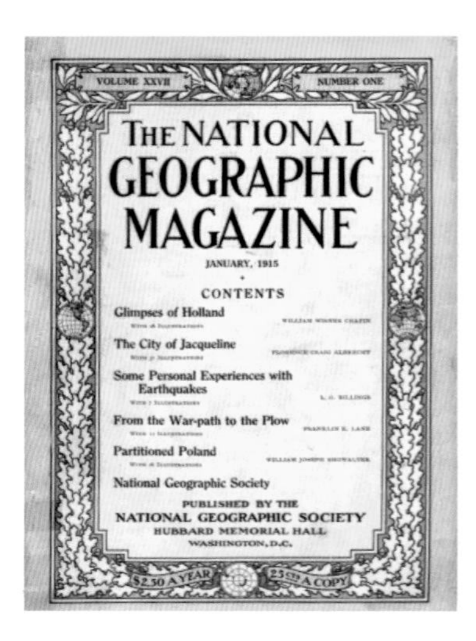

EARLY 20TH CENTURY: *THE NATIONAL GEOGRAPHIC MAGAZINE*, A VERY MUCH TEXT-BASED SCIENTIFIC JOURNAL. 21ST CENTURY: *NATIONAL GEOGRAPHIC ADVENTURE*, FULL-COLOR NATURE – FIT FOR PRINT

BBC *WEIRD NATURE*: SCIENCE MIGHT BE STRANGER THAN MYTH, THE LATTER IS SOMETIMES STRONGER

The landscape is no longer a place where trees grow and wildlife nibbles on the grass. Human culture has invested images of nature with deep symbolic value. Nature, and those who escape to it, are presented as more heroic, spiritual, and authentic than their civilized counterparts.

Nature can be depicted as red in tooth and claw, too mighty to be tamed, a universe entirely removed from the human realm. On the flip side, nature can be seen as delicate, a scarce commodity that is always retreating but never entirely disappearing. These contradictory images of nature saturate popular culture, from magazine ads to documentaries. Environmentalists, filmmakers, and advertisers have turned the landscape into the cornerstone of their brands. Our contemporary visual representations of nature take their cures from a long history of landscape art. In Tang Dynasty China, the landscape became a distinct genre in itself. In Europe, the Golden Age of painting in the Netherlands popularized scenes of picturesque ruins and billowing clouds for the rest of the continent. Visual nature-appreciation gained new vitality as a reaction to the Industrial Revolution. The construction of what poet William Blake called "satanic mills" emphasized the divide between mechanized civilization and unsullied nature. During the Romantic period, artists painted landscapes that were imbued with emotional and philosophical subtext. Their images of nature stood for purity and innocence, for rebellion against society, or for the indifference of the cosmos – all tropes that still underpin our current visual culture of 'nature.' Sean Penn's *Into the Wild* (2007) may have come nearly 200 years after Caspar David Friedrich's *Wanderer Above*

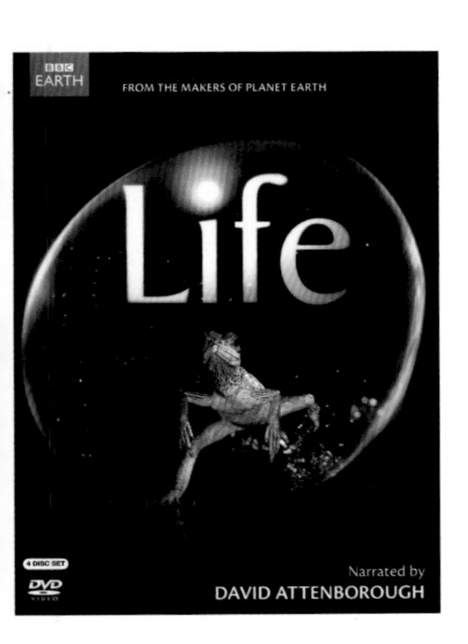

LIFE BY BBC EARTH: NATURE AT ITS HIGHEST RESOLUTION

DISCOVERY CHANNEL'S *SHARK WEEK:* WATCHING DANGEROUS SEA CREATURES ON TV BEING ADVERTISED AS AN 80'S TEENAGE CULT MOVIE

the Sea of Fog (1818), but both present a consistent image of the lone man venturing into exhilarating, dangerous nature. The notion of the wilderness as disappearing, finite, and in need of protection has become an important, even unavoidable aspect of the image of nature in popular culture. 19th century Romantic landscape arts went hand-in-hand with a nascent environmentalist movement.

National Geographic Magazine was instrumental in bringing the aesthetic of conservation to the masses. The 1906 issue featured its first wildlife photography, a decision that proved wildly popular with readers, and marked the magazine's transition from a text-based to an image-based publication. National Geographic's conservationist ethos helped to define nature as a exotic space removed from human incursion, but also easy prey for chainsaws, poachers, and pollution. The magazine's merging of field science and stunning imagery has influenced more than a hundred years of visually rich documentaries from the BBC's *Planet Earth* series to the Discovery Channel's *Shark Week*.

Advertisers, activists, and Hollywood directors all take their cues from an artistic tradition that presents nature not merely as a background for human endeavor, but as a visual language in its own right. The eye of the image maker frames the scene to convey a message, turning icebergs into postcards and forests into billboards. Depending on the context, the image of a snow-capped mountain can be a sign of global warming, a reason to buy hiking gear, or a summit to climb to conquer your fears. The less we live in nature, the more we live in an image of it. | AG

Preservers' Nature

Anthropomorphic Nature

he notion that 'we should save nature' ropagated by countless nature conservation rganizations, inherently imposes an image f nature as a weak and threatened entity hat has to be rescued by people. Although e can all agree that we should cherish iversity and a balanced relationship with ur surroundings, the view of man as the overnor – or even the savior – of 'nature' is t best naïve, undoubtedly paternalistic and rguably arrogant. Are we so self-centered ot to realize that, when considered from he perspective of the entire universe with s infinite amount of planes, supernovas

speck on the crust of one planet?

Why is it that mostly cute animals like pandas, seals, and polar bears have to be rescued from extinction, while at the same time we put extensive effort into extinguishing things like malaria-carrying mosquitoes and HIV? Presumably there is some projection at play and is the real issue not to save nature, but to save ourselves. Nature as a whole wouldn't really care if the living conditions on Earth would be radically altered and become inhabitable for people – evolution would continue nonetheless and for all we know a world without people would have room for lots of

Nature as Something Lost

Although the preservation of nature is typically seen as a progressive and leftish activity – perhaps because of the many clashes with economical powers – it is by definition also deeply conservative. Promoters want to 'conserve' nature. It is their core business to emphasize nature as a declining, weakening and threatened entity. Although nature is associated with primordially, it only appears once it disappears: Nature as a retro effect.

Nature as Commodity

Organizations like the WWF and Greenpeace routinely present nature as a commodity that has become increasingly scarce and will be used up altogether if we don't act. Although this depiction intuitively makes sense, it is in conflict with the idea of nature as pristine and untouched, which is promoted by the same organizations. How is it possible to perceive nature as untouched and consumed at the same time? Arguably, the campaigns have to be interpreted as a critique on corporations and individuals who use up environmental value as a commodity. From that perspective things makes sense again. But there is more. Possibly the tension between nature as untouched versus nature as a commodity, lies at the very root of our environmental crisis.

Exactly the romantic desire to perceive our natural environment as pristine, untouched and undefined, makes it incredibly easy to consume: No ownership, no responsibility. Should we, in order to save the environment, start to define it as in terms of value, ownership, e.g. as a commodity?

POLAR ICE FOR CONSUMPTION?

WWF ADVERTISEMENTS CREATED IN THE IMAGE OF MAN

Google Nature

Imagine you are an intelligent alien from outer space that has just landed on Earth. Before you can mingle with the earthlings you'd need to learn their language. It seemed like a smart idea to start at Google image search. Just type in a word and you'll immediately get a collage of images that show you what it means (by

the way: this is also a helpful tip for the more visually oriented humans among us). Let's start for example with the word dandelion. That teaches you a lot about the different phases of this flower and how it propagates! So far so good, but things are rapidly getting more bizarre. For instance when you try the Beetle, or the Puma: both somewhat confusing. The Lion seems to be fine, that 's the meat-eating animal that jumps on other animals. Jaguar seems to be more schizophrenic again. Better avoid the Apple. The Blackberry, however, is peculiar. It's not certain where they grow. |KVM

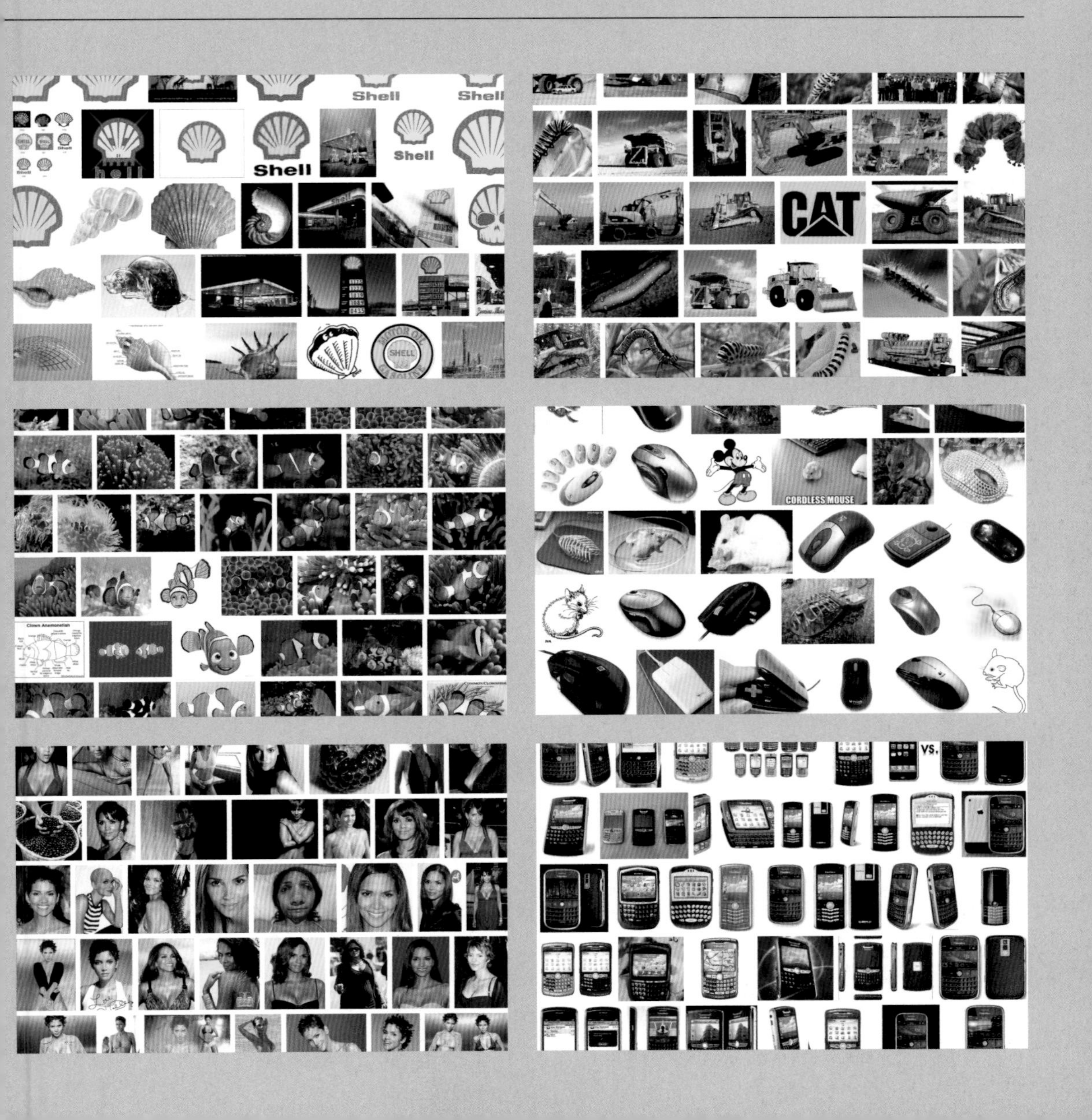

Corporate Nature

Cigarette Country

Tobacco doesn't grow wild in Texas. Cigarettes have no direct tie to the American West. Yet Marlboro has made their brand synonymous with a mythic cowboy lifestyle. A cowherd can drive cattle anywhere, but the cowboy exists only in relation to the American frontier. The Marlboro Man might be the brand's most famous persona, but without Marlboro Country, he would just be a smoker with a thing for horses.

The Marlboro Man and naturalists like Henry David Thoreau have different ends but twin means. The corporate dialogue is one of residence within nature, not dominion over it. The Man does not need to conquer the West because he is a natural extension of it. Only when the lifestyle is the landscape can a cigarette become a vital accessory for outdoor life. AG

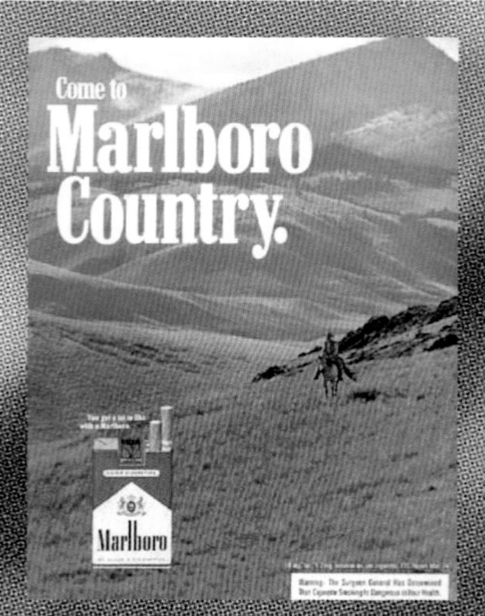

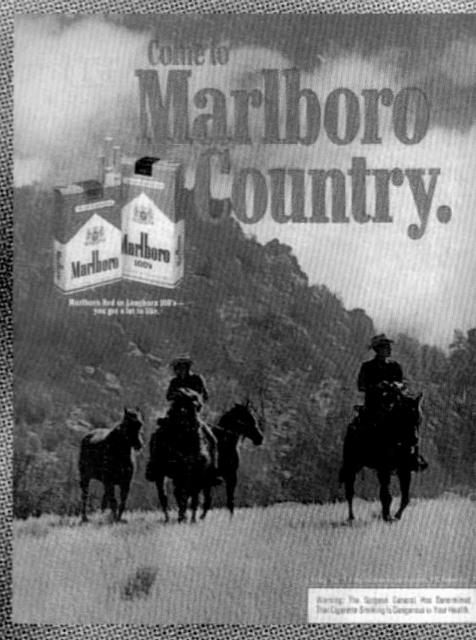

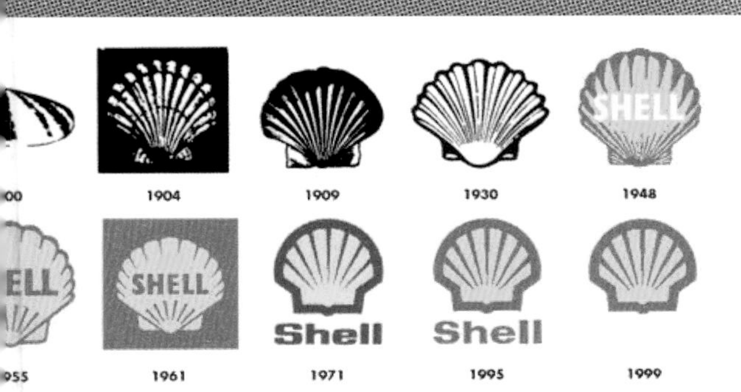

Logo Evolution

Corporate logos constantly have to adapt to survive. Ironically, in the case of the global oil company Shell the result is a logo reminiscent of images from our biology class. The first use of the shell as a logo was in 1891 and stems from the early days of the company. Before oil trading became its core business, Shell primarily shipped antiques, curios and oriental seashells from the Far East. | KVM

Never stop exploring ways to exploit nature

A mountain might be an outcropping of rock and ice, but to The North Face, it is gold. For a brand that's just as focused on subway riders as on climbers, The North Face uses mountains as symbols for the exceptional. The North Face consumer is one of a privileged few able to ascend, literally and figuratively, above the average crowds who can't quite make it past base camp. The mountain is a heroic individual, and so are you.

The North Face sells boots and trades in alpine imagery. The company is named after the 'north face' of a mountain; the logo itself is a stylization of Half Dome in Yosemite. Despite the rugged peaks, corporate nature is strangely tame. The North Face landscape is where you go for athletic self-actualization, not where you go to die of exposure.

The tag line 'never stop exploring' implies that unspoiled and uncharted nature exists for the right band of self-sufficient mountaineers. Nature still remains for the discovering, so long as you're wearing a North Face jacket. But, how convincing do we really find this notion, especially after Mount Everest has been trampled by 2,000 climbers? | AG

Hollywood Nature

Romantic Nature

From the savannah affairs of *Out of Africa* to the coming-out western *Brokeback Mountain*, scenic landscapes serve as a metaphor for romance. Hollywood nature justifies romantic passions repressed in civilization and echoes the raw, untamed personalities of the lovers. The closet door leads to the wilderness.

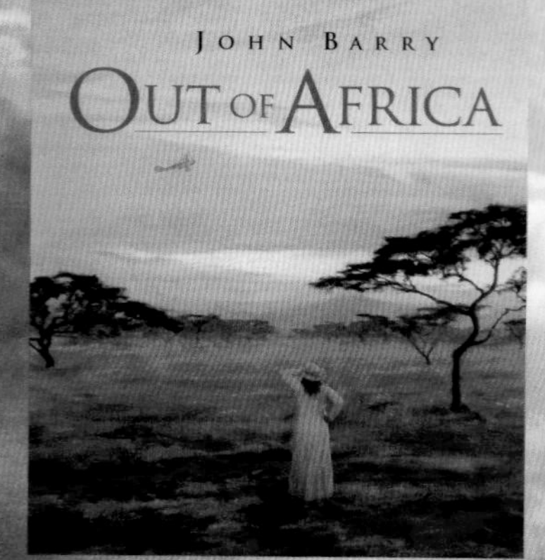

OUT OF AFRICA (SYDNEY POLLACK, 1985)

BROKEBACK MOUNTAIN (ANG LEE, 2005)

AVATAR (JAMES CAMERON, 2009)

Mythical Nature

Derived from mythological animistic traditions, fantasy films turn the natural backdrop into a lead role. In *The Lord of the Rings* trilogy (2001 - 2003), CGI glens and New Zealand's sweeping mountains set the tone of the vast epic. As in Tolkien's books, the landscape itself becomes conscious. Along with sentient beasts, even the trees can pull up their roots and join the action.

Escapist Nature

Into The Wild depicts nature as a place to escape into, where you can find yourself by losing everything else. Here, 'the wild' represents freedom from the messiness of human relationships and the existentially pointless task of making a living in a consumer society. The Alaskan landscape makes the protagonist somehow more authentic than the mainstream family he left behind.

INTO THE WILD (SEAN PENN, 2007)

Imaginature

The mountains in *Avatar* float in computer-generated glory over the bioluminescent rainforest of the moon Pandora. The native people, a blue-hued humanoid species, live in harmony with the planet's flora and fauna thanks to an extension of their brainstem that allows them to telepathically plug into nature. They are spiritual, simple, and half-naked: the ideal Western concept of tribal life. Mankind comes blustering into the scene as part of an industrial-military complex intent on mining 'unobtanium' and blowing up the biggest tree in the universe. The central conflict reflects director James Cameron's fears of unstoppable environmental destruction. In the film, Earth still exists, but as an ecological wasteland. Yet *Avatar* may have trumped its purpose as an environmental parable. Even the most impressive terrestrial vistas seem disappointing after two hours of high-def dragons and glowing plants. | AG

YOUR PIECE OF NATURAL HISTORY

MYPOLARICE.COM

TOGETHER WITH LOCAL INHABITANTS, MYPOLARICE HARVESTS THE BEST ICE LUMPS FROM THE GREENLANDIC ICE CAP. A SOUVENIR FROM THE LAST ICE AGE TO PRESERVE IN YOUR FRIDGE.

WE ARE LIVING IN POSTCARD NATURE

Under the Beach, the Pavement!

City Surrealism

During the riots of 1968, as students in Paris ripped up paving stones and threw them at police, one of the rallying cries was "sous les pavés, la plage" (under the pavement lies the beach).

The beach – the incarnation of a natural, undesignated and non-utilitarian space – was the opposite of the street, a historic relic of a designated, oppressive environment based on private property. Since then, policymakers have learned to better comply with the needs of the public – well, sort of. In various cities throughout the world, every summer a temporary artificial beach is created on the pavement. In one year alone the local government in Mexico City created nine artificial beaches, mostly in poorer parts of the city. In general, the camouflaging of technological infrastructure with 'nature' has proven a successful strategy to provide people with a seemingly more pleasant and acceptable living environment.

LIFE'S A BEACH IN NOTTINGHAM, UNITED KINGDOM (PHOTO: PHOTO BY DARREN STAPLES, © REUTERS/NOVUM)

CITY BEACHES IN (CLOCKWISE) BRATISLAVA (SLOVAKIA), PARIS (FRANCE), BERLIN (GERMANY) AND IN THE NETHERLANDS

UMTS MASTS DISGUISED AS TREES: AS LONG AS THEY MEET OUR IMAGE OF NATURE, WE ARE WILLING TO ACCEPT THEM

The Rise of Cellphone Antenna Trees

They used to be quite rare but over the last few years they have become a dominant species. They especially thrive in nature resorts and pristine or protected landscapes. When the public protests against a telecom company or local government's intention to place a cellphone antenna mast at some location where it would spoil the landscape, consensus is typically reached by disguising the mast as a tree. The rise of cellphone-antenna trees teaches us two things. On the one hand it shows our longing for a natural environment is only skin deep. Apparently it doesn't matter that the tree is made of metal and plastic, as long as it meets our image of nature, we are willing to accept it. It also shows how successful technology makes itself invisible: it becomes part of the fabric of everyday life where you no longer recognize it as technology.

COVER YOUR WASTEBIN UNDER A PLASTIC ROCK

IN CHINA, A PLASTIC TREE STUMP APPARENTLY LOOKS BETTER THAN A TRASH CAN OR A WELL-DESIGNED LID

THIS DUTCH 'KLIKO' WASTEBIN IS DRESSED UP WITH DECORATIVE FLOWER PATTERNS

Nature as Wallpaper

Somehow it provides us with a sense of comfort to drape outlandish phenomena with images of things we are intuitively more accustomed to. The cellphone trees and city beaches show the success of this strategy for policy makers. Also on an individual level, people feel the need to camouflage infrastructure with images of nature. Don't ask why, but somehow it just feels better to have a big rock in your backyard than a trashcan.

FLOWERBEDS IN THE MIDDLE OF THE DESERT. TAKEN IN DUBAI, FROM THE *URBAN DAILY LIFE* SERIES BY REINEKE OTTEN

Camouflaging Irrigation Infrastructure

Sometimes, we even build entire infrastructures for the sole purpose of having a sense of nature in our surroundings. Sophisticated watering systems are installed to ensure the survival of flowers at a location where they would not grow naturally. Luckily in this case the flowers quite elegantly cover up the watering infrastructure so that our illusion of a natural environment remains intact and everyone is happy. As in the end, an untouched illusion of nature is what we care about the most.

ACADEMY ANIMALS BY RICHARD BARNES

Hacking the Zoo

With over 50% of the world's population living in cities, a visit to the zoo is for many people the only opportunity to have contact with live animals.

Unlike the natural history museum, the zoo allows its visitors not merely to see, but also hear, smell and communicate with the exhibits; a favorite summer past time for young and old. It's a place to experience, appreciate and learn about the variety of fellow species inhabiting our planet. In the modern zoo, animals are typically presented in confinements that aim to simulate their natural habitats. It becomes increasingly apparent, however, that these zoo dwellings depict an idealized untouched nature, which hardly exists anymore. Hence the Vienna Zoo in Austria, the oldest existing modern zoo which opened as an imperial menagerie in 1752, decided it was time to add some realism to the habitats of the animals: junk.

In their temporary installation called *Trouble in Paradise* artists Christoph Steinbrener and Rainer Dempf created a new 'natural' context in which to view the animals. In one enclosure, they half-submerged a car in a watering hole used by the resident rhinos. In another enclosure, penguins frolic in the shadow of an oil pump, and in yet another, alligators had to share their modest bayou with a bathtub and a truck tire. The artists hope their intervention forces the viewer to reconsider their idyllic vision of what a natural animal habitat looks like and realize the impact man has had on the environment.

TROUBLE IN PARADISE BY CHRISTOPHER STEINBRENER AND RAINER DEMPF (VIENNA ZOO, AUSTRIA)

A realism of a very different kind can be found at Artis, the Amsterdam City Zoo which besides the extensive collection of animals from around the planet, also houses some local wild species on its premises who immigrated to the zoo of their own accord. The blue heron, an inhabitant of Amsterdam, enjoys the housing conditions at the zoo so much that a whole colony has set up home there. The birds are intruding and squatting in residences of legally immigrated zoo animals, enjoying a good seafood dinner along with the penguins, flamingos and sea lions. Although the government granted the zoo a permit to remove the heron nests, a Dutch fauna protection organization objected on the grounds that the zoo is too intolerant of the wild locals. While this poses questions about the concept of the zoo as a place to present wild life and the dispute over the illegal immigrants continues, the animals in the zoo have already found ways to deal with the birds. According to a zoo spokesman, almost every week a young heron that is learning to fly inadvertently lands on the wrong spot and disappears into the jaws of a lion. Nature is cruel.

INTRUDING HERONS IN ARTIS ZOO, AMSTERDAM

THE BERG BY ARCHITECT JAKOB TIGGES

Man-made Mountain

So you've seen the peak of Mount Everest on tour? Descended the bobsled ride of the faux Matterhorn in Disneyland? Think you've seen it all? Now come and see The Berg in Berlin!

German architect Jakob Tigges explores the outskirts of megalomania with his proposed plan to construct a 1,000-meter tall mountain at the site of the former Tempelhof airport in Berlin – which was refurbished by the Nazis as part of their Germania plan. If realized, 'The Berg' would be the largest man-made icon. A tourist attraction unlike any city has ever served up, providing Berliners and (more importantly) tourists with a convenient location to enjoy a range of activities including hiking, hang gliding, rock climbing and even skiing, as the mountain would collect snow on its peak from September to March offering the perfect winter sport climate in the otherwise slope-less city. The plans for The Berg seem to have spawned out of a severe case of 'peak-envy'. On The Berg's website Tigges wrote: "While big and wealthy cities in many parts of the world challenge the limits of possibility by building gigantic hotels with fancy shapes, erecting sky-high office towers or constructing hovering philharmonic temples, Berlin sets up a decent mountain. Hamburg, as stiff as flat, turns green with envy, rich-and-once-proud Munich starts to feel ashamed of its distant Alp-panorama and planners of the Middle-East, experienced in taking the spell off any kind of architectural utopia immediately design authentic copies of the iconic Berlin-Mountain." Whether the world is gullible or people truly want to see and experience The Berg, the project attracted a lot of

COME SEE THE BERG!

IF REALIZED, *THE BERG* WOULD BE THE LARGEST MAN-MADE ICON. A TOURIST ATTRACTION UNLIKE ANY CITY HAS EVER HAD, PROVIDING BERLINERS AND (MORE IMPORTANTLY) TOURISTS WITH A CONVENIENT LOCATION TO ENJOY A RANGE OF ACTIVITIES INCLUDING HIKING, HANG GLIDING, ROCK CLIMBING AND EVEN SKIING, AS THE MOUNTAIN WOULD COLLECT SNOW ON ITS PEAK FROM SEPTEMBER TO MARCH OFFERING THE PERFECT WINTER SPORT CLIMATE IN THE OTHERWISE SLOPE-LESS CITY.

local media, gathered more than 7000 followers on Facebook and has some promising product endorsements already. Although an uninhabitable monolith of this magnitude might look appealing at first sight, funding for it might be another matter. Not to mention the environmental impact of the gigantic structure. The mountain is so big it would alter the weather surrounding it and attract a wide range of flora and fauna. Nonetheless Berliners are getting behind the project as another tourist-attraction (read: money-making) option for their fair city.

"It's provocative, but not constructive," Tigges told *Der Spiegel* of his proposal. The architect sees his idea as more of a placeholder in the minds of Berliners, a mythical mountain to fire imaginations until an appropriately grand solution is found. In the meantime, Tigges says, he would prefer for Tempelhof to remain untouched as he considers it more interesting for a Sunday walk than your average park landscape. "Tourists would come to the site to take photographs of the mountain that isn't there," said Tigges, who noted that his euphoric mountain renderings serve as a direct critique of the city of Berlin. "The site is much too valuable to sacrifice for mediocre apartment buildings."

Time will tell if Berlin will take on the monumental task of constructing The Berg or leave the plan as a wild speculation. Although the project, if ever realized, would certainly be the largest urban mountain, it would not be the first. Already in the fourteenth century, a monumental artificial mountain was constructed in the Forbidden City of Beijing, China. The creation of 'Prospect Hill' (Jingshan in Chinese) was inspired by Feng Shui principles that dictate it is favorable to site a residence to the south of a nearby hill – it is also practical, gaining protection from cold northern winds. The imperial palaces in the other capitals of previous dynasties were situated to the south of a hill. When the capital was moved to Beijing, no such hill existed at this location, so one had to be built (obviously). The artificial mountain was entirely constructed through manual labor, hence it can be called a truly man-made mountain. The structure is 45.7 meters tall, covers an area of more then 230,000 m² and is visited by thousands of tourists every year. | KVM

JINGSHAN IN BEIJING, SEEN FROM THE NORTH SIDE

Mount Everest on Tour

Mount Everest is the highest mountain on Earth, as measured by the height of its summit above sea level. In May 2005, artist Xu Zhen (known as the maverick of the Chinese art world) claimed to have removed the summit of the mountain, reducing its height by 186cm (Xu Zhen's own height), although this was later proven to be a hoax. The 'summit' of Mount Everest has been touring art exhibits throughout the world ever since.

The British claimed in 1856 that the summit of the Mount Everest was 8,840 meters in height. Various official and independent surveys since have consistently shown that Everest is not as high as had been thought, pointing, perhaps, to evidence of global warming, or a shift in the tectonic plates, though its cause still remains unproven.

ROCKY MESSAGES
THIS SMS-STYLE MESSAGE SCRIBBLED ONTO THE MOUNTAIN BY BELGIAN ARTIST WIM DELVOYE, MAKES US SO HUMBLY AWARE OF THE DISCREPANCY BETWEEN OUR EVERYDAY HUMAN EXPERIENCE AND SOME OF THE LARGER FORCES OF NATURE.

EXPEDITION EVEREST UNDER CONSTRUCTION

Expedition Everest

Expedition Everest is a Himalayan ride in a park stocked with Australian wallabies and African gorillas in subtropical America. Not actually modeled on Mount Everest, Forbidden Mountain joins five other artificial crags in Disney World, and is second highest elevation, real or fake, in the state of Florida.

While Disney's Animal Kingdom is expressly devoted to wildlife conservation, imaginary and actual nature are treated there as equal and interchangeable. In the conceptual space of the park, the animatronic Yeti that menaces the ride-goers is as real as the Komodo dragons and tigers lounging in the 'Asia' section. Forced perspective is used to simulate the appearance of Everest in the background, situating the fictional Forbidden Mountain in actual geographical space. The man-made mountain doesn't trouble visitors with deadly storms or worries of its painted-on glaciers melting. They don't even have to get out to hike. After a trip through the monster-menaced belly of the mountain, the ride ends in true Disney style, by depositing the riders in a gift shop. | AG

EXPEDITION EVEREST ON THE WEBSITE OF DISNEY WORLD

BRAVE NEW WINTER, BY BIANCA ELZENBAUMER AND FABIO FRANZ

BRAVE NEW WINTER, BY BIANCA ELZENBAUMER AND FABIO FRANZ (CONTINUED)

SPOT.
SHOOT.
SHARE.
T-Mobile NL
10:54
Spot!
Recent
Top 50
Cellphone antenna tree
by Koert van Mensvoort
N200
N200
Zeeweg
N200
Wethouder
van Gelukpark
1 / 50
Spotter
Posts
About
NEXT NATURE
SPOTTER FOR
IPHONE
DOWNLOAD THE NEXT NATURE
APP IN THE ITUNES APP STORE
AND SHARE YOUR NEXT NATURE
OBSERVATIONS WITH OTHER
EXPLORERS.

THE WOODS SMELL OF SHAMPOO

CAR FRESHENERS, BY HENDRIK-JAN GRIEVINK

Authentic Artificial Smell

Makers of artificial scents traditionally mimicked existing scents like pine tree, lavender, cherry, peach or strawberry. However, as the whole spectrum of old nature scents has been recreated and commoditized, they are now expanding their product range with the creation of scents that don't refer to an existing theme. New scents like 'Moonshine', 'Mondrian', 'Karma' or 'Nihilism' are artificial, yet at the same time authentic. They provide us with an olfactory experience of phenomena that formerly only existed in the visual or conceptual realm. Whereas with the pine tree forest and cherry scents the simulation was no more than an inferior derivative, it now becomes a cross-sensory creative force. | KVM

Car Freshener
New Car Scent
Car Freshener
Boeing 747
Car Freshener
Big Mac
Car Freshener
Forest Fresh
Car Freshener
Rainbow Rush
Car Freshener
Next Nature

Car Freshener
Nihilism

Car Freshener
Mondrian

Car Freshener
Karma

Highlanders in the Lowlands

At the end of every cold winter there is a debate in the Netherlands about whether the forestry service should feed the oxen, horses and deer grazing in the Dutch nature resorts. The official policy of the Dutch forestry service is to let the ecosystem manage itself, which causes the weaker animals – 24% of the population – to perish because of lack of food: a sight too natural for most 'nature' lovers.

In response to the protests, the initiators of the Dutch 'hands-off' landscape management argue that the protests of hikers, bikers and other tourists merely exemplify how alienated people have become from nature. However, are the premises of these policy makers really valid? Is it defendable to leave the animals in the hands of the elements or is this game getting out of hand?

CONTRADICTING POLICIES

ALTHOUGH THE OFFICIAL POLICY IN THE DUTCH REGENERATED NATURAL AREAS IS TO LET THE ECOSYSTEM MANAGE ITSELF, THE CATTLE DESTRUCTION LAW OBLIGES THE FORESTRY SERVICE TO REMOVE THE CORPSES OF THE HECK AUROCHS. LIKE WITH THE REMOVAL OF (DEAD) WOOD, IMPORTANT NUTRITIOUS ELEMENTS ARE WITHDRAWN FROM THE ECOSYSTEM. DIVERSE SCAVENGER COMMUNITY OF RAVENS, BUGS AND MOULDS ARE DEPENDENT ON THE PRESENCE OF LARGE CORPSES. THE ABSENCE OF THESE LARGE CORPSES IN THE LANDSCAPE IS THOUGHT TO ALREADY HAVE CAUSED THE WITHDRAWAL OF THE RED AND BLACK KITE AND RAVEN FROM THE DUTCH LANDSCAPE.

For the past few decades the policy for nature resorts in the Netherlands has been geared at regenerating the original landscape, as it existed in prehistoric times. In practice this means that land is gained from the ocean or bought from farmers and transformed into the landscape we think existed 8,000 years ago, long before people placed their footprint on it.

Although this policy has been rather successful and resulted in some beautiful, largely self-sustaining resorts, there are also drawbacks. One difficulty is that we don't know exactly what the landscape looked like in prehistoric times and have to make lots of educated guesses about it. Another issue that immediately emerged along with the desire to regenerate an ancient ecosystem is that some elements don't exist anymore. For instance the ancestor of domestic cattle, the aurochs (*Bos primigenius*), a type of large wild cattle that inhabited Europe, Asia and North Africa in prehistoric times, became extinct in 1627. The aurochs were far larger than most modern domestic cattle with a shoulder height of two meters and weight of 1,000 kilograms. Domestication occurred in several parts of the world at roughly the same time, about 8,000 years ago. The aurochs already appeared in Paleolithic European cave paintings – such as those found at Lascaux and Livernon in France – and were regarded as elusive prey. The main causes of extinction of the original aurochs population were hunting, a narrowing of habitat due to the development of farming, climatic changes and diseases transmitted by domestic cattle, which are so different in size and build that they are regarded as separate species. The last recorded live aurochs died from natural causes in 1627 in Poland.

RECREATION IN THE NETHERLANDS: LOWLANDER MEETS HIGHLAND COW

WHO'S THE TOURIST?

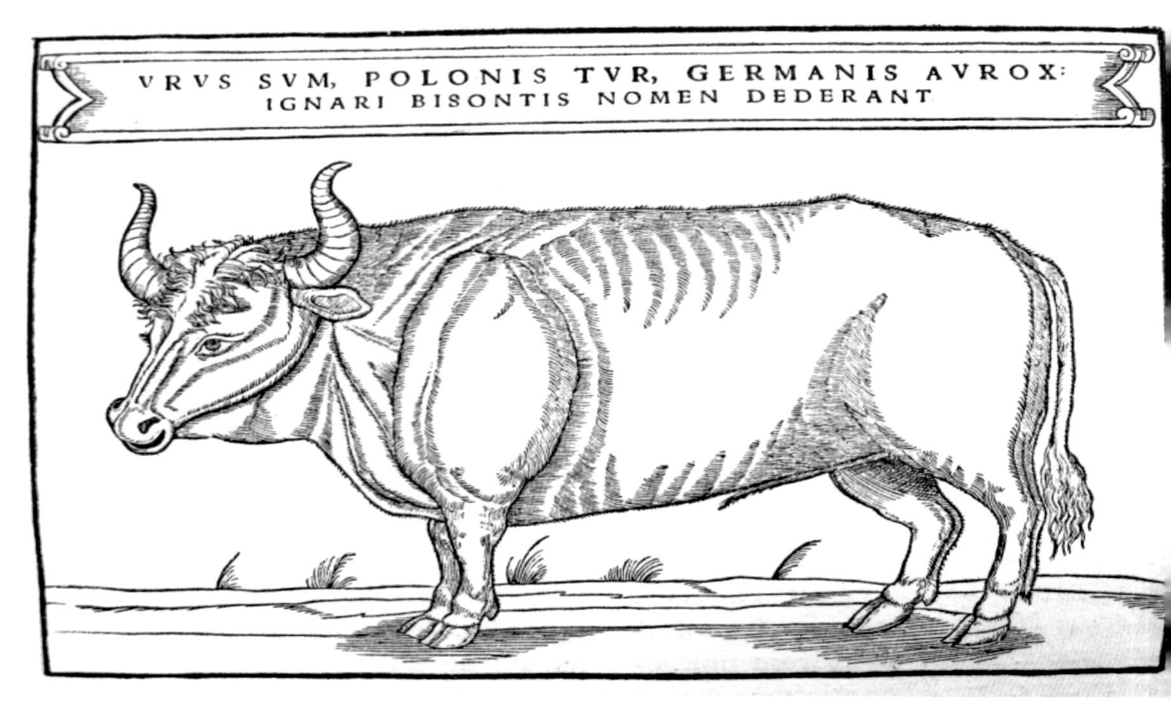

VRVS SVM, POLONIS TVR, GERMANIS AVROX:
IGNARI BISONTIS NOMEN DEDERANT

ILLUSTRATION OF AUROCHS PUBLISHED IN 1556 IN A BOOK BY SIGISMUND VON HERBERSTEIN.

DEPICTIONS OF AUROCHS IN *BREHM'S LIFE OF ANIMALS* (1927). ORIGINATOR: UNKNOWN.

Rebreeding The Extinct Aurochs

In 1920, some 300 years after the aurochs became extinct, the Heck brothers, two German biologists and zoo directors, attempted to recreate aurochs they knew from drawings and paintings. Their idea was that the genetic material of the aurochs was still present scattered over various domesticated animals and that it would be possible to rebreed the original extinct aurochs. The result was called Heck cattle, or reconstructed aurochs that resembled the original aurochs in their ability to live autonomously, although smaller in size. They number in the thousands in Europe today, however, due to their intolerance of humans, regenerated aurochs are less suitable for public nature resorts. At these locations the less aggressive Highland Cow is a stand-in for the extinct aurochs.

Nowadays, at countless locations in the Dutch landscape, Highland cattle appear to be just as strong and independent as the original aurochs – almost like accomplished actors – yet, contrary to the prehistoric aurochs, the Highland cow is a domesticated animal, dependent on people. While in its original setting in Scotland, Highland cattle were brought into the valleys in the wintertime where they could find some shelter in barns in the regenerated prehistoric landscape there is no place for such un-prehistoric shelters. Another difference with the prehistoric aurochs and their contemporary stand-ins is the huge difference in their living environment. In prehistoric times, perhaps a few hundred people lived in an area the size of the Netherlands, today this number is over 16 million, which significantly reduces the freedom of the animals to move around and find a suitable habitat. Arguably, the original aurochs would never have chosen the gated habitat their stand-ins are currently occupying. The absence of predators, like wolfs, only adds to the tragedy.

Regenerated Nature or a Recreational Simulation?

The key question of the debate should revolve around whether we should perceive these so-called nature reserves and the living animals they inhabit, as a truly natural environment, or if we could better consider them as recreational simulations of a prehistoric landscape, made for the pleasure and education of people. Considering the difficulty (read: impossibility) to fully recreate the prehistoric ecology and its inhabitants in such a relatively tiny area already crowded by people, one is inclined to answer the latter. Like the landscape paintings people hang in their houses, these regenerated natural resorts are merely copies of an old nature that is long gone and can never truly be recreated: change happens, evolution goes on. The best we can do is a re-enactment – a simulation of the landscape we believe existed in prehistoric times – and although the realism of this simulation is certainly more profound than in any given landscape painting, in the end it will always remain a derivative.

Regenerated nature or a recreational simulation?

Now don't get it wrong: the Dutch attempts at recreating prehistoric landscapes aren't necessarily bad policy. They are a form of cultural heritage that not only educates people in the history of the region, but also fulfills the need for green zones in an overly urbanized area – which is not only for the benefit of people, but also for wild birds that thankfully spend the winter there.

All things considered, the regenerated prehistoric landscapes of the Netherlands are best understood as theatre: an attempt to recreate the long-gone historic landscape and its actors on a contemporary stage. Like in a theatre play or a movie, we are willing to suspend our disbelief and linger in an imaginary story. Nonetheless, a theatre director should also consider what the actors are enduring for the sake of a good play. In this regard the horrifying and painful deaths of the starving Highland cattle and Heck Aurochs are a form of 'method acting' taken too far. | KVM

Hypernature

PATENTED NATURE
BY 2010, OVER 90% OF ALL SOY IN THE USA WAS GENETICALLY MODIFIED. COMPANIES SUCH AS MONSANTO HAVE PATENTED THE GENOMES OF THEIR CROPS, EXTENDING HUMAN OWNERSHIP TO AREAS PREVIOUSLY ONLY 'OWNED' BY EVOLUTION.

SLICKER, SAFER, MORE CONVENIENT
THE ORIGINAL BANANA IS FULL OF INEDIBLE SEEDS. AFTER MILLENNIA OF CULTIVATION, IT HAS BEEN SELECTED TO BE SWEET, SEEDLESS, AND EASY TO PEEL. DUE TO THE LOSS OF ITS SEEDS, HOWEVER, IT CANNOT PROPAGATE WITHOUT THE FARMER'S EFFORT.

BETTER THAN THE ORIGINAL
HYPERNATURE BRINGS US 'NATURAL' EXPERIENCES THAT COULD NOT EXIST WITHOUT THE HUMAN HAND, BUT CAN BE APPRECIATED NONETHELESS. ARE WE INTUITIVELY FORCED TO BELIEVE RAINBOW-DYED FLOWERS ARE MORE ADVANCED THAN THE ORIGINAL VERSION? NATURE NEVER GOES OUT OF STYLE, YET IT CAN EASILY BE TURNED INTO KITSCH.

The enhanced version of nature

Much of the so-called 'nature' in our lives has taken on an artificial authenticity. Engineered tomatoes are redder, rounder, and larger than the ones from our gardens. Domestic pets could not survive in the wild, but prosper by triggering our empathy. We have made fluorescent fish, rainbow tulips and botanical gardens that contain species from every corner of the globe. Human design has turned nature into hypernature, an exaggerated simulation of a nature that never existed. It's better than the original, a little bit prettier and slicker, safer and more convenient. Hypernature emerges where the born and the made meet. It presents itself as nature, yet arguably, it is culture in disguise.

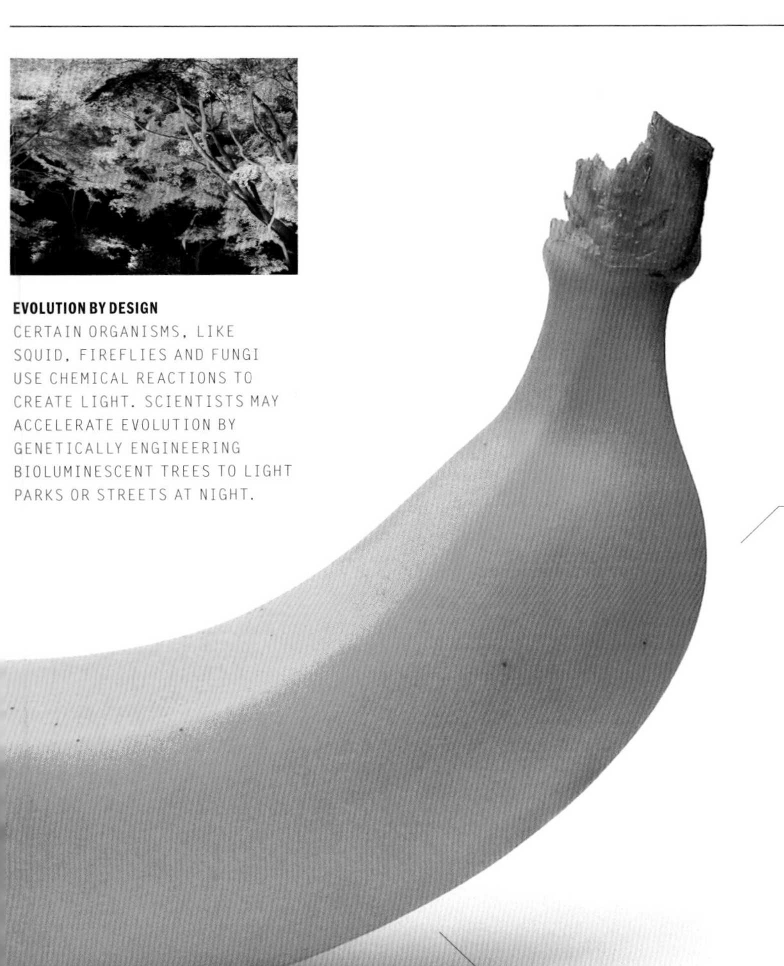

EVOLUTION BY DESIGN
CERTAIN ORGANISMS, LIKE SQUID, FIREFLIES AND FUNGI USE CHEMICAL REACTIONS TO CREATE LIGHT. SCIENTISTS MAY ACCELERATE EVOLUTION BY GENETICALLY ENGINEERING BIOLUMINESCENT TREES TO LIGHT PARKS OR STREETS AT NIGHT.

AESTHETIC SELECTION
THE PINEBERRY REVERSES THE USUAL WHITE-SEED RED-SKIN COLORATION OF A STRAWBERRY, AND HAS A SLIGHT PINEAPPLE FLAVOR. LOOKING GOOD IN THE SUPERMARKET IS NOW PART OF THE SURVIVAL OF THE FITTEST.

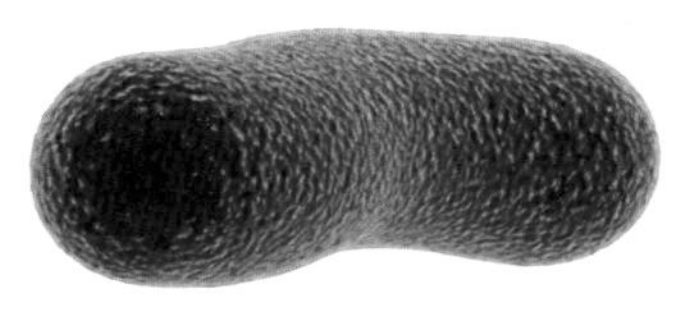

MASS DOMESTICATION
THERE IS A DIRECT LINE BETWEEN THE DOMESTICATION OF CATTLE 10.000 YEARS AGO AND TODAY'S DOMESTICATION OF MICROBES TO TURN WASTE INTO PETROL. SAME PRINCIPLE, SMALLER SCALE.

UPGRADED FOR YOUR CONVENIENCE
THE LABRADOODLE BRINGS US A FINE BLEND OF THE POODLE AND THE LABRADOR. COMBINING SOCIABILITY WITH BEAUTY, THEY ARE THE PERFECT HUMAN COMPANION. LABRADOODLES ARE HYPERNATURE. MOREOVER, THEIR APPARENT ARTIFICIALITY CONCEALS THE FACT THAT LABRADORS AND POODLES ARE HYPERNATURE NO LESS. THE DOGS DISTANT PAST AS WILD GRAY WOLVES IS AN ECHO LONG GONE.

AMPLIFIED & EXAGGERATED
THIS SALMON, GENETICALLY ENGINEERED BY THE COMPANY AQUABOUNTY, GROWS TO FULL SIZE IN HALF THE TIME IT TAKES A WILD SALMON. SUPPOSEDLY IDEAL FOR AQUACULTURE, THE GM SALMON REFLECTS OUR INTEREST IN GETTING FOOD CHEAPER, FASTER, AND EASIER.

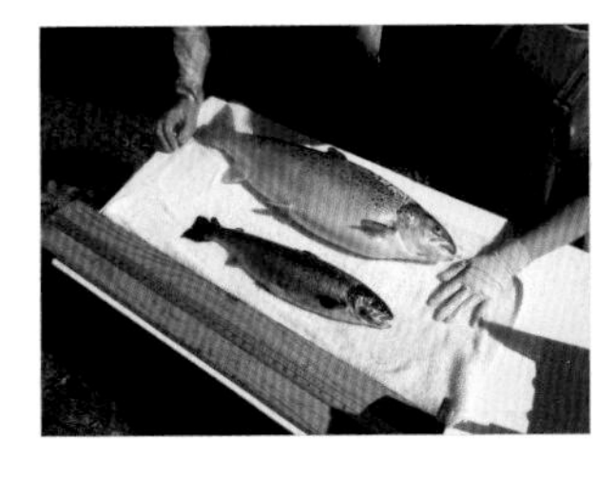

A SOCIETY OF SIMULATIONS

By Koert van Mensvoort

We live in a society where simulations are often more influential, satisfying and meaningful than the things they presumably represent. Media technologies play a fundamental role in our cycle of meaning construction. This is not necessarily a bad thing, nor is it entirely new. Yet, it has consequences for our concepts of virtual and real, which are less complementary than originally thought.

Before you read on, a personal anecdote from my youth: when I was a child, I thought the people I saw on TV were really living inside the box. I wondered where they went when the TV was turned off and I also remember worrying it would hurt the TV, when I switched it off. Obviously, I am a grown man now and I've long learned that the television is just a technological device, created to project distant images into viewers' living rooms and that those flickering people weren't actually living inside the cathode ray tube.

Now I return to my argument. Over the last century or so, the technological reproduction of images has grown explosively. Each of us is confronted with more images every day than a person living in the Middle Ages would have seen in their whole lifetime. If you open a 100-year-old newspaper you will be amazed by the volume of text and the absence of pictures. How different things are today: the moment you are born, covered in womb fluid, not yet dressed or showered, your parents are already there with the digital camera, ready to take your picture. And of course the pictures are instantly uploaded to the family website, where the whole world can watch and compare them with the medical ultrasound photographs already shared before you were born.

Visual Power

Images occupy an increasingly important place in our communication and transmission of information. More and more often, it is an image that is the deciding factor in important questions. Provocative logos, styles and icons are supposed to make us think we are connected to each other, or different from each other. Every schoolchild nowadays has to decide whether he or she is a skater, a jock, a preppie, or whatever. Going to school naked is not an option. But no matter which T-shirt you decide to wear, they are inescapably a social communication medium. Your T-shirt will be read as a statement, which your classmates will use to stereotype you.

I remember the strange feeling of recognition I had when visiting Paris for the first time and saw the Eiffel Tower. There it was, for real! I felt as if I was meeting a long-lost cousin. Of course, you take a snapshot to show you've been there: 'Me and the Eiffel Tower'. Thousands of people take this same picture every year. Every architect dreams of designing an icon like this. Today, exceptional architecture often wins prizes before the building is finished; their iconic quality is already recognized on the basis of computer models.

Picture this!

Does anyone still remember the days when a computer was a complex machine that could only be operated by a highly trained expert using obscure commands? Only when the graphical user interface (GUI) was introduced did computers become everyday appliances; suddenly anyone could use them. Today, all over the world, people from various cultures use the same icons, folders, buttons and trashcans. The GUI's success is owed less to the cute pictures than to the metaphor that makes the machine so accessible: the computer desktop as a version of the familiar, old-fashioned kind. This brings us to an important difference between pictures and *pictures* – it is indeed awkward that we use the same word for two different things. On the one hand, there are pictures we see with our eyes. On the other, there are mental pictures we have in our heads – pictures as in, "I'm trying to picture it".

Increasingly, we are coming to realize that 'thinking' is fundamentally connected to sensory experience. In *Metaphors We Live By*, Lakoff and Johnson[1] argue that human thought works in a fundamentally metaphorical way. Metaphors allow us to use physical and social experiences to understand countless other subjects. The world we live in has become so complex; we continuously search for mental imagery to help us understand things. Thus politicians speak in clear sound bites. Athletic shoe companies do not sell shoes, they sell image. Thoracic surgeons wander around in patients' lungs like rangers walking through the forest, courtesy of head-mounted virtual-reality displays.

You would expect that this surfeit of images would drown us. It is now difficult to deny that a certain visual inflation is present, and yet our irrepressible hunger for more persists. We humans, after all, are extremely visually oriented animals. From cave paintings to computers, the visual image has helped the human race to describe, classify, order, analyze and grow our understanding of the world around us.[2] Perhaps the most extraordinary thing about our visual culture[3] is not the number of pictures being produced but our deeply rooted need to visualize everything that could possibly be significant. Modern life amid visual media compels everyone and everything to strive for visibility.[4] The more visible something is, the more real it is, the more genuine.[5] Without images, there seems to be no reality.

Virtual for Real

When considering simulations, one almost immediately thinks of videogames. Nowadays, the game industry has grown bigger than the film industry and its visual language has become so accepted that it is almost beyond fiction. Virtual computer worlds are becoming increasingly 'real' and blended with our physical world. In some online role-playing games, aspiring participants have to write an application letter in order to be accepted to a certain group or guild. We still have to get used to the fact that you can earn an income with gaming nowadays[6], but how normal is it anyway, that at the bakery round the corner, you can trade a piece of paper – called money – for bread?

Most people would denounce spending too much time in virtual worlds, but which world should be called virtual then? Simply defining the virtual as opposite to physical is perhaps too simple. The word 'virtual' has different meanings that are often entangled and used without further consideration. Sometimes we use the word virtual to mean 'almost real,' while at other times we mean 'imaginary'. This disparity is almost never justified: fantasy and second rank realities are intertwined. It would be naïve to think simulations are limited to video games, professional industrial or military applications. In a sense, all reality is virtual; it is constructed through our cognition and sensory organs. Reality is not so much 'out there', rather it is what we pragmatically consider to be 'out there'. Our brain is able to subtly construct 'reality' by combining and comparing sensory perceptions with what we expect and already know.[7-10]

Even the ancient Greeks talked about the phenomenon of simulation. In the *Allegory of the Cave*[11], Plato describes human beings as being chained in a cave and watching shadows on the wall, without realizing that they are 'only' representations of what goes on behind them – outside of the scope of their sensory perception. In Plato's teaching, an object such as a chair, is just a shadow of the idea Chair. The physically experienced chair we sit on is thus always a copy, a simulation, of the idea Chair and always one step away from reality. Today, the walls of Plato's cave are so full of projectors, disco balls, plasma screens and halogen spotlights that we do not even see the shadows on the wall anymore. Fakeness has long been associated with inferiority – fake Rolexes that break in two weeks, plastic Christmas trees, silicone breast implants, imitation caviar but, as the presence of media production evolves, the fake seems to gain a certain authenticity. Modern thinkers agree that because of the impasto of simulations in our society, we can no longer recognize reality. In *The Society of the Spectacle*, Guy Debord[12] explains how everything we once experienced directly has been replaced in our contemporary world by representations.

Do we still have genuine experiences at all, or are we living in a society of simulations?

Another Frenchman, Jean Baudrillard[13], argues that we live in a world in which simulations and imitations of reality have become more real than reality itself. He calls this condition 'hyperreality': the authentic fake. In summer we ski indoors; in winter we spray snow on the slopes. Plastic surgeons sculpt flesh to match retouched photographs in glossy magazines. People drink sports drinks with non-existent flavors like "wild berry ice zest". We wage war on video screens. Birds mimic mobile-phone ring tones[14]. At times, it seems the surrealists were telling the truth after all. And though you certainly cannot believe everything you see, at the same time, images still count as the ultimate evidence. Did we really land on the moon? Are you sure? How did it happen? Or was it perhaps a feat of Hollywood magic? Are we sure there is no Loch Ness Monster? A city girl regularly washes her hair with pine-scented shampoo. Walking in the forest with her father one day, she says, "Daddy, the woods smell like shampoo." Do we still have genuine experiences at all, or are we living in a society of simulations?

Media Schemas

A hundred years ago, when the Lumière brothers showed their film *L'arrivée d'un train*.[15] people ran out of the cinema when they saw the oncoming train.

Well, of course – if you see a train heading towards you, you move out of the way. Today, we have adapted our media schemas. We remain seated, because we know that the medium of cinema can have this effect.

Media schemas are defined as the knowledge we possess about what media are capable of and what we should expect from them in terms of their depictions: representations, translations, distortions, etc.[10|16|17] The term media schemas stems from the concept of schemas, which in psychology and cognitive sciences is described as a mental structure that represents some aspect of the world.[18] This knowledge enables us to react to media in a controlled way ("Don't be scared, it's only a movie."). A superficial observer might think media schemas are new things. This would be incorrect. For centuries, people have been dealing with developments in media. Think of conducting a telephone conversation, painting with perspective, or composing a letter with the aid of writing technology – yes, even the idea that you can set down the spoken word in handwriting was once new.

Let's face it. Our brains actually have only limited capabilities for understanding media. When our brain reached its current state of evolutionary development in Africa some 200,000 years ago,[19|20] what looked like a lion, actually was a lion! And if you paused to contemplate the nature of reality at that point, you would have made easy feline snack food.[10] Although we do seem to have gained some media awareness over the years, some part of this original impulse – in spite of all our knowledge – still reacts automatically and unconsciously to phenomena, as we perceive them. When we see the image of an oncoming train, we physically still are inclined to run away, even though cognitively we know it is not necessary.

Our media schemas are thus not innate but culturally determined. Every time technology comes out with something new, we are temporarily flummoxed, but we carry on pretty well. We are used to a world of family photographs, television and telephone calls. Imagine if we were to put someone from the Middle Ages into a contemporary shopping street. He would have a tough job refreshing his media schemas. But to us it is normal, and a lucky thing, too. It would be inconvenient indeed if with every phone call you thought, "How strange – I'm talking to someone who's actually far away." We are generally only conscious of our media schemas at the moment when they prove inadequate and we must refresh them, as those people in the 19th century had to do when they saw the Lumière brothers' filmed train coming at them.

Media Sphere

I once took part in an experiment in which I was placed in an entirely green room for one hour. In the beginning everything seemed very green, but after some time the walls became grey. The green was not informative any more and I automatically adjusted. Something similar seems to be going on with our media. Like the fish, who do not know they are wet, we are living in a technologically mediated space. We have adjusted ourselves, for the better because we know we will not be leaving this room any time soon. Today, media production has expanded by such leaps and bounds that images and simulations are often more influential, satisfying and meaningful than the things they simulate. We consume illusions. Images have become part of the cycle in which meanings are determined. They have bearing on our economy, our judgments and our identities. In other words: we are living the simulation.

Our brains actually have only limited capabilities for understanding media.

A disturbing thought, or old news? In contrast to Plato, his pupil Aristotle believed imitation was a natural part of life.[21] Reality reaches us through imitation (Aristotle calls it mimesis): this is how we come to know the world. Plants and animals too, use disguises and misleading appearances to improve their chances of survival (think of the walking stick, an insect that looks like a twig). Now then, the girl that says that "the woods smell like shampoo", should we consider this a shame and claim that this young child has been spoiled by media? Or is this child merely fine-tuning herself with the environment she grows up in? In the past, the woods used to smell like woods. But how interesting was that anyway?

Our Interfaced World-View

Four centuries ago, when Galileo Galilei became the first human being in history to aim a telescope at the night sky, a world opened up to him. The moon turned out not to be a smooth, yellowish sphere but covered with craters and mountains. Nor was the sun perfect: it bore dark spots. Venus appeared in phases. Jupiter was accompanied by four moons. Saturn had a ring. And the Milky Way proved to be studded with hundreds of thousands of stars. When Galileo asserted, after a series of observations and calculations, that the sun was the center of our solar system, he had a big problem. No one wanted to look through his telescope to see the inevitable.

While some dogs have such limited intelligence that they chase their own tails or shadows, we humans like to think we are smarter; we are used to living in a world of complex symbolic languages and abstractions. While a dog remains fooled by his own shadow, a human being performs a reality check. We weigh up the phenomena in our environment against our actions to form a picture of what we call reality. We do this not only individually, but also socially.[22] Admittedly, some realities are still rock solid – just try kick a stone to feel what I mean. However, this is not in conflict with the point I am trying to make, which is that the concepts of reality and authority are much more closely related to one another then most people realize. Like the physical world, which authority is pretty much absolute, media technologies are gradually but certainly attaining a level of authority within our society that consequently increases their realness.

Today the telescope is a generally accepted means of observing the universe. The earth is no longer flat. We have long since left the dark ages of religious dogma and have experienced great scientific breakthroughs, and yet there are still dominant forces shaping our world-view. As we are descending into the depths of our genes, greet webcam-friends across the ocean, send probes to the outskirts of the universe, find our way using car navigation, inspect the roof of our house with Google Earth and as it is not unusual for healthy, right-minded people to inform themselves about conditions in the world by spending the evening slouched in front of the television, we come to realize that our world-view is fundamentally being shaped through interfaces. As media technologies evolve and are incorporated within our culture, our experience of reality changes along. This process is so profound – and one could argue, successful – it almost goes without notice, that to a large extent, we are living in a virtual world already.

REFERENCES

1 LAKOFF, G. & JOHNSON, M. (1980). METAPHORS WE LIVE BY, THE UNIVERSITY OF CHICAGO PRESS, CHICAGO AND LONDON, 1980.

2 BRIGHT, R. (2000). 'UNCERTAIN ENTANGLEMENTS' IN SIAN EDE (RED.), STRANGE AND CHARMED SCIENCE AND THE CONTEMPORARY VISUAL ARTS, CALOUSTE GULBENKIAN FOUNDATION, LONDON 2000, P.120-143

3 MIRZOEFF, N., (1999). AN INTRODUCTION TO VISUAL CULTURE, ROUTLEDGE, LONDON, ISBN: 0415158761.

4 WINKEL VAN, C. (2006) HET PRIMAAT VAN DE ZICHTBAARHEID, NAI UITGEVERS, 2006, ISBN 90-5662-424-5

5 OOSTERLING, H. (2003). ACT YOUR ACTUALITY, IN GERRITZEN, M. ET AL. 2004. VISUAL POWER: NEWS. GINGKO PRESS, CORTE MADERA, CA, 2004, ISBN: 9063690568.

6 HEEKS, R. (2008). CURRENT ANALYSIS AND FUTURE RESEARCH AGENDA ON "GOLD FARMING": REAL-WORLD PRODUCTION IN DEVELOPING COUNTRIES FOR THE VIRTUAL ECONOMIES OF ONLINE GAMES. WORKING PAPER SERIES, DEVELOPMENT INFORMATICS GROUP, UNIVERSITY OF MANCHESTER.

7 DENNETT, D. (1991). CONSCIOUSNESS EXPLAINED. BOSTON: LITTLE, BROWN, AND CO.

8 GREGORY, R L. (1998) EYE AND BRAIN: THE PSYCHOLOGY OF SEEING (5TH EDITION). OXFORD UNIVERSITY PRESS.

9 HOFFMAN, D.D. (1998). VISUAL INTELLIGENCE: HOW WE CREATE WHAT WE SEE. NEW YORK: W.W. NORTON AND CO.

10 IJSSELSTEIJN, W. A. (2002). ELEMENTS OF A MULTI-LEVEL THEORY OF PRESENCE: PHENOMENOLOGY, MENTAL PROCESSING AND NEURAL CORRELATES. IN PROCEEDINGS OF PRESENCE 2002 (PP. 245-259). PORTO, PORTUGAL.

11 PLATO (360 B.C.) THE REPUBLIC, TRANSLATED BY ROBIN WATERFIELD, OXFORD UNIVERSITY PRESS, USA

12 DEBORD, GUY (1967). SOCIETY OF THE SPECTACLE, ZONE BOOKS, 1995, ISBN 0942299795

13 BAUDRILLARD, J. (1981). SIMULACRA AND SIMULATION. TRANSLATED BY SHEILA FARIA GLASER, UNIVERSITY OF MICHIGAN PRESS (1995), ISBN: 047206521

14 ATTENBOROUGH, D. (1998). LIFE OF BIRDS. DVD FROM THE BBC SERIES, 2 ENTERTAIN VIDEO, ASIN: B00004CXKJ

15 LUMIÈRE, AUGUSTE & LUMIÈRE, LOUIS. (1895). L'ARRIVÉE D'UN TRAIN À LA CIOTAT

16 MENSVOORT VAN, K. & DUYVENBODE VAN, M. (2001). HET BOS RUIKT NAAR SHAMPOO, DOCUMENTARY. VPRO TELEVISION, APRIL 2001.

17 NEVEJAN, C. (2007) PRESENCE AND THE DESIGN OF TRUST. PHD. DISSERTATION, UNIVERSITY OF AMSTERDAM, 2007.

18 PIAGET, J. (1997). JEAN PIAGET: SELECTED WORKS. ROUTLEDGE, 1997, ISBN 9780415168892

19 HEDGES, S. B. (2000) HUMAN EVOLUTION: A START FOR POPULATION GENOMICS, NATURE 408, 652-653 (7 DECEMBER 2000)

20 GOODMAN M., TAGLE D.A., FITCH D.H., BAILEY W., CZELUSNIAK J., KOOP B.F., BENSON P., SLIGHTOM J.L. (1990) PRIMATE EVOLUTION AT THE DNA LEVEL AND A CLASSIFICATION OF HOMINOIDS. JOURNAL OF MOLECULAR EVOLUTION 1990;30(3):260-6.

21 ARISTOTLE, (350 BC). POETICS, TRANSLATED BY STEPHEN HALLIWELL. CHAPEL HILL: UNIVERSITY OF NORTH CAROLINA PRES, 987.

22 SEARL, JOHN (1995). THE CONSTRUCTION OF SOCIAL REALITY, FREE PRESS (1997), ISBN: 0684831791.

FAKE FOR REAL MEMORY GAME

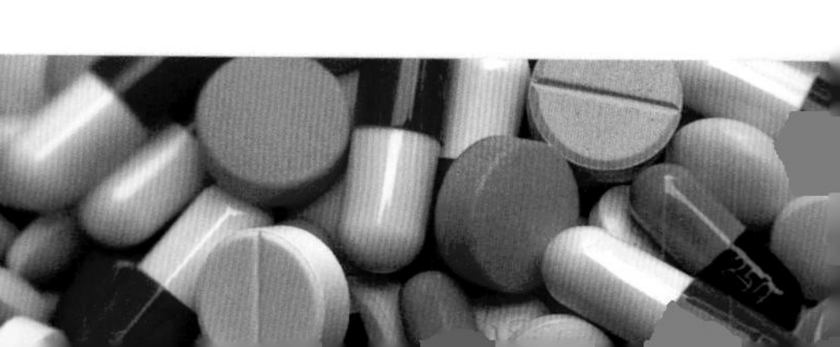

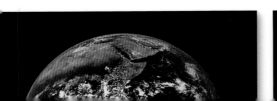

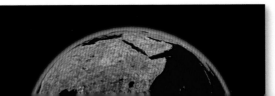

#2 RECREATION

wwww.nextnature.net/
themes/recreation

NEXT NATURE

ISBN 978-84-92861-53-8
9 788492 861538

WILD SYSTEMS

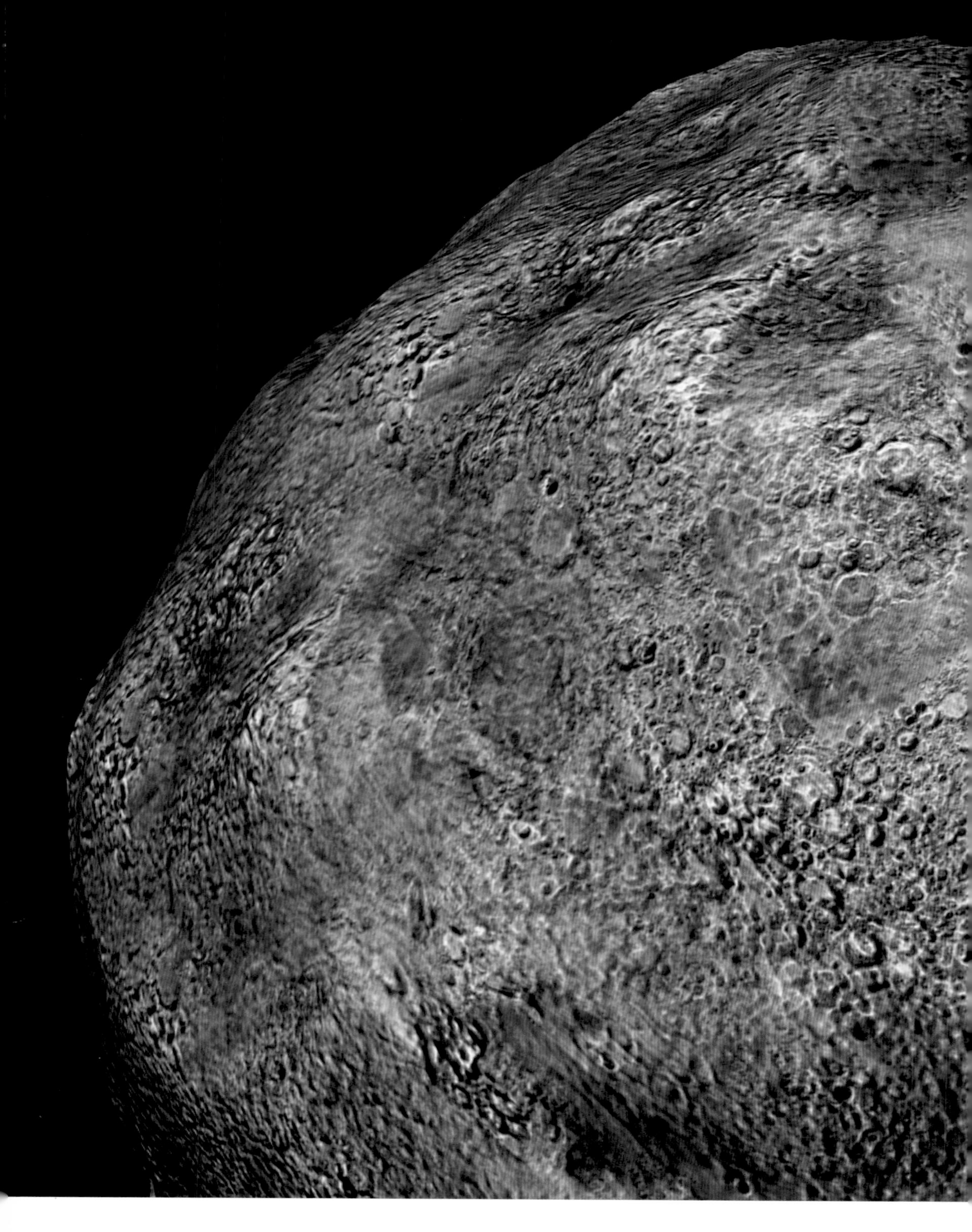

MOST PLANETS HAVE A GEOSPHERE.

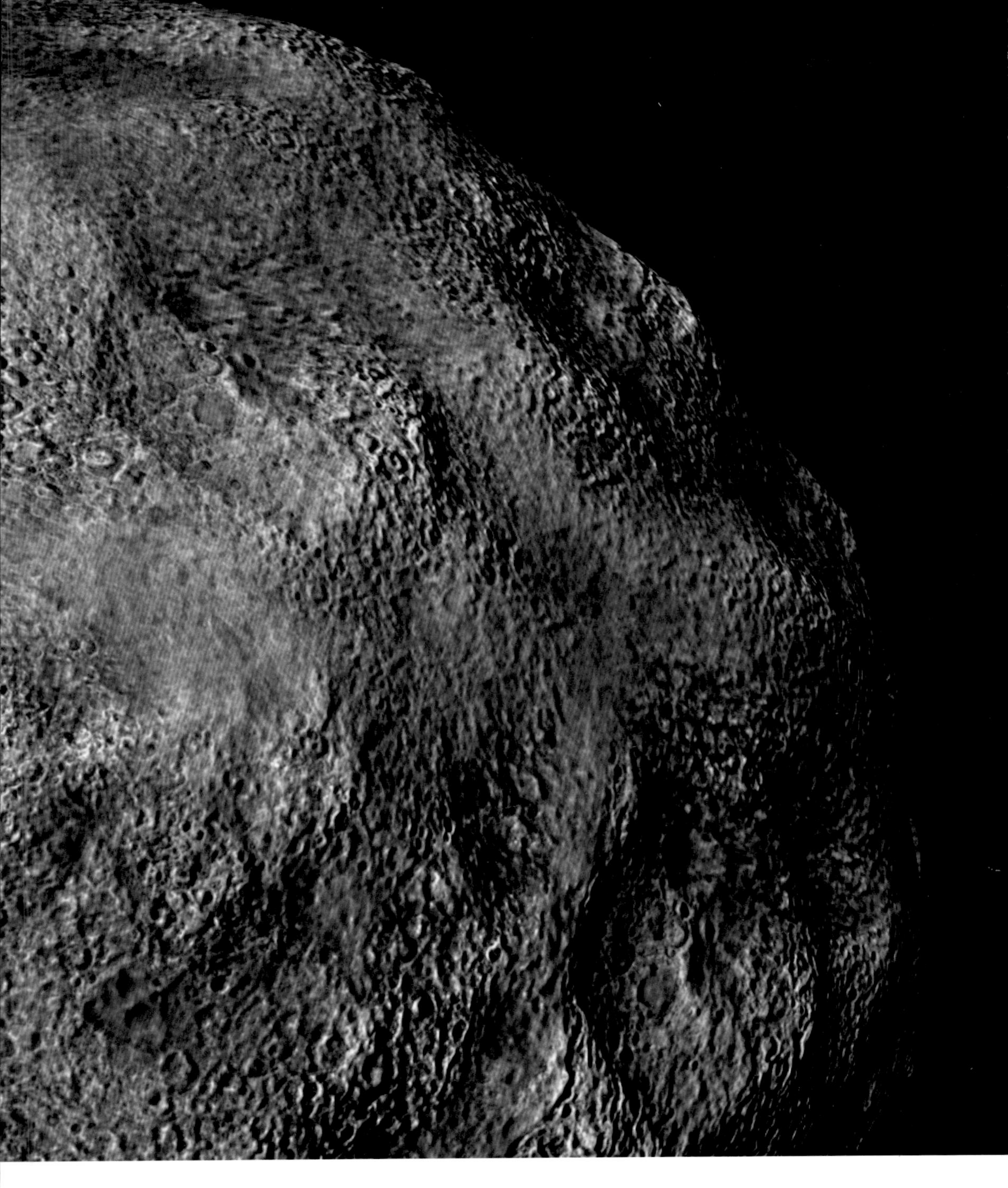

THE GEOSPHERE CONSISTS OF THE ROCK, WATER, AND SOIL OF A PLANET. ON EARTH, THE GEOSPHERE ENCOMPASSES ALL SOLID, INANIMATE MATTER. THE ROCKY EARTH FORMED 4.5 BILLION YEARS AGO AND EXISTED AS A LONELY LUMP FOR ANOTHER ONE BILLION YEARS BEFORE THE ADVENT OF LIFE.

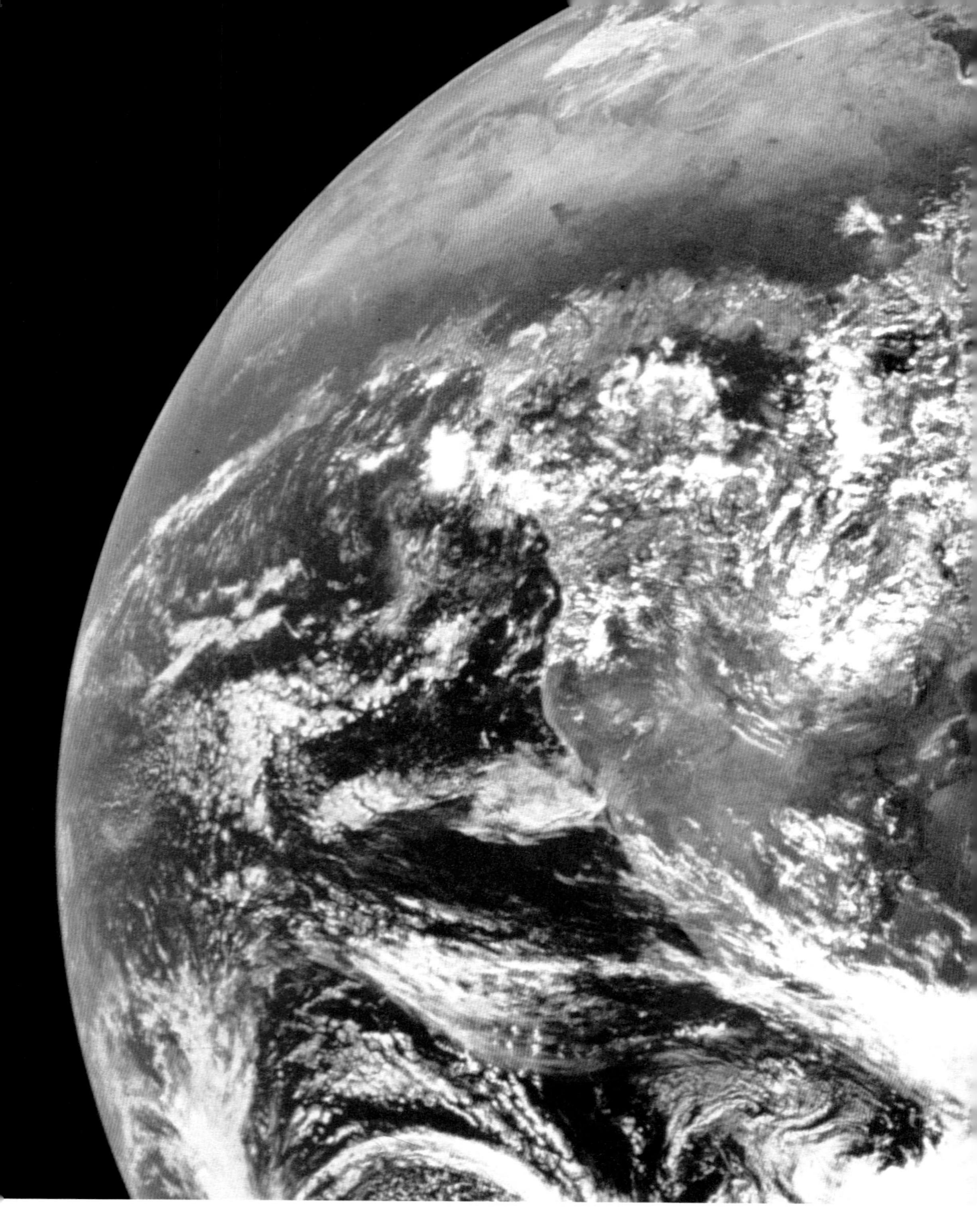

SOME PLANETS ALSO HAVE A BIOSPHERE.

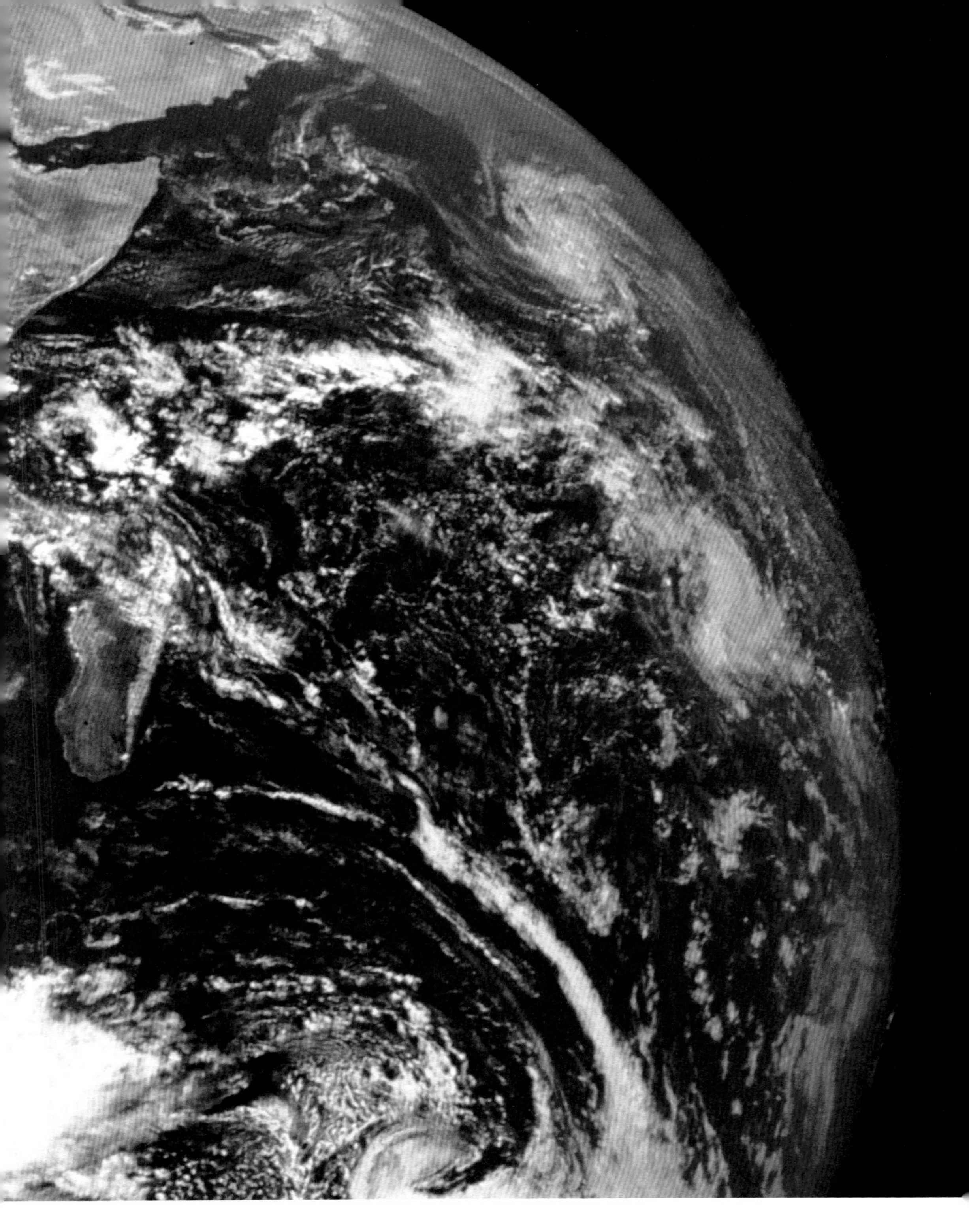

THE BIOSPHERE IS THE LAYER OF LIFE ON EARTH, THE SUM TOTAL OF EVERY ECOSYSTEM. THE BIOSPHERE HAS RADICALLY TRANSFORMED THE GEOSPHERE SINCE LIFE FIRST BLINKED INTO EXISTENCE MORE THAN 3.5 BILLION YEARS AGO. ALGAE FIRST CREATED THE OXYGEN-CONTAINING ATMOSPHERE; BACTERIA CYCLE NITROGEN FROM THE AIR INTO THE GEOSPHERE. EARTH IS THE ONLY PLANET WE KNOW OF WITH A BIOSPHERE, ALTHOUGH WE KEEP LOOKING FOR A REPLACEMENT.

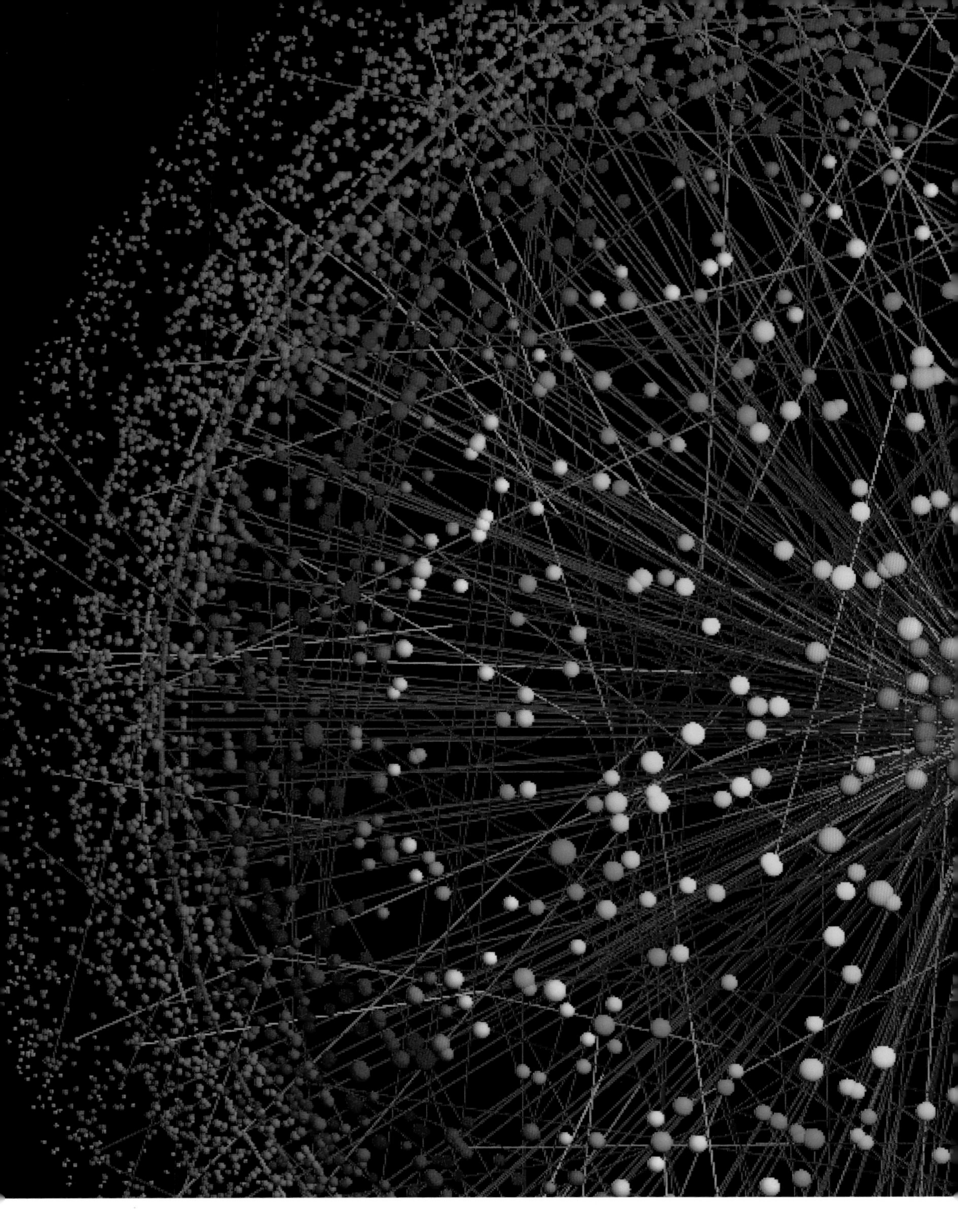

PLANET EARTH NOW HAS A NOOSPHERE.

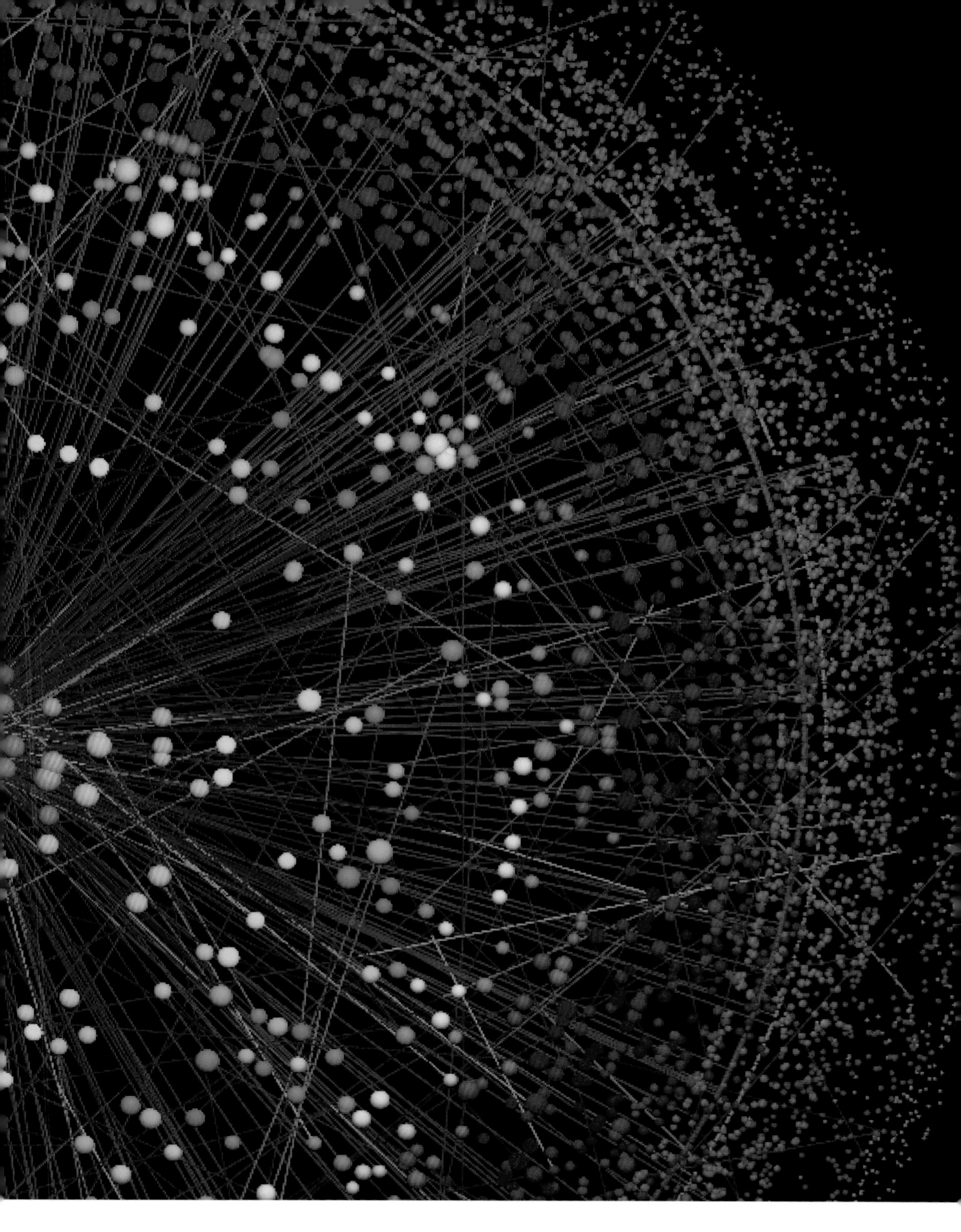

THE NOOSPHERE IS THE REALM OF HUMAN THOUGHT. JUST AS THE BIOSPHERE HAS HAD AN IRREVOCABLE IMPACT ON THE GEOSPHERE, THE NOOSPHERE HAS IRREVERSIBLY IMPACTED THE PREVIOUS TWO SPHERES. HUMAN ENDEAVOR HAS CARPETED THE EARTH IN CITIES AND FARMS, LEVELED MOUNTAINS FOR MINES, AND TRANSFORMED THE CHEMICAL MAKEUP OF THE ENTIRE ATMOSPHERE.

THE MOST WILD AND UNPREDICTABLE SYSTEMS ARE MAN MADE.

IF WE COMPARE THE VARIOUS ECOLOGIES OF OUR PLANET, IT IS STRIKING TO LEARN THAT ANCIENT, BIOSPHERE RELATED ECOLOGIES LIKE RAINFORESTS AND CORAL REEFS, ARE BEING THREATENED IN THEIR EXISTENCE, WHILE THE NEWER, NOOSPHERE RELATED ECOLOGIES LIKE THE FINANCIAL SYSTEM, ARE GROWING ERRATICALLY AND ARE ARGUABLY MORE THREATENING THAN THE LARGEST HURRICANE.

THE TECHNO-LOGICAL SUBLIME

By Jos de Mul

The sublime is an aesthetic concept of 'the exalted,' of beauty that is grand and dangerous. Through 17th and 18th century European intellectual tradition, the sublime became associated with nature. In the 20th century, the technological sublime replaced the natural sublime. Has our sense of awe and terror been transferred to factories, war machines and the unknowable, infinite possibilities of computers and genetic engineering?

When we call a landscape or a piece of art 'sublime,' we express the fact that it evokes particular beauty or excellence. Note that the 'sublime' is not only an aesthetic characterization; a moral action of high standing or an unparalleled goal in a soccer game may also be called 'sublime.' Roughly speaking, the sublime is something that exceeds the ordinary. This aspect of its meaning is expressed aptly in the German word for the sublime: the 'exalted' (das Erhabene). In the latter term we also hear echoes of the religious connotation of the concept. The sublime confronts us with that which exceeds our very understanding.

The notion of the sublime goes back a long way. It stems from the Latin *sublimis*, which – when used literally – means 'high up in the air,' and more figuratively means 'lofty' or 'grand.' One of the oldest essays on the sublime dates back to the beginning of our calendar. It is a manuscript in Greek entitled Περὶ ὕψους (*On the Sublime*), long ascribed to Longinus, though probably incorrectly so. In this treatise, the author does not provide a definition for 'the sublime,' and some classicists even doubt whether 'the sublime' is even the correct translation of the Greek word used – *hypsous*. Using a number of quotes from classical literature, the author discusses fortunate and less fortunate examples of the sublime. For one, the sublime must address grand and important subjects and be associated with powerful emotions. For pseudo-Longinus, the sublime landscape even touches upon the divine. Nature "has implanted in our souls an unconquerable passion for all that is great and for all that is more divine than ourselves".[1]

Longinus' essay was hardly noticed by his contemporaries and, in the centuries that followed, we rarely find references to this text. The essay was printed for the first time as late as 1554 in Basel. But only after the French translation by Boileau (1674) and the English translation by Smith (1739) did the text begin its victory march through European cultural history. From the Baroque period onward, which culminated in Romanticism, the sublime grew to become the central aesthetic concept, at which time it was often associated with the experience of nature. In the eighteenth century, we find it predominantly in the descriptions of nature of a number of British authors, portrayals of their impressions collected on Grand Tours through Europe and the Alps (a common practice in those days among young people from prosperous families). These authors use the term to render the often fear-inducing immensity of the mountain landscape in words.

The sublime refers to the wild, unbounded grandeur of nature, which is thus contrasted starkly with the more harmonious experience of beauty. In *A Philosophical Inquiry into the Origin of Our Ideas of the Sublime and Beautiful*, Edmund Burke defines the sublime as a "delightful terror".[2] That the forces of nature may nevertheless leave the viewer in a state of ecstasy is connected with the fact that the viewer observes these forces from a safe distance.

In the era of converging technologies, it is technology itself that gains a confounding character in its battle with nature.

In German Romanticism, however, the sublime loses its innocent character. The work of Immanuel Kant has been of particular critical importance in this respect. In the *Critique of Judgement* (*Kritik der Urteilskraft*, 1790), Kant, following Burke, makes an explicit distinction between the beautiful (das Schöne) and the sublime (das Erhabene). Beautiful are those things that give us a pleasant feeling. They fill us with desire because they seem to confirm our hope that we are living in a harmonious and purposeful world. A beautiful sunrise, for instance, gives us the impression that life is not that bad, really. The sublime, on the other hand, is connected with experiences that upset our hopes for harmony. It is evoked by things that surpass our understanding and our imagination due to their unbounded, excessive, or chaotic character.[3]

Kant makes a further distinction between the mathematical sublime and the dynamic sublime. The first, the mathematical sublime, is evoked by that which is immeasurable and colossal, and pertains to the idea of infinitude. When we view the immensity of a mountain landscape or look up at the vast night sky, we are overcome by a realization of our insignificance and finitude. Kant associates the second, the dynamic sublime, with the superior forces of nature. The examples he uses include volcanic eruptions, earthquakes and turbulent oceans. Here, too, we experience our insignificance and finitude, but in these cases this understanding is supplemented by the realization that we could be destroyed by the devastating power of these forces of nature. The dynamic sublime evokes both awe and fear; it induces a 'negative lust'.[3] in which attraction and repulsion melt into one ambiguous experience.

Since the sublime remains primarily an aesthetic category in Kant's work, he maintains the idea that 'safe distance' characterizes the experience of the sublime. When viewing a painting of a turbulent storm at sea, one can contemplate the superior force of nature while remaining comfortably assured that one is safely in a museum and not at sea! Friedrich Schiller, in contrast, takes things one step further and 'liberates' the sublime from the safe cocoon of aesthetic experience. The political terror under Jacobin rule following the French Revolution had deeply impressed him and shaped his view of the sublime, as elaborated in a series of essays.

In order to accomplish this liberation, Schiller rephrases Kant's distinction between the mathematical sublime and the dynamic sublime. In a 1793 text called *On the Sublime* (*Vom Erhabenen*), Schiller argues that the mathematical sublime ought to be labeled the theoretical sublime. The immeasurable magnitude of the high mountains and the night sky evoke in us a purely reflexive observation of infinitude. When nature shows itself to be a destructive force, on the other hand, we experience a practical sublime, which affects us directly in our instinct for self-preservation. Still, in Schiller's view, we need to make yet another distinction. When we view life-threatening forces from a safe distance – for instance, by observing a storm at sea from a safe place on land – we might experience the grandeur of the storm, but not its sublime character. An experience can only be truly sublime when our lives are actually endangered by the superior forces of nature. And yet, for Schiller, even that is not enough. Human beings have an understandable urge to shield themselves both physically and morally from the superior forces of nature. He who protects his country by building dykes attempts to gain 'physical certainty' over the violence of a westerly gale; he who believes his soul will live on in heaven after death protects himself by means of 'moral certainty.' He who manages to truly conquer his fear of the sea, or of death, shows his grandness, but loses the experience of the sublime. According to Schiller, truly sublime is he who collapses in a glorious battle against the superior powers of nature or military violence. "One can show oneself to be great in times of good fortune, but merely noble in times of bad fortune" ("Groß kann man sich im Glück, erhaben nur im Unglück zeigen").[4] Schiller's work transforms the sublime from an ambiguous aesthetic category into a no less ambiguous category of life.

History doesn't stop, however. Over the nineteenth and twentieth centuries, the main site for the ambiguous experience of the sublime has gradually shifted from nature to technology. Our current period is viewed as the age of secularization. God is retreating from nature and nature is gradually becoming 'disenchanted' in the process. Nature no longer implants in us, as was the case in Longinus's time, "an unconquerable passion for all that is great and for all that is more divine than ourselves," but invites technical action and control. Divine rule has become the work of man. The power of divine nature has been transferred to the power of human technology. In a sense, the sublime now returns to what it was in Longinus' work: a form of human technè. However, these days it no longer falls into the category of the alpha technologies, such as rhetoric, but rather, we find ourselves on the brink of the age of sublime beta technologies. Modern man is less and less willing to be overpowered by nature; instead, he vigorously takes technological command of nature.

As David Nye has documented in great detail in his book, *American Technological Sublime*[5], Americans initially embraced the technological sublime with as much enthusiasm as they had embraced the natural sublime. The admiration of the natural sublime, as it might be experienced in the Grand Canyon, was replaced by the sublime of the factory, the sublime of aviation, the sublime of auto-mobility, the sublime of war machinery, and the sublime of the computer.[5] The computer in particular discloses a whole new range of sublime experiences. In a world in which the computer has become the dominant technology, everything – genes, books, organizations – becomes a relational database. Databases are onto-logical machines that transform everything into a collection of (re)combinatory elements.

Twenty-first century man has been denied the choice to not be technological.

As such, the database also transforms our experience of the sublime, and the sublime as such. The mathematical sublime in the age of computing manifests itself as a combinatorial explosion. As Borges has shown in *The Library of Babel*, the number of combinations of a finite number of elements – in his story, 25 linguistic symbols – is hyper-astronomical.[6] Borges' library, consisting of books of 410 pages, each having 40 lines of 80 characters – contains no less than $25^{1.312.000}$ books. The number of atoms in the universe (estimated by physicists to be roughly 10^{80}) is negligible compared to the unimaginable number of possible (re)combinations in the 'Database of Babel.' And the number of possible (re)combinations of the three billion nucleotides of the human genome is even more sublime.[7]

Moreover, by actively recombining the elements of the database (by genetic manipulation or synthetic biology, for example), we unleash awesome powers and, in so doing, transform the dynamic sublime. In our (post)modern world it is no longer the superior force of nature that calls forth the experience of the sublime, but rather, the superior force of technology. However, with the transfer of power from divine nature to human technology, the ambiguous experience of the sublime also nests in the latter. In the era of converging technologies – information technology, bio-technology, nanotechnology and the neurosciences – it is technology itself that gains a confounding character in its battle with nature. While technology is an expression of the grandeur of the human intellect, we experience it more and more as a force that controls and threatens us. Technologies such as atomic power stations and genetic modification, to mention just two paradigmatic examples, are Janus-faced: they reflect, at once, our hope for the benefits they may bring as well as our fear of their uncontrollable, destructive potentials.

According to David Nye, this explains why enthusiasm for the technological sublime has transformed into fear in the course of the twentieth century. This is also why it is often said, in relation to such sublime technologies, that we 'shouldn't play God.' At the same time, twenty-first century man has been denied the choice to not be technological. The biotope in which we used to live has been transformed, in this (post)modern age, into a technotope. We have created technological environments and structures beyond which we cannot survive. The idea that we could return to nature and natural religion is an unworldly illusion. In fact, because of its Janus-faced powers, technology itself has become the sublime god of our (post)modern age. Assessments regarding the fundamental transformation from the natural to the technological sublime may vary; however, no one can deny that technology is a no less inexhaustible god.

REFERENCES

1 LONGINUS. 1965. ON THE SUBLIME. IN CLASSICAL LITERARY CRITICISM. EDITED BY T.S. DORSCH. BALTIMORE: PENGUIN BOOKS.

2 BURKE, E., AND D. WOMERSLEY. 1998. A PHILOSOPHICAL ENQUIRY INTO THE ORIGIN OF OUR IDEAS OF THE SUBLIME AND BEAUTIFUL: AND OTHER PRE-REVOLUTIONARY WRITINGS. PENGUIN CLASSICS. LONDON/NEW YORK: PENGUIN BOOKS.

3 KANT, I. 1968. KRITIK DER URTEILSKRAFT. VOL. X, THEORIE-WERKAUSGABE. FRANKFURT.

4 SCHILLER, F. 1962. VOM ERHABENEN. IN SÄMTLICHE WERKE. MÜNCHEN: HANSER.

5 NYE, D.E. 1994. AMERICAN TECHNOLOGICAL SUBLIME. CAMBRIDGE, MASS.: MIT PRESS.

6 BORGES, J.L. 1962. FICCIONES. NEW YORK,: GROVE PRESS.

7 BLOCH, W.G. 2008. THE UNIMAGINABLE MATHEMATICS OF BORGES' LIBRARY OF BABEL OXFORD: OXFORD UNIVERSITY PRESS.

OUR TECHNOLOGICAL ENVIRONMENT IS A NATURE OF ITS OWN

Climate (Un)Control

Climate change is real and caused by people. Yet, it is not new: climate change occurs constantly throughout the history of the earth. It is crucial to realize climate change has both winners and losers. It is a political problem, rather than a natural problem.

LONDON

DOGGERLAND
A SPECULATIVE SATELLITE RENDERING OF WESTERN EUROPE AS IT (PRESUMABLY) WAS [illegible]0,000 YEARS AGO.

AMSTERDAM

So You Think Climate Change is New?

Thousands of years ago our ancestors took the easy route from London to Amsterdam. And, oh yeah, they walked...

Doggerland is the name of a vast plain that joined Britain to Europe for nearly 12,000 years, until sea levels began rising dramatically after the last Ice Age. Taking its name from a prominent shipping hazard - Dogger Bank - this immense land bridge vanished beneath the North Sea around 6000 B.C.

Like all land bridges, Doggerland seems to have been a pretty busy thoroughfare for ancient hunters and gatherers. But archaeologists hardly gave it a second thought until 2002, when a small group of British researchers laid hands on seismic survey data collected by the petroleum industry in the North Sea. It's thought that the sea level rose no more than about one or two meters per century, and that the land would have disappeared in a series of punctuated inundations.

According to marine archaeologist Nic Flemming, from the National Oceanography Centre of University of Southampton, "It was perfectly noticeable in a generation, but nobody had to run for the hills". Although hunter-gatherers usually didn't have much sense of ownership, land would have become an increasingly precious resource as the sea rose. According to Vince Gaffney, a landscape archaeologist at the University of Birmingham, who established the mapping project to outline the terrain of Doggerland, the transformation of Doggerland from harsh tundra into a fertile paradise and eventually into contemporary northern European landscape in only a few thousand years, "put human adaptability to the test". Doesn't that sound familiar? | KVM

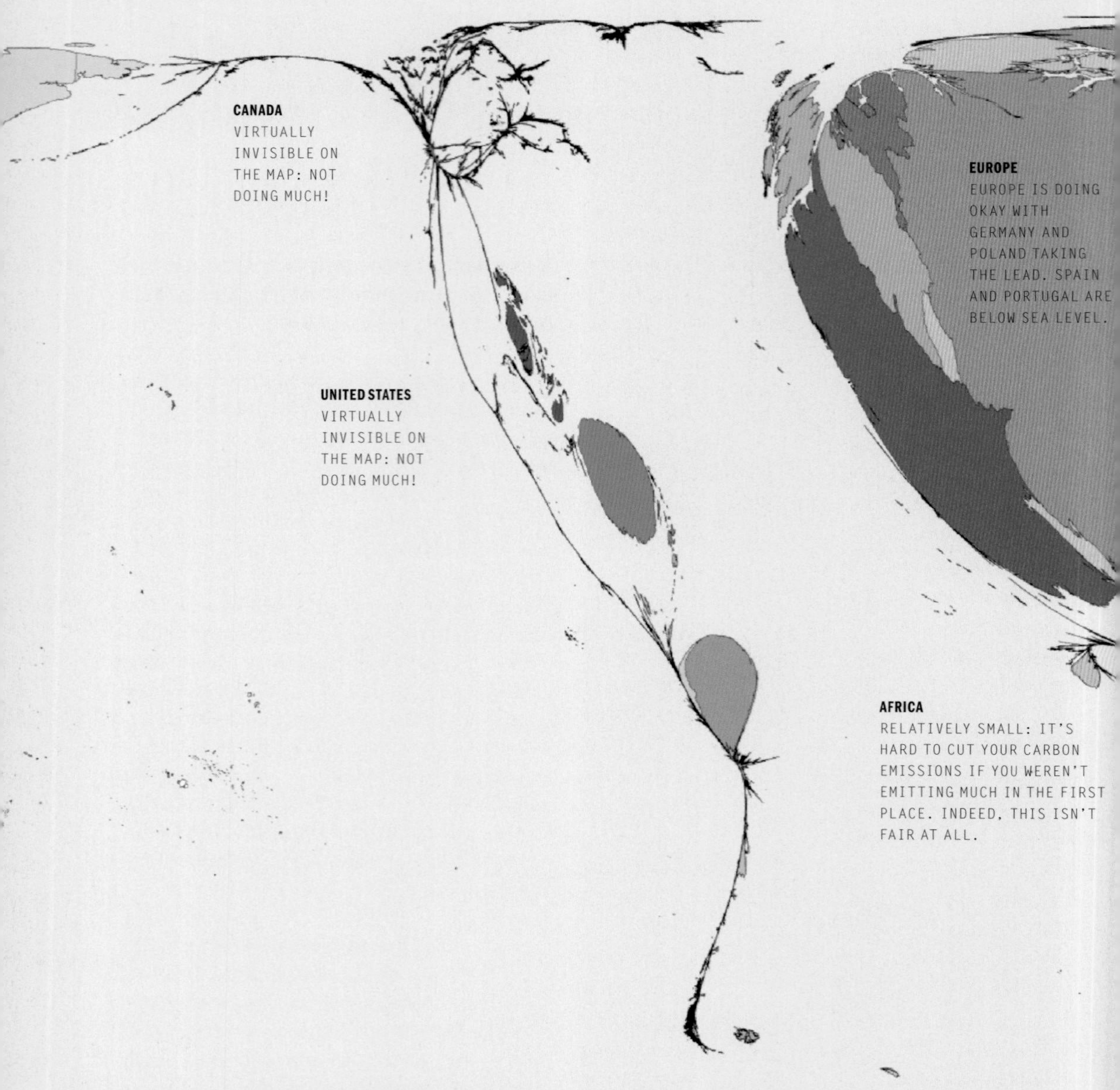

CANADA
VIRTUALLY INVISIBLE ON THE MAP: NOT DOING MUCH!
UNITED STATES
VIRTUALLY INVISIBLE ON THE MAP: NOT DOING MUCH!
EUROPE
EUROPE IS DOING OKAY WITH GERMANY AND POLAND TAKING THE LEAD. SPAIN AND PORTUGAL ARE BELOW SEA LEVEL.
AFRICA
RELATIVELY SMALL: IT'S HARD TO CUT YOUR CARBON EMISSIONS IF YOU WEREN'T EMITTING MUCH IN THE FIRST PLACE. INDEED, THIS ISN'T FAIR AT ALL.

Imagine if the Effects of Global Warming were Fair

Contrary to popular belief, global warming is not only a bad thing: there are winners and losers. While low-lying countries, like Bangladesh, are expected to suffer enormously from rising temperatures and sea levels, countries situated at the top of the Northern Hemisphere, like Canada and Russia, might gain large regions of pristine exploitable farming ground, as temperatures rise. Global warming is not a natural disaster: it is a political disaster. The countries that cause the global warming effect aren't necessarily the countries that suffer the consequences. National political agendas hardly align with their globally felt consequences. Imagine if the effects of global warming were fair. The information graphic above shows a distorted world map, in which the landmass of countries is scaled according to carbon emission reductions between 1980–2000. Life would be so simple, if polluting countries just disappeared into the sea. | KVM

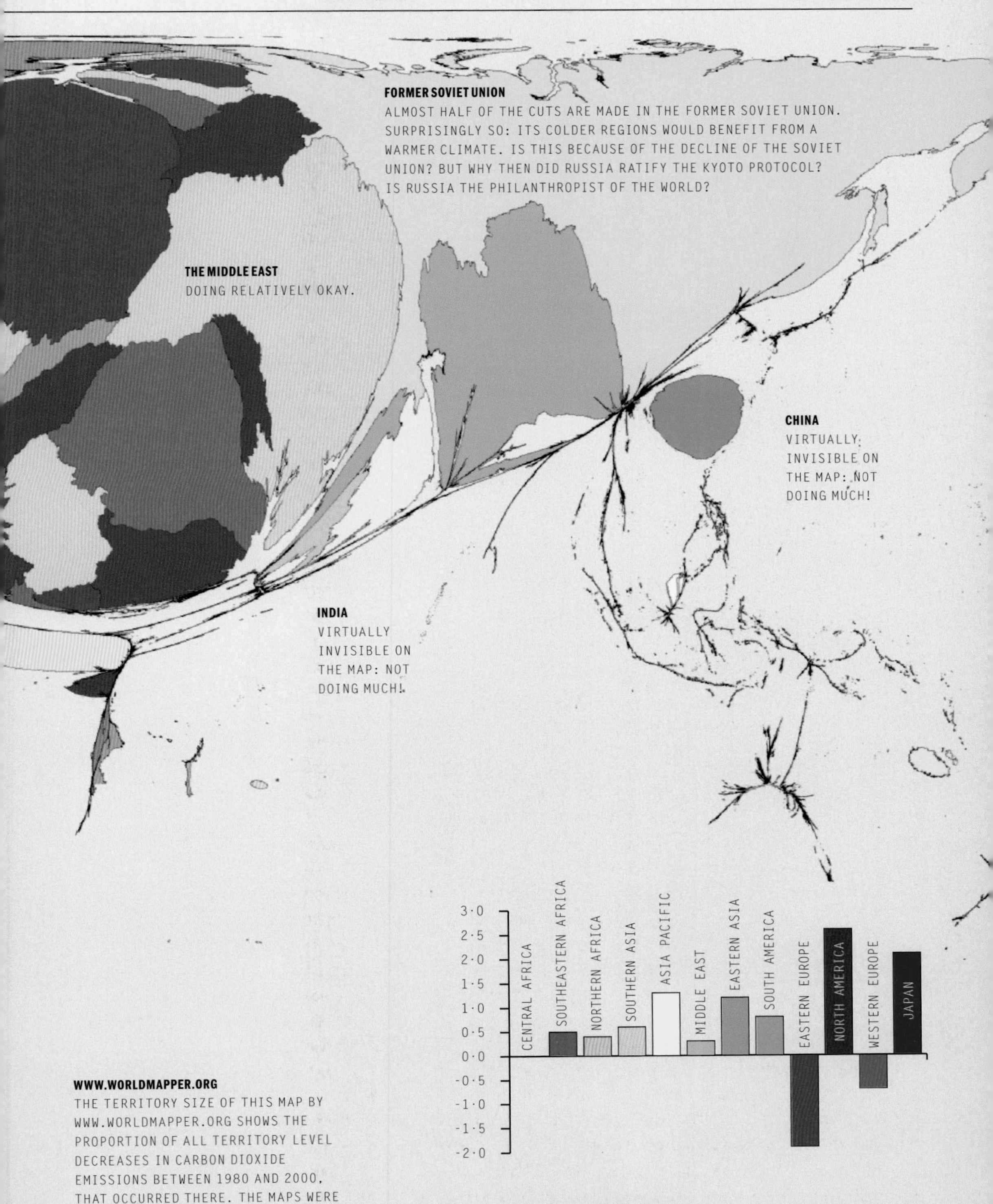

FORMER SOVIET UNION
ALMOST HALF OF THE CUTS ARE MADE IN THE FORMER SOVIET UNION. SURPRISINGLY SO: ITS COLDER REGIONS WOULD BENEFIT FROM A WARMER CLIMATE. IS THIS BECAUSE OF THE DECLINE OF THE SOVIET UNION? BUT WHY THEN DID RUSSIA RATIFY THE KYOTO PROTOCOL? IS RUSSIA THE PHILANTHROPIST OF THE WORLD?
THE MIDDLE EAST
DOING RELATIVELY OKAY.
CHINA
VIRTUALLY INVISIBLE ON THE MAP: NOT DOING MUCH!
INDIA
VIRTUALLY INVISIBLE ON THE MAP: NOT DOING MUCH!
3·0
2·5
2·0
1·5
1·0
0·5
0·0
-0·5
-1·0
-1·5
-2·0
CENTRAL AFRICA
SOUTHEASTERN AFRICA
NORTHERN AFRICA
SOUTHERN ASIA
ASIA PACIFIC
MIDDLE EAST
EASTERN ASIA
SOUTH AMERICA
EASTERN EUROPE
NORTH AMERICA
WESTERN EUROPE
JAPAN

WWW.WORLDMAPPER.ORG

THE TERRITORY SIZE OF THIS MAP BY WWW.WORLDMAPPER.ORG SHOWS THE PROPORTION OF ALL TERRITORY LEVEL DECREASES IN CARBON DIOXIDE EMISSIONS BETWEEN 1980 AND 2000, THAT OCCURRED THERE. THE MAPS WERE CREATED BY DANIEL DORLING, MARK NEWMAN AND ANNA BARFORD AND PUBLISHED IN THE BOOK *THE ATLAS OF THE REAL WORLD: MAPPING THE WAY WE LIVE.*

CHANGES IN CARBON EMISSIONS

CHANGE IN ANNUAL CARBON EMISSIONS, IN EXTRA OF TONNES PER PERSON PER YEAR, FROM 1980-2000. ALL DATA AND MAPS VIA WWW.WORLDMAPPER.ORG

LIVE

U.S. PAYS MEXICO TO ALLOW HURRICANE ON ITS TERRITORY

BREAKING NEWS CNN

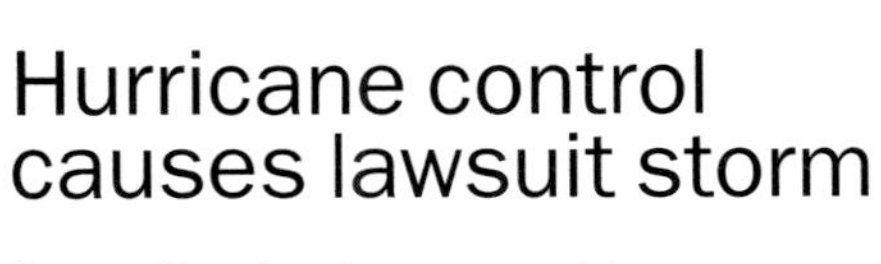

Hurricane control causes lawsuit storm

Controlling hurricanes could save many lives and dollars. According to a study conducted by climate physicist Daniel Rosenfeld – adding dust to Hurricane Katrina's base could have weakened the storm and sent it spinning away from New Orleans.

However, few scientists believe these new ideas will be tried outside the computer lab anytime soon. The problem isn't science. It's lawyers. A manipulated storm could destroy towns that otherwise might not have been hit – leading to liability issues regardless of whether the storm was weakened, or pushed away from a major city. Even if 'Hurricane Control' technology was totally robust, who decides where to direct a storm?

'Hurricane Control' perfectly illustrates how the cultivation of old nature leads to unexpected new menace: the natural disaster ceases to exist, but it is replaced by a political threat. As scientific papers on the subject are being published, conspiracy theories will soon follow. The idea that Hurricane Katrina was secretly guided away from Florida towards New Orleans is rather tempting if you look at the pictures of its path. We don't believe this theory holds; science just isn't that advanced yet.

Nonetheless, a newspaper story in which one of the hurricane scientists described how in an early experiment, lawyers advised them to keep silent about their cloud-seeding activities after a storm they had tinkered with swerved and battered South Carolina, is rather disturbing. Will 'natural disasters' soon be exclusive to the poor and powerless? | KVM

Hack the Planet

The 1991 eruption of mount Pinatubo is thought to have caused a certain level of global cooling (amongst other serious side effects). If a volcano can do that, why not humans?

Increasingly, scientists are recommending planetary 'geo-engineering' to avoid climate change. Different scenarios, ranging from space mirrors to planting reflective crops or injecting a huge cloud of ash into the atmosphere, have been suggested and politicians are listening.

Geo-engineering might be a viable solution to an overheated planet. For instance, it could work as an emergency measure to slow a melting ice cap. Yet, before we dive in the game of deliberately manipulating the Earth's climate to counteract the effects of global warming, we must realize that *maakbaarheid** is never finished: every cultivation of nature typically causes the rise of a next nature that is wild and unpredictable as ever. Just like the inventor of the fridge did not anticipate a hole in the ozone layer, we should be bracing ourselves for some serious side effects of geo-engineering. Thus, if we get in the geo-engineering game our methodology should be one of 'guided growth' rather than 'maakbaarheid'.

Arguably, rather than desperately attempting to stop all changes in the climate, we should as a culture gain more of a flexibility towards a constantly mutating environment. After all, change happens. | KVM

MAAKBAARHEID

A DUTCH WORD THAT DEFIES A DIRECT ENGLISH TRANSLATION. DESCRIBED LITERALLY AS 'MAKEABILITY' IT IS A COMBINATION OF 'MANUFACTURED', 'CONSTRUCTED' AND 'MOLDED'

Geo-engineering Strategies
Space Mirrors
Pump liquid CO_2 into the deep sea
Cloud Seeding
Reflective Crops
Aerosols in stratosphere
Iron fertilization of sea
Giant reflectors in orbit
Wrap Greenland
Grow more trees
Green the deserts
Pump liquid CO_2 into rocks
Grow genetically engineered crops
Artificial Trees
Cloud Seeding
Genetically engineer crops

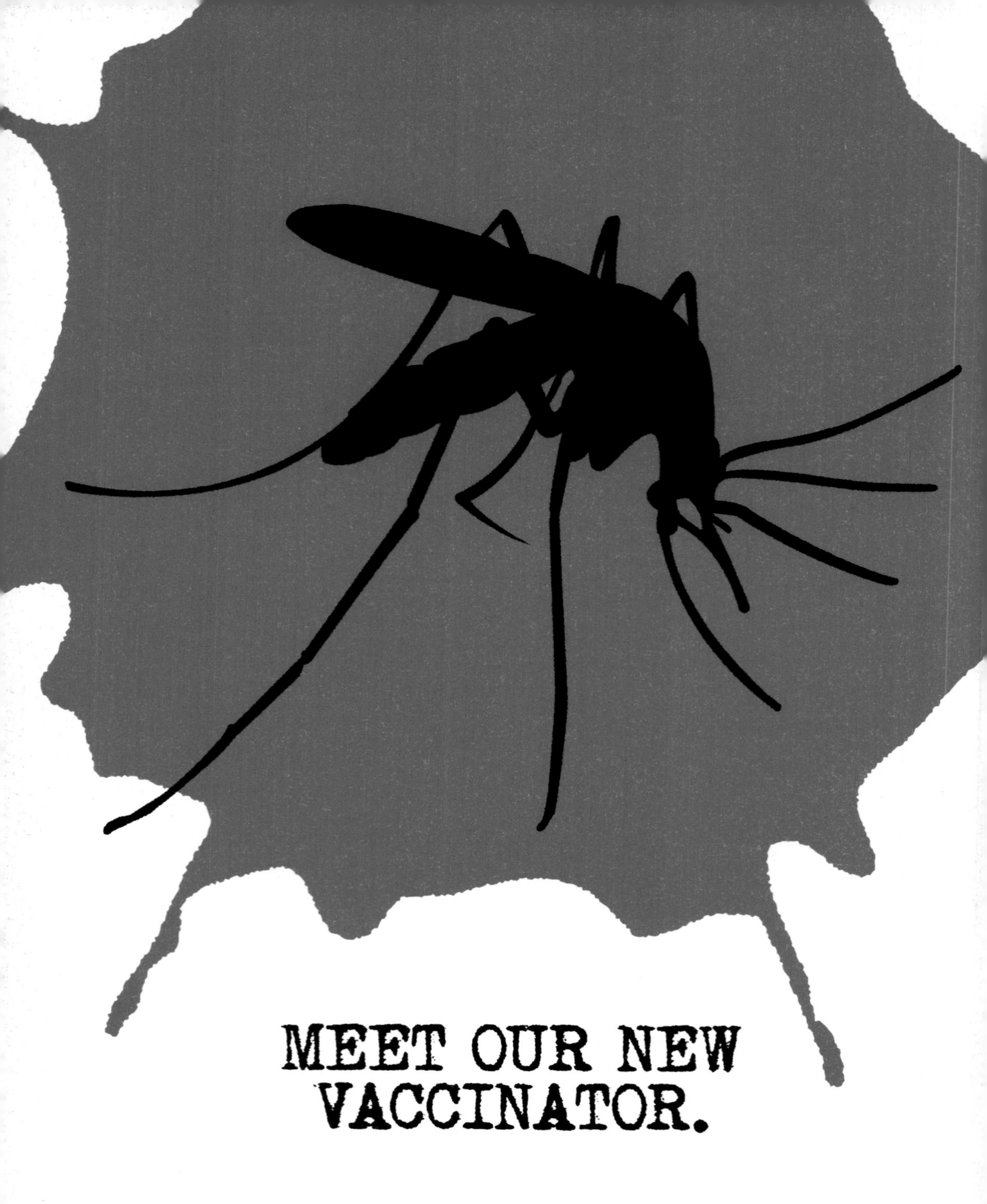
MEET OUR NEW
VACCINATOR.

VACCINATORS
WITHOUT BORDERS

USING MOSQUITOES AS OUR FLYING DOCTORS, WE CAN NOW DISTRIBUTE OUR VACCINES TO THE MOST REMOTE REGIONS OF AFRICA. THANKS TO THE RESEARCH BY MOLECULAR GENETICIST SHIGETO YOSHIDA OF JICHI MEDICAL UNIVERSITY IN TOCHIGI JAPAN THE FORMERLY NOTORIOUS, MALARIA-SPREADING INSECTS HAVE BEEN TRANSFORMED INTO VACCINE-CARRYING SYRINGES. THEY CARRY VACCINES FOR A WIDE VARIETY OF DISEASES. SO NEXT TIME YOU ARE BITTEN BY ONE OF OUR FLYING DOCTORS, YOU KNOW IT IS FOR YOUR OWN GOOD.

WE HAVE TO EMBRACE COMPLEXITY

Guided Growth

Contrary to the 20th century top-down approach of modular construction and engineering on the basis of detailed blueprints, guided growth is a design methodology that focuses on the steering of processes as a production principle. The outcome of the process may be partly unpredictable, as it is dependent on the context in which it takes place. This allows for a certain constrained randomness to enter in the production process. The methodology can be applied in various domains ranging from product design, to architecture to tissue engineering. It is particularly suitable where 'the made' and 'the born' meet. Although the methodology might seem like a futuristic response to the linear and rigid engineering methodologies that have been dominant since the industrial revolution, it is in fact rooted in ancient design principles and traditions.

Root Bridges

A spectacular example of the guided growth principle is found in the depths of north-eastern India, one of the wettest places on earth, where bridges aren't built – they're grown. The living bridges of Cherrapunji, India are made from the roots of the *Ficus elastica* tree. This tree produces a series of secondary roots from higher up its trunk and can comfortably perch atop huge boulders along the riverbanks, or even in the middle of the rivers themselves. In order to make a rubber tree's roots grow in the right direction – say, over a river – the Khasis use betel nut trunks, sliced down the middle and hollowed out, to create root-guidance systems. The thin, tender roots of the rubber tree, prevented from fanning out by the betel nut trunks, grow straight out. When they reach the other side of the river, they're allowed to take root in the soil. Given enough time, a sturdy, living bridge is produced. |KVM

THE MAPLE OF RATIBOR
ACCORDING THE ORIGINAL DESCRIPTION IN *THE PICTURE MAGAZINE* 1893, TWO FLOORS THAT COULD ACCOMMODATE 20 PEOPLE ARE GROWN INTO THIS 80-YEAR-OLD MAPLE.

LIVING ROOT BRIDGE, CHERRAPUNJI (INDIA)

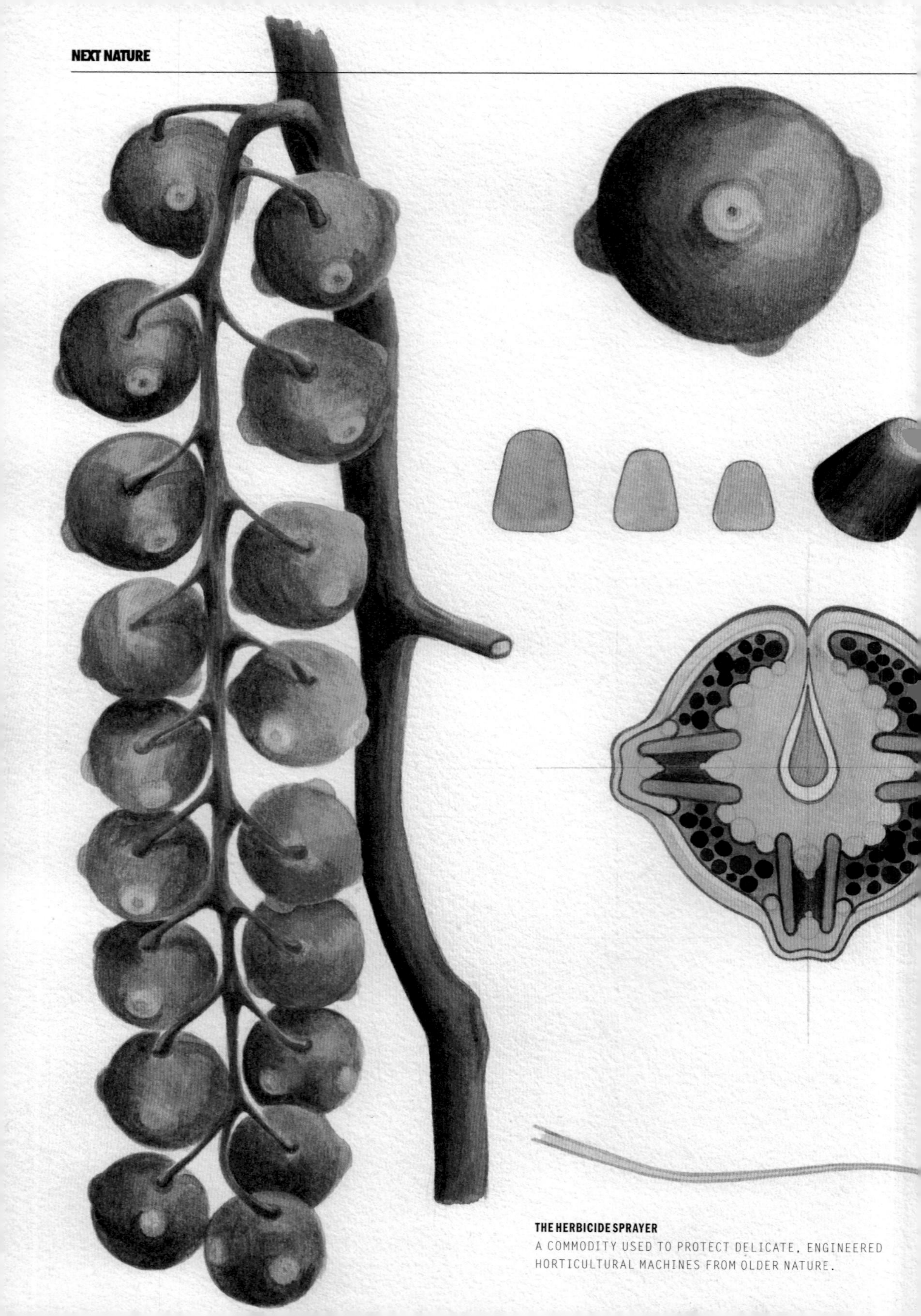

THE HERBICIDE SPRAYER

A COMMODITY USED TO PROTECT DELICATE, ENGINEERED HORTICULTURAL MACHINES FROM OLDER NATURE.

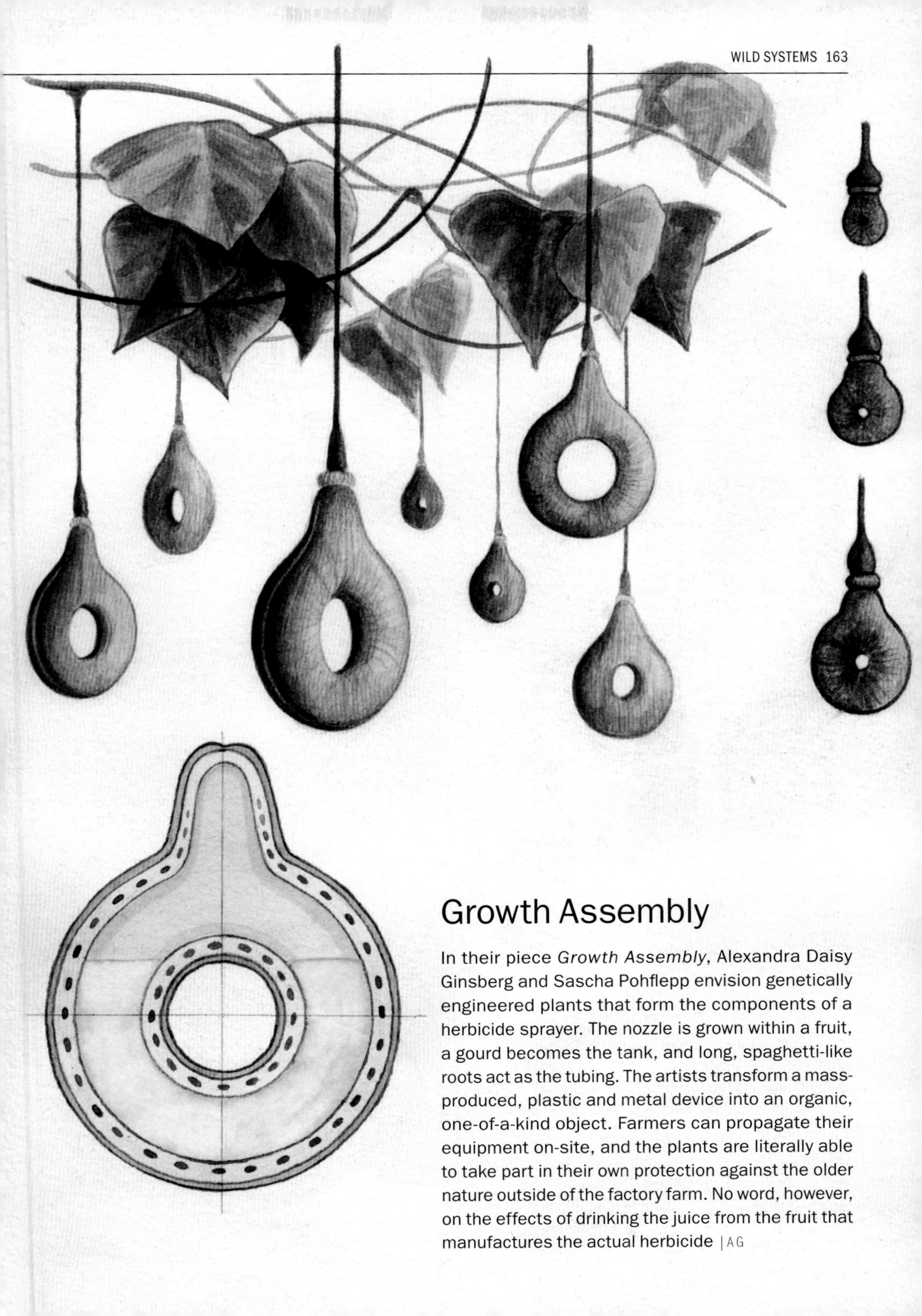

Growth Assembly

In their piece *Growth Assembly*, Alexandra Daisy Ginsberg and Sascha Pohflepp envision genetically engineered plants that form the components of a herbicide sprayer. The nozzle is grown within a fruit, a gourd becomes the tank, and long, spaghetti-like roots act as the tubing. The artists transform a mass-produced, plastic and metal device into an organic, one-of-a-kind object. Farmers can propagate their equipment on-site, and the plants are literally able to take part in their own protection against the older nature outside of the factory farm. No word, however, on the effects of drinking the juice from the fruit that manufactures the actual herbicide | AG

FORTRESS OF SOLITUDE, THE PLACE SUPERMAN CALLS HOME AFTER A HARD DAY'S WORK

CAVE OF CRYSTALS DISCOVERED BY INDUSTRIAS PEÑOLES MINERS A 1,000 FEET (300 METERS) BELOW NAICA MOUNTAIN IN THE CHIHUAHUAN DESERT.

Learning from Superman's House

Superman already knew it: guided growth is the future of architecture. The spaceship in which Superman arrives on Earth from his home planet Krypton carries an intelligent crystal that, when thrown in an arctic ice field, melts into the ice and grows into a huge crystalline building. Rather than constructed, Superman's fortress is grown. A remarkably similar architecture was discovered at the Industrias Peñoles mine a thousand feet (300 meters) below Naica mountain in the Chihuahuan desert. The cave contains some of the largest natural crystals ever found. Researchers have been studying the structures to learn out how they could grow to such gigantic proportions. The secret recipe for cooking up such staggeringly huge, pillar-like crystals is fairly simple: Start out with a warm subterranean broth. Add the right pinch of minerals at the right times for a few million years. Wait and wait as your tree trunk-thick gypsum crystals grow.

Obviously our current technologies aren't advanced enough to reproduce the house of Superman. Yet, on a smaller scale successes have been made. The production process of the Crystal Chair by Japanese designer Tokujin Yoshioka revolves entirely around the guidance of growth: a substrate made of polyester elastomer forms the skeleton and is submerged in a tank, after which the crystals grow on that substrate. A more metaphorical interpretation of the guided-growth methodology is explored by architects Aranda & Lasch, who with their project *Rules of Six* envision an unpredictable, self-generating landscape of interlocking hexagons that could represent rooms, buildings or entire urban neighborhoods. The growth element is not so much in the construction of the buildings, but focuses on the planning element. The work explores self-assembly and modularity across scales. Using an algorithm that mimics the growth patterns of microscopic structures, they create a sprawling matrix of three-dimensional structures that can multiply indefinitely without sacrificing stability. The result is an organic-algorithmic city. |KVM

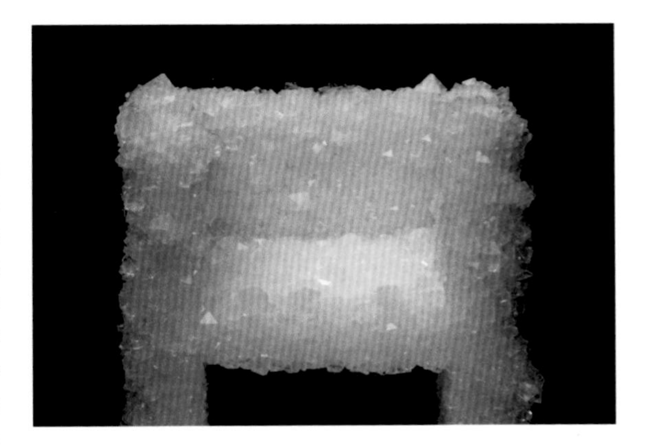

VENUS, NATURAL CRYSTAL CHAIR BY TOKUJIN YOSHIOKA

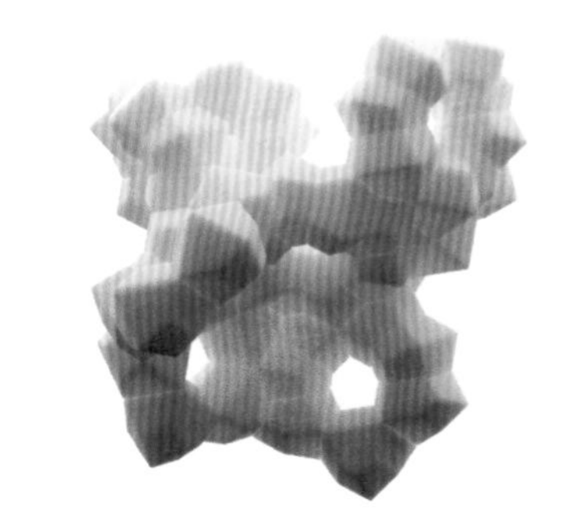

SELF-GENERATING ARCHITECTURE BY ARANDA & LASCH

VENICE RENDERING BY ARCHITECTS GMJ

Self-repairing Architecture

By Rachel Armstrong

All buildings today have something in common: They are made using Victorian technologies. This involves blueprints, industrial manufacturing and construction using teams of workers. All this effort results in an inert object, which means there is a one-way transfer of energy from our environment into our homes and cities. This is not sustainable. I believe that the only possible way for us to construct genuinely sustainable homes and cities is by placing them in a constant conversation with their surroundings. In order to do this, we need to find the right language.

Metabolic materials are a technology that acts as a chemical interface or language through which artificial structures such as, architecture, can connect with natural systems. I am developing this technology in collaboration with scientists working in the field of synthetic biology and origins of life sciences whose model systems of investigation are materials that belong to a new group of technologies being described as 'living technology',[1] which possess some of the properties of living systems but are not considered 'alive'. The characteristic of metabolic materials is that they possess the living property of metabolism, which is a set of chemical interactions that transform one group of substances into another with the absorption or production of energy. This transfer of energy through chemical exchange directly couples the environment to the living technology and embeds it within an ecosystem. Metabolic materials work with the energy flow of matter and systems using a bottom up approach to the construction of architecture.

Metabolic materials need water to chemically participate in an ecological landscape since they have not, developmentally speaking, reached the origins of life transition through which they are able to leave the water and adapt to 'life' on the land, bringing with them all the necessary support systems for survival on air. Currently metabolic materials can be thought of as architectural symbionts since they coexist alongside

ARTIFICIAL REEF UNDER VENICE FOUNDATIONS. COMPUTER RENDERING BY CHRISTIAN KERRIGAN.

structural materials and offer a medium through which a chemical dialogue between the classical architectural framework and the environment can take place. Metabolic materials may also be thought of as the next generation of architectural skins that are more than just decorative cladding but living integuments designed to give biological-like functionality to building exteriors. With further technological development metabolic materials may become autonomous structures and not dependant on existing infrastructures for 'survival'. These continually recycling, auto-cannibalizing architectures would emerge from derelict building sites being shaped by their environmental context and responsive to changing urban land use. Metabolic materials are able to carry out their dynamic functions without the need for DNA, which is the information processing system that biology uses. One specific example of agents that are capable of generating functional metabolic materials is protocells. These are dynamic oil in water droplets that are chemically programmable and exhibit some of the properties of living systems. Protocell oil droplets are able to move around their environment, sense it, modify it and undergo complex behaviors, some of which are architectural. The architectural properties of protocells include the shedding of skins, altering the chemistry of an environment through their 'waste' products, the precipitation of solids, population based interactions, light sensitivity and responsiveness to vibration.

Protocells can be 'programmed' chemically to achieve particular outcomes. For example, is possible to create

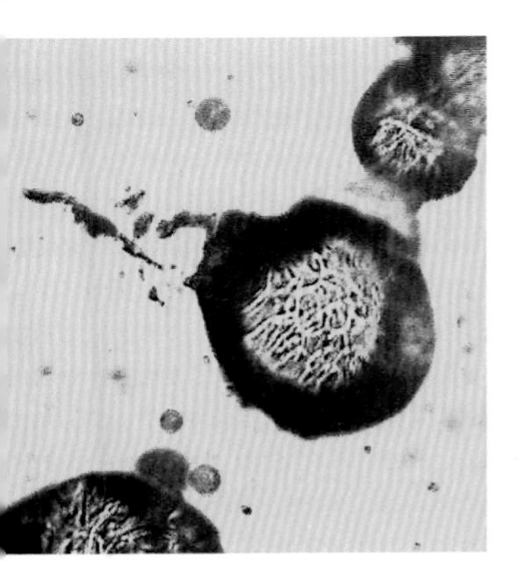

PROTOCELLS

PROTOCELL OIL DROPLETS ARE ABLE TO MOVE AROUND THEIR ENVIRONMENT, SENSE IT, MODIFY IT AND UNDERGO COMPLEX BEHAVIORS, INCLUDING THE SHEDDING OF SKINS, ALTERING THE CHEMISTRY OF AN ENVIRONMENT THROUGH THEIR 'WASTE' PRODUCTS, THE PRECIPITATION OF SOLIDS, POPULATION-BASED INTERACTIONS, LIGHT SENSITIVITY AND RESPONSIVENESS TO VIBRATION. THE CHALLENGE IS TO CHEMICALLY 'PROGRAM' THE PROTOCELLS TO ACHIEVE PARTICULAR OUTCOMES THAT CONTRIBUTE TO THE LARGER ARCHITECTURAL SYSTEM.

a 'carbonate' shell from insoluble carbonate crystals that are produced by protocells when they come in contact with dissolved carbon dioxide. Protocells can therefore produce a limestone like substance and artificially extend the development of this material (created by the accretion of the skeletons of tiny marine organisms), which can continue to grow, self-repair and even respond to changes in the environment. We are developing a coating for building exteriors based on this principle. A practical example of how the first protocell based metabolic materials may inform architectures was developed for a series of collaborations with architect Philip Beesley where active protocells were engineered to be accessible for public display. Sargasso Sea (CITA collaboration for 'Architecture and Climate Change' exhibition, Royal Danish Academy, December 2009), Hylozoic Grove, (Quebec, February, 2010) and Hylozoic Field (Mexico City, Festival of Mexico, March-April 2010) featured protocell 'incubators' that took the form of flasks of modified protocells reaching several centimeters in diameter. A propositional relationship was created between the soft technology and the synthetic framework of the cybernetic field suggesting that living materials in the incubators would replace the inert scaffolding materials of the main exhibit.

A more intricate chemical landscape was designed to exist within a similar cybernetic framework at the Canadian Pavilion for the Venice Biennale, which was exhibited from September to November 2010 in Venice, Italy. The proposed chemical systems within this installation performed a functional and dynamic relationship both to the cybernetic installation and the human visitors. The metabolic materials 'breathed in' carbon dioxide that was naturally dissolved in the water drawn from Venice's canals and were able to demonstrate a carbon fixation process where the waste gas was recycled it into millimeter scale building blocks. In this way metabolic materials turned products of human activity into bodily components for the construction of Beesley's giant synthetic 'life form'. Metabolic materials will challenge the assumptions that we have about architectural building processes and since they require water for their development they are likely to be useful in areas with repeated flooding or in urban areas that are lower than sea level or, as in the case of Venice, have a complex relationship with the sea. Protocell technology could stop the city of Venice sinking on its soft geological foundations by generating a sustainable, artificial reef under the foundations of Venice and spreading the point load of the city.

Metabolic materials will challenge the assumptions we have about architectural building processes

The speculative technology underpinning the construction of an artificial reef under Venice employs a species of carbon-fixing species of protocell technology that is engineered to be light sensitive. The protocell system would be released into the canals, where it would prefer shady areas to sunlight. Protocells would be guided towards the darkened areas under the foundations of the city rather than depositing their material in the light-filled canals, where they would interact with traditional building materials and turn the foundations of Venice into stone. With monitoring of the technology, the woodpiles would gradually become petrified and at the same time, a limestone-like reef would grow under Venice through the accretion and deposition of minerals.

The issues involved with the reclamation of Venice are complex and this particular protocell-based approach addresses just one aspect of a large range of factors that threaten the continued survival of the city. However, other metabolic materials besides the protocell technology may have further potential to address other significant issues in this multifactorial situation, such as the very pressing problem of rising damp in the fabric of Venice's buildings where functional 'seaweed wraps' may be able to extract water from waterlogged traditional building materials and attenuate the ongoing significant damage caused by this process.

Metabolic materials may even be able to regenerate problematic areas within urban environments and contribute to regeneration by revitalizing poor areas through carbon fixation methods. Not only would the buildings thrive on the carbon emissions from pollution

COMPUTER VISUALIZATION OF THE VENICE REEF BY CHRISTIAN KERRIGAN

but would add value to the buildings by recycling carbon into the fabric of the buildings the where metabolic materials would function as synthetic 'lungs' on building exteriors. The regenerating buildings would become an integral part of the carbon and construction economies since the buildings would be able to perform useful functions and 'grow' as a result of sinking the waste gases into their substance and transforming these formerly toxic and undesirable environments into useful and even desirable locations. In the next 10 years additional functionality to these urban metabolic materials will go beyond carbon capture and storage so that these interfaces provide a site through which it is possible to recycle the captured carbon and produce fuels and other materials that have been created by further metabolic processing of the chemical systems. The recycled fuel could then be collected through systems within the 'breathing organs' (like air sacs within a lung) and reused within the architecture, consequently making more efficient use of oils and combusted substrates and providing further basis for a thriving economy.

Ongoing developments and engineering of metabolic materials even suggest that they will have a restorative effect on the environmental chemistry where the most effective way to 'heal' a stressed ecology may be to construct living buildings. In this case metabolic materials could be thought of as performing the role of environmental pharmaceuticals. These architectural interventions may not intend to provide housing for human inhabitants and merely exist in an environmentally restorative capacity where they would be difficult to distinguish from natural materials and accepted as an inherent part of our biological landscape. Metabolic materials and living buildings will not only be able clean up the pollutants that we pump into the environment but will have the capacity to serve as a first line of defense against climate change and unpredictable environmental events since their sensors, intelligence and efforts are embedded in real environmental event not ones that are simulated using traditional computers. Moreover, metabolic materials possess a language that is found everywhere on planet earth in the physics and chemistry of matter and this new approach to constructing architecture could benefit developing countries as much as First World nations. In this scenario our architectures would be able to serve as an early warning system for catastrophe in a manner similar to the potential of animals to sense impending disaster. In the advent of adversity, living buildings would be the first to respond to damage or detect human life within collapsed frameworks and in many ways they may come to be regarded as our architectural 'best friends'.

REFERENCES

1 BEDAU, M. 2009. LIVING TECHNOLOGY TODAY AND TOMORROW. TECHNOETIC ARTS. VOLUME SEVEN, NUMBER TWO, INTELLECT JOURNALS, PP 199-206.

GROW YOUR SNEAKER

www.rayfishfootwear.com

RAYFISH©
FOOTWEAR

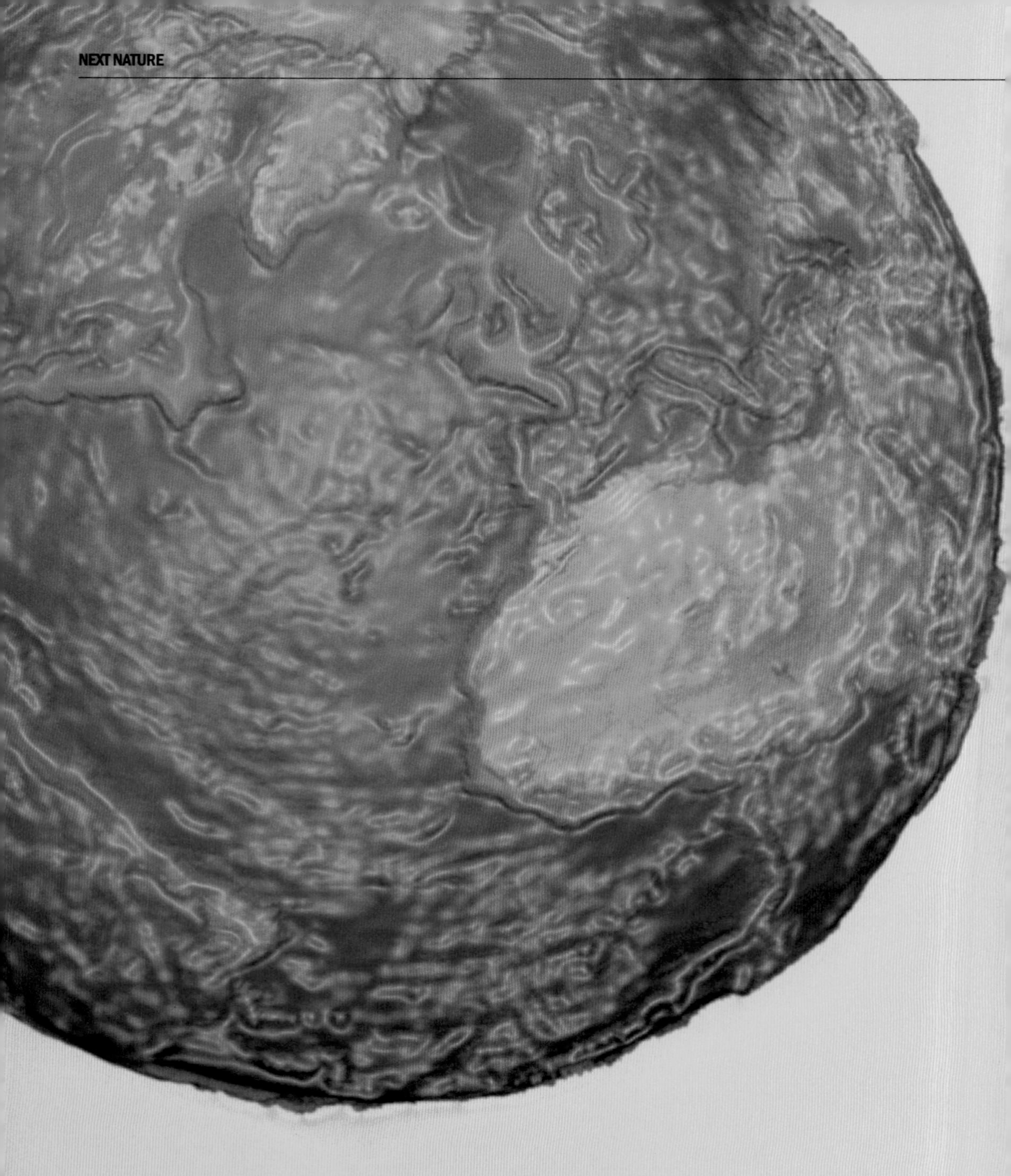

Everybody's Plastic, But I Love Plastic. I Want To Be Plastic.

Andy Warhol

Plastic Planet

We tend to think of plastic as a cheap, inferior and ugly material used to make children's toys, garden furniture and throwaway bottles. But as an experiment, imagine for a moment a world in which plastic was extremely rare, like gold or platinum, and plastic objects were devastatingly expensive to produce. One would encounter plastic objects only at special occasions; one would see and touch very few plastic objects throughout one's lifetime. I know it's a challenge, but try to imagine, for the sake of our experiment, that plastic was scarce, available only to the happy few, and the masses lived in a world of wood, pottery and metals. Ready?

Now look around you and grab the first plastic object you see. Look at the object. Study it. It doesn't matter if it's a coffee cup, a cigarette lighter, a pen or a plastic bag. This is a special moment. You are now holding one of the few, delicate pieces of plastic you will ever get to touch. Feel how durable it is. Feel how light it is considering its volume. Feel how strong and rigid it is, or how very flexible. Get a sense of how easy it must have been to mold. Understand that it could be molded into something else again. If plastic weren't such an omnipresent material, we would realize that it is beautiful. And that it is a disgrace that we throw so much of it away.

The word plastic stems from the Greek word *plastikos*, meaning capable of being shaped or molded. It refers to the malleability, or plasticity during manufacture, that allows it to be cast, pressed or extruded into almost any shape. Before the first synthetic plastics were produced, substances occurring in old nature – gutta-percha, shellac and the horns of animals – were used as plastic material. Bakelite, the first plastic based on a synthetic polymer, was invented in 1907. It was molded into thousands of forms, such as cases for radios, telephones and clocks, and billiard balls. After the Second World War, improvements in chemical technology led to an explosion in new plastics – among them polypropylene and polyethylene – which rapidly found commercial application in a wide spectrum of products, from coffee cups, to shampoo bottles, to bags, eyeglass frames, medical instruments and, well, almost everything and anything that surrounds us.

Oceans of Plastic

In 1997, a Californian sailor named Charles Moore came across an enormous stretch of floating debris while traveling across the edge of the North Pacific Subtropical Gyre (a region often avoided by seafarers). Throughout the week it took him to traverse the area, there was always some piece of plastic bobbing by: a bottle cap, a toothbrush, a cup, a bag, and a torrent of unidentifiable pieces of plastic bits. Moore sensed there was something wrong. He had identified what is now called the Great Pacific Garbage Patch, an area in the central North Pacific Ocean larger than France or Texas, which contains exceptionally high concentrations of marine trash.

The fact that plastic hardly breaks down is well known, but rarely talked about. Plastic does not biodegrade, as microbes haven't evolved to feed on it. It can photodegrade however, meaning that sunlight causes its polymer chains to break down into smaller and smaller pieces, a process catalyzed by friction, as when pieces are blown across a beach or rolled by waves. The same process is in play when pieces of rock are rounded by ocean waves. It is this type of frictional erosion that accounts for the majority of unidentifiable flecks and fragments making up the massive plastic soup at the heart of the Pacific.

Captain Moore's research revealed six times more plastic in the area than plankton. Further, it was discovered that 80% of the debris had initially been discarded on land – a finding later confirmed by the United Nations Environmental Program. Wind blows the plastic through streets and from landfills. It makes its way into rivers, streams and storm drains, then rides the tides and currents out to sea, finally ending up in an ocean gyre. And the trash-vortex Moore discovered isn't the only one – the planet has six additional major tropical oceanic gyres, all of them swirling with debris.

Mermaids' Tears

Nearly all the plastic items in our lives begin as these little manufactured pellets of raw plastic resin, known in the industry as nurdles. They are made from the natural gas portion of our petroleum resources. Their pelletized form – typically under five millimeter in diameter – represents the most economical way to ship large quantities of a solid material. Over 100 billion kilograms of nurdles are shipped each year to processing plants, where they are used as the raw material that is heated up, stretched and molded into familiar plastic products and packaging. Nurdles are small and light enough to become airborne in a strong wind and float wonderfully in water. The most common source for nurdles is industrial spills from trucks and container ships. Because nurdles are so small, they are hard to contain and effortlessly slip away from containers into waterways or into the ocean directly. You can find nurdles on virtually every beach; hence their nickname: mermaids' tears. These days, there is no such thing as a pristine sandy beach anymore. The ones that look pristine are usually groomed, yet if you look closely you will always find plastic particles that have been dropped by the tides. All this plastic has accumulated in less than a century. It's as if plastic fell on the world, like a tiny drop of water, with ever expanding ripples. The long-term impact of mermaids' tears on the oceans and the planets ecosystem is hard to predict. We know that plastic is extremely durable, but will it last long enough to enter the fossil record? Millions of years from now will geologists find the fossilized imprints of your garden furniture imbedded in conglomerates formed in seabed depositions? Chances are, they will – assuming geologists are also still around.

NURDLES, PLASTIC FOUND ON THE BEACH

STYROFOAM GRANULES ON A BEACH ROCK, PHOTO BY RALPH HOCKENS

A New Material in the Earth's Ecosystem

Plastic is now part of our planet's food chain; the problem is that there is nothing in the food chain that can digest it. It ends up in the stomachs of many sea creatures from fish, to turtles and albatrosses. According to the United Nations Environment Program, plastic is killing a million seabirds a year and 100,000 marine mammals and turtles. Next to the deaths from entanglement thanks to six-pack rings and discarded synthetic fishing lines and nets, a common cause of death in animals is choking and clogged digestive tracts, leading to fatal constipation. One wonders what Darwin would have thought of albatross babies being fed plastic by their parents, who soar out over the vast polluted ocean foraging for what looks like food to bring back to their young. We know by now that every second nature typically stresses a first nature, which in effect deteriorates, after which the victorious second nature becomes the first. But are we ready for a plastic planet? Cleaning up the huge accumulation of plastic in the ocean is basically impossible, although it is feasible to collect the larger bits of plastic debris. Most biologists are focused on beach cleanup, and on the reduction of garbage, which could end up in the ocean. Obviously we need to change our act: put a stop to the throwaway society, increase awareness of the environmental impact and produce biodegradable forms of plastic. We can do that. And we should. Yet, that bit of mindful recycling you are urging yourself to turn into a habit, might work as a mantra, but won't undo the damage that has already been done. The proliferation of mermaids' tears could continue to hurt marine organisms for thousands of years, even if we stopped all plastic production today. Plastic is a new material in the Earth's ecosystem and we've caused it.

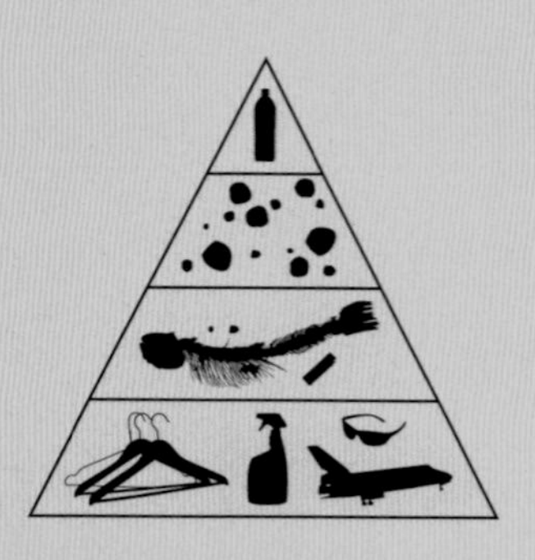

HOW PLASTIC ENTERS THE ECOSYSTEM

BIODEGRADABLE PLASTIC BOTTLE

Bugs that Eat Plastic

Some geologists have already suggested the current period in the Earth's history will be remembered as the anthropocene; a geological time frame characterized by the global impact of human activities on the Earth's ecosystem. Since our planet came into existence, it took about a billion years to form a biosphere of life around its geosphere. Some three and a half billion years later humankind emerged. We in our turn created a noosphere – a sphere of human thought and activity – that now influences both the biosphere and the geosphere. Plastic is our poisonous gift to the planet. We took the oil out of the ground, transformed it into plastic and brought it to the oceans. Ironically, like plastic today, oil too was once waste, created by the remains of vegetation that died millions of years ago, sunk in the sediments and due to the long-term geological pressure was transformed into oil where it remained until people discovered it was of use to them and started pumping it up. Who knows, in due time, some other organism or intelligence will appreciate plastic as a valuable material and mine or feed on it, like we have done with oil. But how many sea creatures will have to perish before this happens?

The only sensible way to think of plastic nowadays is to consider it as a raw material within the ecosystem of the earth. The apparent problem is that there is no species, process or actor that can feed on it: it is a next nature material, with its balancing counterpart yet to evolve. Perhaps some future-evolving microbe, able to digest plastic, could thrive on the vast amount of plastic 'food' available in the ecosystem. It might take a million years, however, for such a plastic eating microbe to evolve. Yet, it would certainly have enough food to proliferate.

But why wait for evolution? 16-year-old high school student prodigy Daniel Burd developed a micro-organism in 2008 that can rapidly biodegrade plastic. Daniel realized something that even the most renowned scientists had not considered: Although plastic is one of the most indestructible manufactured materials, it does eventually decompose. This means there must be microorganisms out there to do the decomposing. Daniel wondered whether those microbes could be bred to do the job faster and tested this by a simple yet clever process of immersing ground plastic in a yeast solution that encourages microbial growth. After which he only had to isolate the most productive organisms – a sort of speedy evolution. His first results were encouraging, so he went on, selecting out the most effective strains and interbreeding them. After some weeks of tweaking and optimizing temperatures he achieved a 43% degradation of plastic in six weeks. Daniel presented his results at the Canada-Wide Science Fair in Ottawa where he won the first price for his study. Meanwhile a 16-year-old girl from Taiwan had discovered a microbe able to break down Styrofoam.

Despite the excitement of having these young geniuses creating plastic eating microbes, we should be extremely careful before applying them. It may sound brilliant to have a colony of plastic-eating microbes clean up the oceans at first sight, yet we must not be naïve about the potential side effects. One of the main advantages of plastic – and why we use it everywhere – is its resistance to biodegradation. Plastic is used in hospitals, vehicles, homes, industrial settings, etc. One can easily imagine the potential dangers of having a plastic eating bug out in the wild. The risk of having microorganisms devour your garden furniture is perhaps acceptable; having them enter a hospital setting would be more problematic. Imagine the mayhem an attack of plastic eating microbes would cause in that precariously sterile environment, causing dangerous drugs, viruses and fluids to run loose. Imagine a plastic eating bug colony gobbling up the coatings of electric cables, causing our communication networks to break down.

The dilemma we face is that in our ignorance towards our environment we have introduced a new material into the Earth's ecosystem that – like a giant meteor from outer space – has radically altered its equilibrium. If we do nothing, sea life will continue to suffer for ages. At the same time, letting plastic eating microbes clean up our mess is not like pushing the "undo" button and reverting to the ecological balance previously enjoyed. Things will be different. There will be side effects. Nonetheless, as catalysts of evolution, it seems sensible to steer towards a balance that is considerate of our own interests and those of our fellow species. Designing plastic eating microbes, if we must. | KVM

PLASTIC STYLE. LEATHER BAG.

THE DISPOSABLE LOOK, THE SUISTAINABLE PRINCIPLE.

NEXT NATURE SERVICES

By Bas Haring

Intentionality separates culture from nature. A dog is intentional, a fox is not; a park is intentional, a forest is not. Since trash, ruined buildings, and automated computer programs are unintentional, they are also a type of nature. Nature provides human society with valuable 'ecosystem services' such as water purification or erosion control. Next nature provides ecosystem services of its own, although they might not be what we expect.

2010 was the International Year of Biodiversity. The United Nations introduced the concept as a way to draw attention to the decline of nature. Advocating on nature's behalf, a relatively new argument emerged, 'ecosystem services': useful things nature does, unbeknownst to us. Forests filter dust from the air, scrub prevents erosion, and insects pollinate our crops. Incidentally, nature provides us with services that would otherwise have cost a fortune. Leaving aside the question of where they could be purchased. Is it conceivable that one day there will be next nature services, delivered in passing and unintentionally by new, future ecologies?

A rainforest is nature, a park is not. Foxes are nature, dogs are not.

But what makes nature nature? What makes it so valuable and special? Perhaps seeing nature in exaggerated and simplified terms, I can start to think about its future. Is spontaneity not the essence of nature? Put differently, the absence of conscious planning is the essence of nature. A rainforest is nature, a park is not. Foxes are nature, dogs are not. And the ocean is nature, but an oceanarium is not. Parks, dogs and oceanariums have been thought up – we intentionally created and designed them. Nature, by contrast, is not a result of intention. Nature just is. At most it's a consequence of a 'natural process'. The very phrase 'natural process' illustrates the essence of nature: 'that's just the way it is' or 'of itself'. The absence of this deliberation or intention is also the source of nature's charm. Nature is surprising. It can be surprising, because no one has thought about it in advance. Nature humbles us in all her beauty. Beauty that we had no part in. Ferns, ibises and dragonflies are magnificent, but we didn't create them or think them up.

The distinction I draw between the intended and unintended shows that there is still a place for 'real' nature in the manufactured nature of the park and the oceanarium. Grass stubbornly creeps between the paving stones in the park, and millions of unintended and uninvited plant and animal species live in the water at the oceanarium. Even sheepdogs, shining examples of obedience in the animal world, will occasionally, unintentionally go against their character by chasing after rabbits. Parks, oceanariums and dogs are less natural than forests, oceans and wolves, because they are deliberately designed rather than having simply evolved.

I wonder how to interpret the statement, 'Meadow birds belong in the Netherlands'. Or other pronouncements about what nature is supposed to be like: 'Lions belong in Africa' and 'Oranges belong on orange trees.' I don't think the sentence 'Meadow birds belong in the Netherlands' is a strange one. I might even think it's true. But if I believe nature is a product of random circumstance, then what do I mean by that sentence? Can something belong somewhere without intent? I believe so. Even if everyone knows meadow birds are indigenous to the Netherlands – they are simply there – one can still believe they belong there. Despite the fact that oranges were not invented or intended (humans did not invent oranges to grow) to grow on orange trees, it's not strange to argue that they belong there. Something that is intentional should be as it was intended to be. But something unintentional can evidently also belong somewhere. There is a difference between 'belonging to' and 'belong'. An orange may belong on an orange tree, but that does not directly imply that the orange tree is supposed to be that way. But enough about the difference between intention and belonging. Let's get back to nature.

Is nature green per se, made up only of organic molecules and living cells? I don't believe it is. Mountains are nature too. They came into being through a natural, unplanned process. And mountains are not composed of organic molecules but of materials like silicon dioxide and limestone, as are streams and salt flats. These things are not green or made of organic material, and yet as far as I'm concerned, they're part of nature.

Picture yourself in Iceland, walking on top of a volcano with a friend. Around you are bare rocks as far as the eye can see, and to your left is a mountain stream. At one point your friend says, 'Isn't nature spectacular?' You probably won't be surprised – 'But this isn't nature; nature's made out of organic material!' Instead, you will agree with your friend – 'Yes, it's spectacular'. Following this line of thought, it's possible that nature can consist of other materials too. If lime and salt are okay, then why not plastic and electronics? As long as something is unintentional, it can be natural, or perhaps it is even natural by definition.

Incidentally, nature provides us with services that would otherwise have cost a fortune.

Near the Dutch city of Almere is an unfinished modern castle, it was originally intended as a luxury hotel, but it was never completed and will never be. Instead of a modern replica of a medieval castle, there is a rough castle-shaped block of building materials – nowhere near the original intention. This modern ruin in the middle of the forest is more natural than the surrounding woods. The trees were planted, intentionally; the castle's current form is an accident. The unintentional, chaotic organization of large companies could perhaps also be understood as next nature – marketing departments redoing the work of communication departments; little groups of people who don't know what the others are doing and may even be working against each other, unknowingly. And then there are the messages generated by Twitter bots, automatic tweet-generating programmes. No one creates these random tweets (if you don't include the programmer) – another new kind of nature. In the future, maybe Twitter bots will have brief conversations with each other, without any human intervention: 'How are you?' 'Fine, thank you. How are you?' These unintentional conversations can be considered a new kind of nature.

The world is becoming increasingly planned and thus increasingly unnatural. The more people there are, taking up more space, the more we think about that space. Unplanned, natural space turns into planned, unnatural space. But I believe the unintentional will keep creeping up in between all those intentions, like grass between the paving stones in the park. It may happen in odd places – inside computers, on building sites, in organizations – but the unintentional will stick around. Will this new nature potentially be of value? When it comes to value in nature, the following paradox applies: plants and animals hold value for us mainly in manufactured sense. The value of agricultural crops is obvious, but maize and grain fields are not nature. The most valuable trees grow in planted forests, not in 'real' nature. And the animals we eat are rarely wild, natural ones.

The term 'value' is a complicated one. There are 'intrinsic value', 'aesthetic value' and 'economic value', and probably many other kinds too, but to reduce the complexity somewhat, I will refer here mainly to economic value – not because I believe it is the only kind of value that matters but because it is the easiest to grasp and the least debatable. The plants and animals that possess the most value to us – maize, grain, vegetables, oak, pigs, grass, cows and chickens – no longer have value in nature. They are cultivated, planned and controlled, in fields, barns and planted forests. It is non-nature, lifted out of nature through intention that has obvious value.

In one town, an old rubbish dump was transformed into an indoor piste, giving new value to something that once had none.

But what about the value of genuine nature – the virgin forests of Siberia, the gulls in the Wadden Sea? Don't they still hold value, even if it is unintended? And it is precisely here that we find the invisible ecosystem services: that nature provides. Worms, along with millions of species of bacteria and single-celled organisms, keep the soil fertile so that we have maize and grain to harvest; forests filter dust from the air; insects pollinate our crops. These are invisible, valuable services provided by nature – incidental services from unintentional nature. And they are much more exciting than the value of intentional animals and plants in parks, barns and oceanariums. Those are intended, here for a reason, and so, logically they have value. But the fact that unintentional nature has value too might come as a surprise.

Ecosystem services supply nature conservationists with a timely argument for their cause. And ecosystem services are one piece of evidence the U.N cites in its defense of nature. If nature contributes incidental value, then it would seem logical that unintended new nature can too. If the essence of nature is its lack of planning, if nature has various unintentional kinds of value, generated in passing, and if nature is not made of organic material per se but could also consist of plastic, buildings and software in the future, then this suggests that new nature will also have new kinds of value in the future.

As long as something is unintentional, it can be natural, or perhaps it is even natural by definition.

Is this really conceivable? Is it possible that tweeting robots, chaotic organizations, modern ruins and other forms of new unintentional nature secretly have value, without it being intentional, and without us knowing it yet?

It just might be, and I have already seen the first indications. The Netherlands is a flat country and this is of value. It makes a big difference to the cost of agricultural labor. But a hill here and there can also be valuable, even if it's just used for skiing. In one town, an old rubbish dump was transformed into an indoor piste, giving new value to something that once had none. It is true that the site was built intentionally and according to plan, but as a dump, not a ski slope. Its value as a hill only became apparent later. Shipwrecks and sunken drilling platforms are another example (can be warm or cold ocean, doesn't matter to fish). Without intention, they lie rusting and rotting on the seabed. Yet they have turned out to be of great value. Fish and other forms of life gather around these wrecks. Divers swim there, and fishermen make extraordinary catches. These unplanned wrecks have unintentional value: a service is provided accidentally by a new, next nature. The fibers in wrecked cars from wiring insulation and upholstery are a final example. These fibers are a byproduct of modern car salvage. After the steel and other valuable materials have been removed, rubber and fibers remain. People found no use for these fibers until it was discovered that they could be used in water purification. Certain pollutants bind to them perfectly. Perhaps even the plastic island – the enormous accumulation of synthetic material floating near Hawaii that is larger than France – secretly has value, as an island that was not planned and is therefore nature. It is not inconceivable that this plastic mountain will turn out to have incidental value. In any case, we must continue to look at possible new natures with a fresh eye. Nature is spontaneous, and therefore it is also unexpected. Next nature could manifest itself in many unexpected ways, with many unexpected kinds of value.

ECONOMY
= ECOLOGY

10,000 BC

TWO GOATS FOR A COW

10,000 YEARS AGO COWS WERE USED AS MONEY. SHEEP ARE USED AS SUBSIDIARY UNITS AND SMALL CHANGE IS GIVEN IN LAMBSKINS.

Virtual Money is a Pleonasm

A few years ago the first virtual millionaire was announced, yet there have also been reports of people enslaved to virtual gold farms. The Chinese government recently announced a limit on the use of 'virtual' currencies.

Firstly, we should realize the term virtual money is a pleonasm, a redundant expression. Money is, by definition, virtual. And it always has been. Well, maybe not during the time people used cows and goats to barter. A cow is a living creature, and useful too. You can drink its milk and when it no longer produces milk you can always slaughter it for meat. We may not think about it, but it is actually a miracle that we can exchange a piece of paper for a rump steak at the butcher.

If we are to fully understand how money came into being, we must go back to ancient China where, as in many other places, a lively trading system arose: an apple for an egg, two goats for a cow, a hammer for a bucket. Because they were durable and universally useful, many tools were traded such as knives and ploughs, and in particular the shovel became a popular barter object. A rich man had a whole row of shovels on his land that were, in fact, not used to work the land but served purely as a means of trade. Blacksmiths made additional shovels that became progressively smaller: these were not handy for digging, but extremely handy for bartering. At a given moment, the shovels had become as small as a present-day coin. Then finally somebody came up with the idea of making them round. The abstraction was complete.

Authority = Reality

In order to emphasize their value, coins were frequently made from precious materials such as silver or gold. This brings us to a following step in the ongoing 'virtualization' of money. In 1973, the American government decided to dispense with the gold standard.

1,000 – 500 BC

SPADE CURRENCY

BECAUSE THEY WERE DURABLE AND UNIVERSALLY USEFUL, TOOLS WERE TRADED AND THE SHOVEL BECAME A POPULAR BARTER OBJECT. A RICH MAN HAD A WHOLE ROW OF SHOVELS ON HIS LAND THAT WERE NOT USED TO WORK THE LAND BUT SERVED SOLELY AS A MEANS OF TRADE. BLACKSMITHS MADE ADDITIONAL SHOVELS THAT BECAME PROGRESSIVELY SMALLER OVER TIME.

Paper money had been introduced earlier: beautifully designed and printed notes but the material used, however, in contrast to coins, did not represent any particular value. The value of paper money exists purely thanks to the bank guarantee, by which a bank note can always be exchanged for the equivalent value in coins. Much handier than having to lug round a bag of coins! And thus the gold standard was relinquished: Money was henceforth only based on trust. This is, in fact, a collective illusion, but as long as everybody believes in the value, it works excellently. Can it become even more virtual? Agreed, you can introduce debit cards and credit cards, which means that your money is stored digitally somewhere on the bank's server. Because these methods of payment have only recently been introduced we still have to grow accustomed to the idea and consider these monetary systems as 'virtual', which incorrectly suggests that the bank note I use to pay for my rump steak is actually 'real'. We conveniently, or perhaps it is better to say pragmatically, ignore the fact that money is actually by definition virtual: it only has a symbolic value that is ingeniously constructed as replacement for the unhandy and less than precise barter of commodities. The step from the bank note to the digital administration of your possessions is only a baby step in comparison to the enormous symbolic leaps which had been made in the course of centuries: the replacement of valuable living creatures and commodities (such as cow, goat or tool) with valuable materials (such as gold and silver) to valueless representations of valuable materials (bank notes), to simple faith in a government that claims to guarantee the value of your bank note without having it backed up by a fort filled with gold... now that's progress! Using your direct debit card for a pair of sports shoes (total bill: 50 Euro) is much easier than exchanging them for two chickens and an egg. It makes very precise payments, complex financial construc-

600 BC

SPADES BECOME COINS
FOR PRACTICAL REASONS, THE SHOVELS HAD BECOME AS SMALL AS A PRESENT-DAY COIN. BEING MADE OF BASE METAL THE CHINESE COINS WERE OF RELATIVELY LOW VALUE AND THEREFORE INCONVENIENT FOR EXPENSIVE PURCHASES. IN OTHER REGIONS, COINS WERE MADE OF VALUABLE METALS LIKE BRONZE, SILVER AND GOLD.

tions and money that 'flashes' round the world at the speed of light possible. But here is also a downside: although it may be physically rather more convenient (because you no longer have to cart around a herd of cows or a basket of eggs) the introduction of money has created a cognitive burden: you have to think about it! You can't simply leave your money in a corner and be sure it will retain its value, you have to deposit it in the right account or invest it sensibly, keep track of inflation and watch out that the bank or government that guarantees its value doesn't collapse.

Sometimes things can go terribly wrong, as they did in China in the 15th century, where a bank note worth 1,000 coins dropped in value, as a result of hyper inflation, to just three coins, and this subsequently caused the government to do away with paper money altogether. Incidentally, the Chinese bank note was also dispensed with in the 11th century and later reintroduced. Can that happen again? Of course. We shouldn't expect bank notes, credit cards and debit cards to last forever.

Paying with cell phone minutes

An interesting banking innovation has taken off in Africa. In various parts of the continent, cows were still used as currency until the 1960s and overgrazing caused by cattle that were only used as 'value deposits' was responsible for environmental problems until the eighties. Necessity, they say, in the mother of invention, and this is certainly the case in Africa where people have no access to a stable banking infrastructure and therefore have started using phone minutes as currency. It is estimated that last year there were 3.8 billion mobile subscriptions in the developing world – that's 73% of global subscriptions. They are used to download music, to send text messages and for playing games, but also as a wallet. The vast majority

100 BC

BANKNOTES REPLACE HEAVY COINS

IN CONTRAST TO COINS, BANKNOTES DID NOT REPRESENT ANY MATERIAL VALUE. THEIR VALUE EXISTS ONLY THANKS TO A BANK GUARANTEE, BY WHICH THE NOTE CAN BE EXCHANGED FOR THE VALUE IN COINS: THE SIGNIFIER BECOMES THE SIGNIFIED.

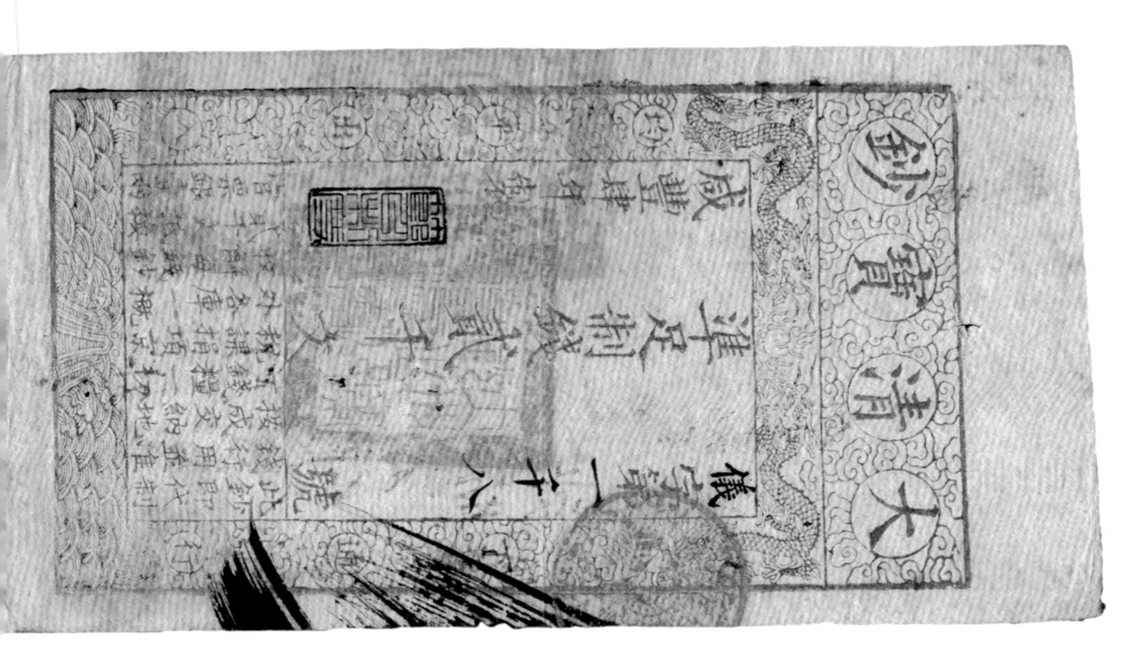

of telephone owners do not have a subscription, but purchase phone minutes from one of the many telephone shops.

A mobile network in Kenya has an innovative payment technology called M-Pesa (M is for 'mobile', pesa is Swahili for 'money'), which allows people to transfer money to one another using airtime resellers or outlets. At the start of 2008, the political instability in Kenya led to violence and many telecom shops were forced to close their doors; telephone cards became scarce. People started sending each other phone minutes and - you know where this is going - these minutes were not used for telephone calls: they were used as currency. Very quickly, telephone cards became more valuable than cash. Family members could send each other phone minutes over great distances; that was much more reliable than an envelop with bank notes. Even aid organizations started distributing telephone cards which people then used to purchase food and other basic needs.

Telecoms followed this development with considerable interest. They are, after all, the ones who can create phone minutes at the press of a button. Certainly in the current situation, in which the financial system is under pressure throughout the world and traditional bankers have, with their unconcealed greed, shown their worst sides, mobile network providers believe they have a good chance of becoming the new bankers. Perhaps in time it will no longer be the government that guarantees your money, but companies like Vodafone, AT&T and T-Mobile. |KVM

20TH CENTURY

CAN IT BECOME EVEN MORE VIRTUAL?

IN 1974, U.S PRESIDENT NIXON ABANDONED THE GOLD STANDARD. THE DOLLAR NO LONGER HAD AN INTRINSIC VALUE IN GOLD, BUT BECAME A FIAT CURRENCY, WHOSE VALUE IS MERELY BASED ON ITS DECLARATION BY THE GOVERNMENT AS A LEGAL TENDER. THE FRENCH, BRITISH AND JAPANESE HAD ALREADY DECOUPLED THEIR GOLD STANDARD EARLIER.

CAN IT BECOME EVEN MORE VIRTUAL?

THE USE OF CREDIT CARDS ORIGINATED IN THE UNITED STATES DURING THE 1920S, WHEN INDIVIDUAL FIRMS, SUCH AS OIL COMPANIES AND HOTEL CHAINS, BEGAN ISSUING THEM TO CUSTOMERS. IN 1950, DINERS CLUB ISSUED THEIR FIRST CREDIT CARD, INTENDED TO PAY RESTAURANT BILLS, IN THE UNITED STATES. AMERICAN EXPRESS ISSUED THEIR FIRST CREDIT CARD IN 1958. LATER THAT SAME YEAR, BANK OF AMERICA LATER ISSUED THE BANK AMERICARD (NOW VISA) CREDIT CARD.

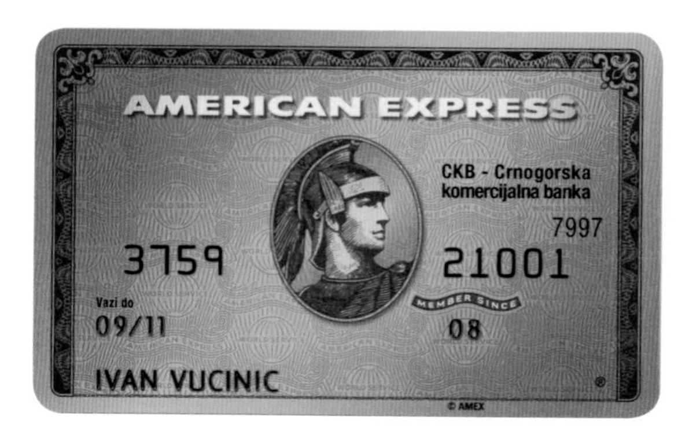

Paper Money is in Fact Just the Poor Man's Credit Card

Marshall McLuhan

21ST CENTURY

TELECOM PROVIDERS BECOME BANKS

IN KENYA, SAFARICOM SETS UP THE FIRST SUCCESSFUL MOBILE BANKING SERVICE. DOES THIS SIGNAL A NEW FUTURE FOR TELECOM COMPANIES?

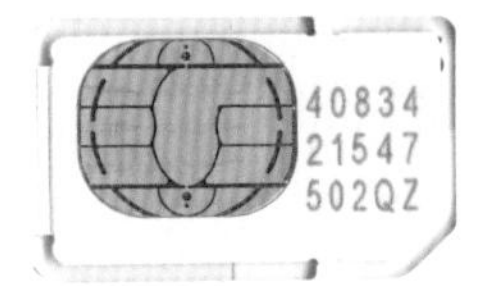

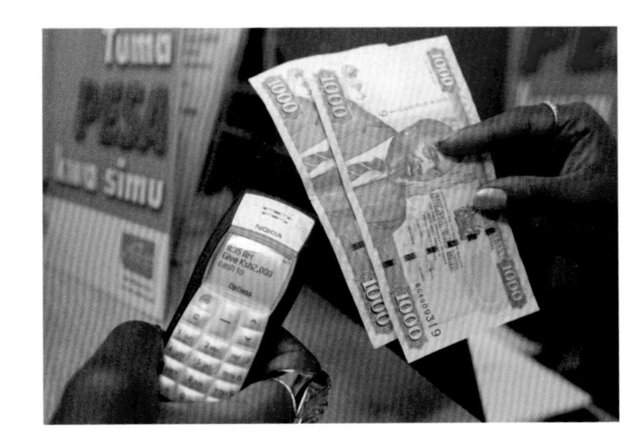

Datafountain

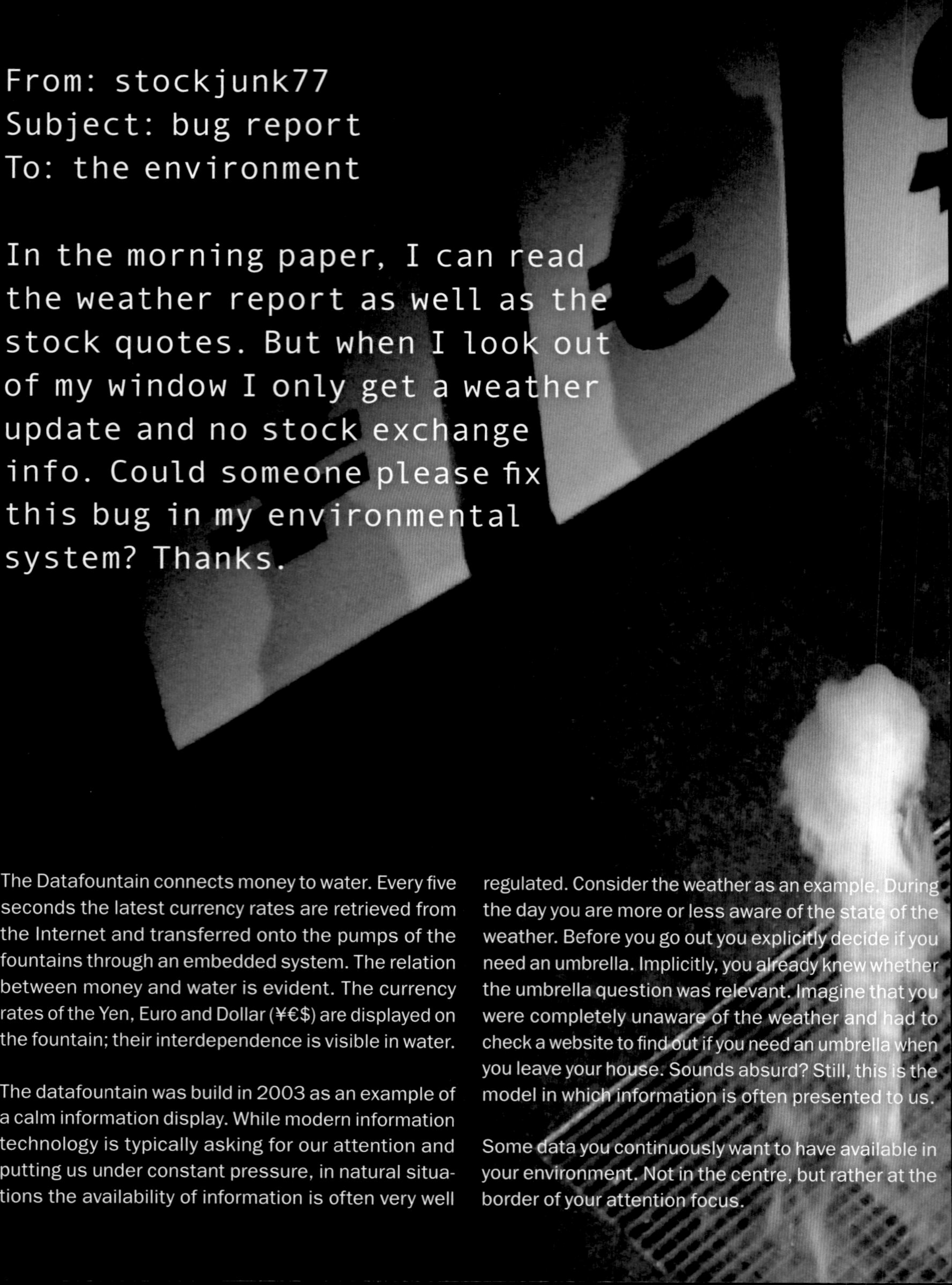

From: stockjunk77
Subject: bug report
To: the environment

In the morning paper, I can read the weather report as well as the stock quotes. But when I look out of my window I only get a weather update and no stock exchange info. Could someone please fix this bug in my environmental system? Thanks.

The Datafountain connects money to water. Every five seconds the latest currency rates are retrieved from the Internet and transferred onto the pumps of the fountains through an embedded system. The relation between money and water is evident. The currency rates of the Yen, Euro and Dollar (¥€$) are displayed on the fountain; their interdependence is visible in water.

The datafountain was build in 2003 as an example of a calm information display. While modern information technology is typically asking for our attention and putting us under constant pressure, in natural situations the availability of information is often very well regulated. Consider the weather as an example. During the day you are more or less aware of the state of the weather. Before you go out you explicitly decide if you need an umbrella. Implicitly, you already knew whether the umbrella question was relevant. Imagine that you were completely unaware of the weather and had to check a website to find out if you need an umbrella when you leave your house. Sounds absurd? Still, this is the model in which information is often presented to us.

Some data you continuously want to have available in your environment. Not in the centre, but rather at the border of your attention focus.

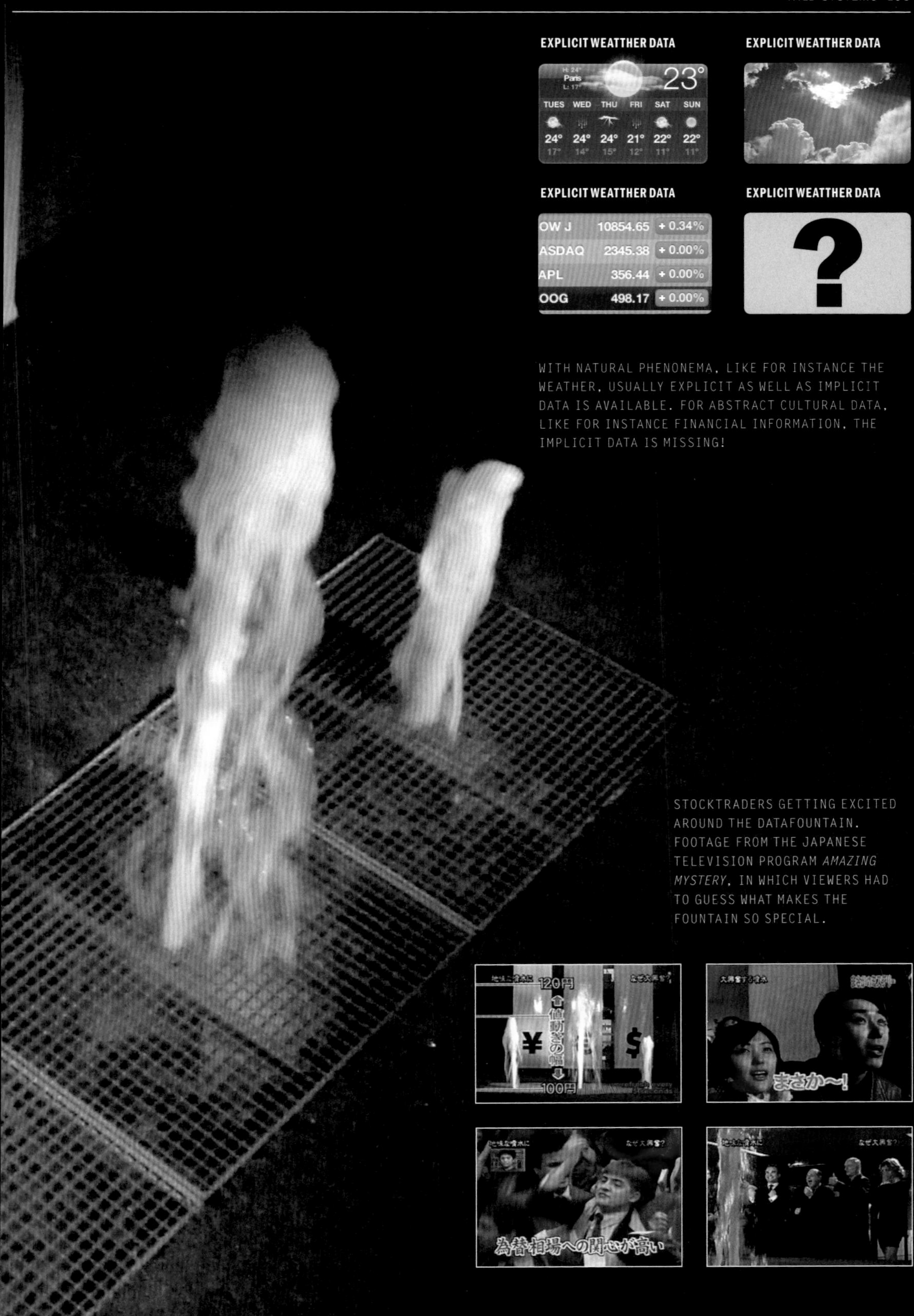

WITH NATURAL PHENONEMA, LIKE FOR INSTANCE THE WEATHER, USUALLY EXPLICIT AS WELL AS IMPLICIT DATA IS AVAILABLE. FOR ABSTRACT CULTURAL DATA, LIKE FOR INSTANCE FINANCIAL INFORMATION, THE IMPLICIT DATA IS MISSING!

STOCKTRADERS GETTING EXCITED AROUND THE DATAFOUNTAIN. FOOTAGE FROM THE JAPANESE TELEVISION PROGRAM *AMAZING MYSTERY*, IN WHICH VIEWERS HAD TO GUESS WHAT MAKES THE FOUNTAIN SO SPECIAL.

OLD ECOLOGY
The rainforest: a
stable and sustainable
yet threatened ecology,
that feeds on sunlight.

Economy is Ecology

Imagine an alternative currency for environmental value. So instead of destroying rainforests, farmers could earn an ECO currency for maintaining them. Can strengthening the link between economy and ecology help circumvent environmental disaster?

The starting point of the ECO-currency* project is the hypothesis that an important factor in the ongoing environmental crisis is the disconnect between the economical and environmental ecologies. With the latter we mean the ecology of plants, trees, animals, and other organic material, whereas the economical ecology is defined by our financial system of market, money, goods and other economical exchange. Our second working hypothesis is that we could address environmental issues by strengthening the link between the economical and the environmental spheres.

Before diving into our analysis we should address the question of whether it is necessary to be critical of addressing ecology as metaphor or structure vis-à-vis ecology as an organic threatened living environment that we are part of? Traditional environmentalists might object to describing both spheres as 'ecologies' and argue it is inconsiderate to use the same term for a man-made system as well as for the older and deeper organic ecology of nature. Surely they have a point here: the environmental ecology is not only much older than the economic, it also rests on the premise that the economical ecology cannot exist without the environmental ecology.

On the other hand it's important to note, that while the environmental ecology is threatened, the economical ecology is currently the most threatening one. Hence it should be taken seriously and not be waved aside as simply a man-made structure. Once we agree that both systems are ecologies in their own right, we can move on and study the structure of both ecologies and their interdependence. The environmental ecology has been a dynamic and stable system for centuries and its proportions are closely linked to the size of the Earth. The economical ecology, on the contrary, is much more unstable and has been growing explosively for the last 200 years. For any ecology to grow so rapidly it has to be fueled by an outside entity: in this case the environmental ecology. So what we call a growth economy is in fact a process of an economic sphere feeding on the environmental sphere. For a long time the environmental sphere seemed to be an infinite resource, because it was so much bigger than the economic sphere. In the last few decades, however, we have reached a point where the impact on the environmental sphere becomes painfully visible.

We can consider both the rainforest and the financial system as ecologies. The most important question is: what kind of ecologies are they?

As described in the introduction, the core of the problem is that environmental values – although so crucial for the existence and wellbeing of humans, animals and plants on the planet – can hardly be expressed in terms of economical value. As this may sound a bit abstract, let's imagine a Brazilian farmer, who cuts trees in the rainforest, sells the wood and transforms the soil into farming land for corn. Through his actions he gains economic value – he supports his family with the money he makes. Obviously, the downside of his work is that another piece of the forest is lost, the amount CO_2 of in the atmosphere increases and global warming is perpetuated. The righteous environmentalist response would be to call on the farmer's moral obligation NOT to destroy the environment. But then again, who are we to make a moral call upon that farmer? After all, he is only adding a tiny droplet to the bucket of global warming which is almost full, thanks to all of us – yes, that includes you! Especially for people from industrialized developed countries it would be hypocritical to make a moral call on the Brazilian farmer, who might simply respond, "Okay, so you destroyed your own prehistoric woods a long time ago for your own economical benefit and now you're telling me not to cut mine? No, thank you." Shouldn't we, rather than patronizing the farmer on his moral obligation towards the environment, search for ways to economically compensate the farmer to leave the rain forest untouched?

The ECO Currency Project explores the idea of introducing a separate currency for environmental value. Besides the U.S Dollar, Euro, Japanese Yen, Chinese Yuan, British Pound, Brazilian Real, there would also be the ECO, a currency that expresses environmental value. The purpose of the ECO is to counter the lack of representation of environmental values in the economical sphere. For the case of our Brazilian farmer this would mean his piece of untouched rainforest would have a value in ECOs.

Rather than facing the dilemma of cutting down trees and earning money or not cut and contribute to a better environment, the farmer could decide to actively steward his piece of rainforest, protect the trees and earn ECOs, which can be traded for real money, goods and economical benefit for his family.
The funding will come from a tiny tax on the financial system. This tax on banks – not on people – can start as low as 0.005%, but when levied on the funds flowing round the global finance system every day via transactions such as foreign exchange, derivatives trading and share deals, it has the power to raise hundreds of billions every year.

The value of the Brazilian farmer's work can be determined through a collaborative online 'wisdom of the crowds' type platform that distributes the ECOs among the participants: people who conduct activities that contribute to steward the environment. There will be room for a wide variety of jobs, ranging from him sustaining a piece of rainforest to a scientist studying biodiversity. People can offer any job on the platform and when the community decides it contributes to the environment is will be granted ECOs. The value of the ECO elegantly moves along with both the strength of the economy – stronger economy means more pressure on the biosphere, but also more funds to support the ECO – as well as the urgency of the environmental crisis – if thousands of people are laboring to support the environment, the value of the eco will decrease along with the lowered need for more activity in this domain.

Strengthening the link between the economical and the environmental sphere might contribute to a resolution of our current environmental crisis. Potentially, the introduction of the ECO currency can contribute to this link; yet, further study is needed for the strengths, weaknesses, opportunities and threats of the ECO concept. As described, the idea of the ECO certainly isn't the most romantic route towards a better environment, as it asks for an explicit description of environmental values and it remains to be seen if people are ready for this. The implementation of the ECO currency will require a substantial re-cultivation of our financial system, which will certainly not be welcomed by some of the current players. Tough decisions will have to be made, both politically, economically as well as in the moral domain. The implementation of this proposed linkage between the economical and environmental spheres via the ECO currency will surely be far from trivial. Nonetheless we can be sure of one thing: the Amazon rainforest is of enormous value. Are we willing and able to make this value explicit? | KVM

NOTE ON THE ECO-CURRENCY PROJECT

* THE CONCEPT OF THE ECO CURRENCY EMERGED AT THE PARALELO CONFERENCE 2009 IN BRAZIL. THE CONFERENCE BROUGHT TOGETHER ARTISTS AND DESIGNERS FROM BRAZIL, THE NETHERLANDS AND THE UNITED KINGDOM. THE AIM WAS TO DISCUSS WAYS IN WHICH INTER-CULTURAL AND INTER-DISCIPLINARY COLLABORATIONS CAN ENABLE RESEARCH AND NEW INSIGHTS INTO GLOBAL AND LOCAL ECOLOGICAL PROBLEMS. PARTICIPANTS INCLUDED: KOERT VAN MENSVOORT, LUNA MAURER, EDO PAULUS, TACO STOLK, ESTHER POLAK, JAMES BURBANK, KATHERINE BASH. SUBSEQUENTLY THE ECO-CURRENCY PROJECT WAS RESEARCHED IN THE NEXT NATURE LAB AT THE EINDHOVEN UNIVERSITY OF TECHNOLOGY BY MARCEL VAN HEIST, BILLY SCHONENBERG, JOP JAPENGA, WHO PROPOSED THE INITIATION OF A COLLABORATIVE PLATFORM TO DISTRIBUTE THE ECOS AMONGST PEOPLE WHO CONTRIBUTE TO A BETTER ENVIRONMENT.

NEW ECOLOGY
The economy: an ecology that cannot sustain itself and feeds on our biosphere.

The Eco Currency

How does it relate to other currencies?

The ECO currency will be a separate currency that expresses environmental value. Like other currencies its value can fluctuate depending on the strength of the world economy, the scarcity of environmental value and the number of laborers in the environment.

What counts as environmental value?

We have to make environmental value explicit in order to preserve it, although this is almost impossible – there are so many profound environmental qualities we are yet to understand. Think of deep-sea microbes that play a balancing role in the oceans or medicines in undiscovered plants in the rainforest – not environmental value itself – but the active labor of people who steward these, is valuable.

How is environmental labor valued in this project?

Through an online platform where people can offer jobs and other members can give credits to those jobs. The wisdom-of-the-crowds-based communities have been successful as a form of co-creation. See kickstarter.com, a platform that allows anyone to offer and request support for projects.

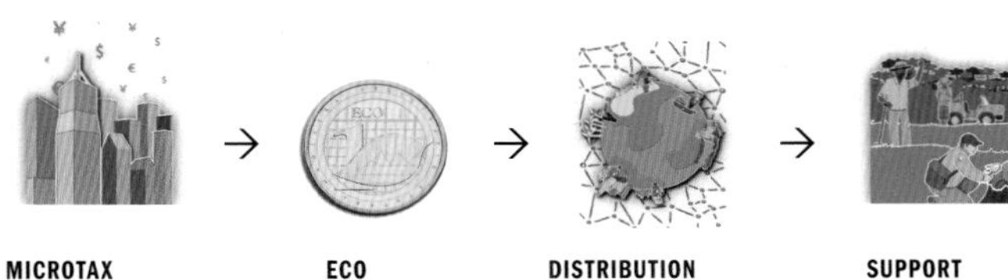

How can I earn ECOs?

There will be room for a wide variety of jobs. People can offer any job on the platform and as long as the community feels it contributes to the environment it will be granted ECOs. Participants can match their skills to an activity. This could range from a farmer sustaining a piece of rainforest to deep-sea biologist doing research, to an administrator guarding against corruption by inspecting participants' jobs.

Are the incentives moral or economical?

Both. As the ECO is a real currency that has a conversion rate in Dollars or Euros, there will be a reasonable economical incentive to earn ECOs. At the same time, one might feel morally righteous and proud of earning ECOs, as this means you are contributing to a better environment.

How does the ECO relate to existing initiatives?

An existing system to balance economy and ecology is the CO_2 credit trading system introduced following the Kyoto Protocol (1997), however a major difference is that this is run as a technocratic scheme excluding a broader public audience. Another related initiative is the U.N program for Reducing Emissions from Deforestation and Forest Degradation that rewards countries who preserve their forests by valuing the amount of carbon the standing forest represents in the global carbon market. | KVM

WORLD

THE WORLD'S MOST PRESTIGIOUS AND

WORLD CARD WOOD

For those who were always rich and demand only the best of what life has to offer, the exclusive Visa World Card Wood is for you. The World Card Wood is not just another piece of plastic. Made from burled wood, it is the ultimate buying tool.

The World Card Wood is not for everyone, even not for the 'nouveau rich'. In fact, it is limited to only 1% of U.S. residents to ensure the highest caliber of personal service is provided to every card member.

Become a World Card Wood member today and enjoy our 24-hour world class Concierge Service ready to assist you with all your business, travel and leisure needs.

WORLD CARD GOLD

Show your class , and let Gold access everything you need.

WORLD CARD SILVER

For those who want comfort everyday.

#3 WILD SYSTEMS

wwww.nextnature.net/
themes/wildsystems

NEXT NATURE

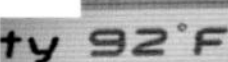

OFFICE GARDEN

ISBN 978-84-92861-53-8

WITH PRIMITIVE PEOPLE, WORK AND LEISURE WERE INTEGRATED.

IN THE NOT-TOO-DISTANT PAST, MOST PEOPLE EARNED THEIR LIVING AS SMALL-SCALE AGRICULTURAL WORKERS. FARMERS HAD LITTLE MOTIVATION TO WORK BEYOND WHAT WAS NECESSARY FOR COMFORTABLE SURVIVAL. SLEEP, MEALTIMES, AND SOCIALIZATION COULD COME WHEN CONVENIENT OR BIOLOGICALLY NECESSARY. YET, IN A LITTLE OVER 200 YEARS, INDUSTRIALIZED SOCIETY HAS REDEFINED LABOR AND LEISURE AS ENTIRELY SEPARATE PURSUITS. FOR MOST PEOPLE, WORK IS WHAT YOU DO, AND LEISURE IS WHO YOU ARE.

MODERNITY DOMESTICATED MAN INTO A SYSTEM WORKER.

DURING THE INDUSTRIAL REVOLUTION, LABORERS WHO ONCE WORKED AND SLEPT IN ACCORDANCE WITH THE SUN WERE STARTLED AWAKE BY THE SOUND OF THE FACTORY WHISTLE. ELECTRICITY PROLONGED THE DAY FAR INTO THE DEPTHS OF THE NIGHT, AND ACTED AS A SUBSTITUTE SUN BEFORE DAWN. GROWING MOBILITY IN THE SEARCH FOR EMPLOYMENT SHRANK TRIBAL-STYLE KIN NETWORKS DOWN TO THE NUCLEAR FAMILY. IN THE ASSEMBLY LINE AND THE CUBICLE FARM, MODERNITY HAS CREATED NEW MACHINES, NOT JUST MECHANICAL AND DIGITAL, BUT ALSO HUMAN.

BUT THE SYSTEMS ARE NOT ALWAYS GEARED TOWARDS THE LIFE POTENCIES OF PEOPLE.

A FACTORY ROBOT CAN PERFORM THE SAME TASK FOR HOURS. A FACTORY WORKER, HOWEVER, CAN SUFFER FROM REPETITIVE-STRESS INJURIES, NOT TO MENTION MENTAL STRESS FROM ISOLATING OR NUMBING TASKS. UNLIKE MACHINES, HUMANS REQUIRE FREEDOM, SECURITY, AND CLOSE SOCIAL BONDS TO REALIZE OUR LIFE POTENCIES. SYSTEM WORK IS EXCELLENT AT CREATING MONETARY WEALTH – AT LEAST FOR A LUCKY FEW – BUT IT OFFERS NO GUARANTEE OF HAPPINESS.

HOW NATURAL IS IT TO WORK IN AN OFFICE FOR FIVE DAYS A WEEK, WEARING A SUIT AND TIE?

HUMANS ARE THE BEST QUICK STUDIES IN THE ANIMAL KINGDOM, BUT THERE MAY BE LIMITS TO OUR ADAPTATIONS. WE ARE STILL TRIBAL PRODUCTS AT THE CORE, ATTUNED TO THE SUN AND OUR SOCIAL GROUPS. THE OFFICE MAY BE AN 'ENVIRONMENT,' BUT IT IS NOT A PROPER HUMAN HABITAT. THE VISUAL MONOTONY AND SOCIAL ISOLATION OF THE CUBICLE, WITH ITS SCHEDULED WORKING AND FEEDING TIMES, IS AS ARTIFICIAL AS OVERHEAD LIGHTING.

WE NEED MORE HUMANE TECHNOLOGIES.

THEORIST MARSHALL MCLUHAN ONCE SAID 'THE MEDIUM IS THE MESSAGE.' MAYBE THE MEDIUM NEEDS TO BE THE MASSAGE. OUR MOST BASIC BIOLOGY AND INSTINCTS SHOULD NOT BE TWISTED TO FIT OUR MACHINES. RATHER, OUR TECHNOLOGIES SHOULD SERVE US BY BECOMING MORE INTUITIVE, MORE SOCIAL, AND MORE SENSITIVE TO THE STRENGTHS AND LIMITATIONS OF THE HUMAN MIND AND BODY. TECHNOLOGY HAS BEEN A MERE MACHINE FOR LONG ENOUGH. IT'S ABOUT TIME IT GRADUATES TO FULL HUMANITY.

TIME BETWEEN EMERGENCE AND DESIGN

By Caroline Nevejan

Previously, experiences of time emerged from nature as given – offering seasons, the rhythm of humans, plants and animals. Nowadays, people integrate nature-time, body-time, inner-time, clock-time, and global 24/7 systems-time. Human beings, in past, current and next natures, have to deal with emergence and design of time in order to survive.

To think about how future new worlds are visualized, assumes that these images reveal how life in decades to come will be shaped. These visualizations offer insight into today's imagination of next natures and next cultures to come. However, in these visualizations 'time' as a process of emergence and design, is often forgotten. This essay argues that time design is distinct in any next nature that will emerge.

Witnessing Spatiotemporal Trajectories

At the end of his life, American philosopher Thomas Kuhn[1] concluded that in communities of practice human beings' need to recognize other beings' spatio-temporal trajectories to be able to share concepts and thereby develop language. In this statement he suggests that without understanding other beings' movements through time and space no communication will be possible. This statement challenges today's experience of global systems-time of millions of people who manage to communicate with people they do not know or see in the online world. Nevertheless in today's experience the feeling of having 'no time' has become a common good. Reaching out to anyone anywhere seems to generate 'no time' as a result. Will human beings be able to overcome the loss of sharing spatiotemporal trajectories and share concepts in next natures to come? What time design requirements would be needed to facilitate a time design that will foster the emergence of communication and possible new language as well?

In the past 15 years systems-time has invaded and restructured many professional practices the world over and people have developed a variety of time designs to make the 24/7 economy work for them. Without formulating it as such, a widespread knowledge and experience of time design has emerged in businesses, organizations and personal practices too. In current interdisciplinary research at the Delft Technical University, four features have surfaced as being crucial in time design for human beings involved: integrating rhythm, synchronizing performance, moments to signify and duration of engagement. Hereunder these four dimensions are outlined with the awareness that more research in any of these will benefit future time design.

Integrating Rhythms

When working in distributed teams, organizing a shared rhythm is crucial for keeping communication and business processes in flow.[2] Simple things, like one well-structured online meeting a week, generate trust and well being for all involved. When working in different time zones, adaptation to others at the expense of personal time has to be taken into account. In small businesses people benefit from the fact that distributed work on a day-to-day basis facilitates personal life styles for those involved. Finding the ultimate rhythm between people's personal time given the work that has to be done, is crucial for success. Global 24/7 systems-time has expanded human experience of time fundamentally. It offers immediate connections to other places anywhere facilitating interaction and transaction anytime and affects social structures of finance, law, business and family life profoundly. Human beings, through a methodology of trial and error, find solutions to integrate different rhythms they are confronted with. Different kinds of time merge necessarily in personal, social and collective experience of time: nature-time, body-time, inner-time, clock-time and systems-time.

Human beings have to deal with emergence and design of time in order to survive.

Nature-time has a huge diversity of scale in time designs. Long eras and short time spans, stretched rhythms and instant events are deeply interwoven. This is the environment in which human presence exists. Human bodies can only exist in one place and therefore human beings have partial perspective on nature-time as a whole. Human biological existence, the holder of body-time, is dependent on rhythms like day and night, heartbeat and breath. Human existence also contains a sense of psychological inner-time, which has hardly been investigated and yet underlies processes of growth and transformation and defines how social situations and events are perceived.[3]

Many centuries ago clock-time was introduced to mechanically structure shared social time. In the variety of clock-times, nature-time was integrated. Whether the clock was made by use of the sun, by smaller and smaller radars or by digits in contemporary design; clocks made it possible to socially anticipate what will happen next. Clock-time always offers a local perspective on time because it is fundamentally connected to a specific region or place. Places are defined by nature-time offering seasons, climates and specific ecological systems that characterize a place. Clock-time and nature-time are integrated in local agendas take that into account the context in which the human body survives.

Integrating rhythm is part of any next nature that will emerge

Today's systems-time, based on algorithms operating on a global scale, is changing the planetary landscape profoundly. Where before systems were built on principles of mandate and delegation, systems have become participants in communities of people in their own right.[4] Systems need clock-time to synchronize, but they are detached from nature-time. Like climate and weather, systems-time can also only be known through partial perspective, but unlike climate and weather, human beings can communicate in systems-time and many mitllions do so everyday. Above all the use and impact of systems-time is its immediacy. Human beings can travel to expand their experience and mental map of the place they live. Systems-time offers an expansion of connection in an instant, any place anytime. It fosters the experience of being in one place while bodies involved reside in different places. Just as nature-time profoundly challenges human existence, so does systems-time.

Nature-, body-, inner- and clock- time offer rhythms that are shared and structure social life. Rhythms cannot not integrate.[5] Over several centuries humankind developed a conscious integration of rhythms, inventing work hours, school hours, lunch breaks, agendas, holidays and more. Systems-time is challenging the integration of rhythms, since it does not seem to have a rhythm of its own. In day-to-day experience individuals integrate systems-time to their benefit, but for organizations this is more problematic. Research into beneficial systems-time design has not been taken up yet. Integrating rhythm is part of any next nature that will emerge, even though it is not clear which rhythm will dominate human life in the end. Human beings need to recognize and integrate rhythms to survive: nature-time, body-time, clock-time, inner-time. Especially systems-time, which gains importance day by day, is hard for human beings to recognize even though systems participate in human society more and more.

Synchronizing Performance

In seeking wellbeing and survival human presence judges and anticipates what will come next. In meeting a new person there is a moment when the encounter starts. Bodies reach out through perception and from the first instance a careful tuning of presence emerges. Lots of tacit knowledge is exchanged in such moments of exploring doubt and hesitation. Granular perception offers instant negotiation resulting in synchronizing the performance of presence to establish common ground upon which interaction may proceed.

The tuning of body rhythms in this process is profound; already a piece of glass between two people sitting at the same table breaks synaesthesia between them.[5] Sensory perceptions, simple emotions and more complex feelings influence processes of synchronization fundamentally. To facilitate synchronization social structures have invented gestures of encounter. The handshake is such an example. Body language is distinct in these moments; the possible recognizing of each other's spatiotemporal trajectories is at stake. Mediating granular perception is complex. Collaborating distributed teams cannot communicate a simple phenomenon like color, for example.[6] Nevertheless, human beings do synchronize in mediated communication in the variety of media they use. In a phone call – where bodies are not present but the voice is – this negotiation happens through a switch between talking at the same time and silences that are just too long before conversation continues smoothly. SMSes need to arrive just in time and so on. On the Internet, digital handshakes have the character of 'pitching one's presence' after a period of investigating an online environment.[7]

And even during participation, the process of synchronization is continuously ongoing in social networks and mailing lists because community members correct each other all the time to protect the 'tone of voice' they have agreed upon. When not sharing physical interaction people synchronize through engagement in time, through pitching and judging performance, through social control. Synchronization of performance of presence will remain a feature as long as human beings want to interact in any next nature that may emerge. Synchronization between human beings and animals, ecosystems and larger technology systems is indispensible for interaction to take place.

Moments to Signify

Part of human existence is that meaning and signification are continuously generated in personal lives and in social structures that emerge through time. Emphasizing specific moments of transformation, of passage of time, highlights the process of time. It helps people to deal with time. Human societies have invented rituals and celebrations for specific moments in time through which meaning emerges for those involved.

Just as nature-time profoundly challenges human existence, so does systems-time.

In personal lives signifying moments play an important role. Be it a private experience of becoming aware, or a collective celebration in which one partakes, these signifying moments produce identity and are fundamental for cultures to survive. Through orchestrating signifying moments, shared experience emerges and offers participants a perspective on their individual position in context of the biological, ecological, technological or social whole. In offering a perspective, it also produces this perspective, which is how cultures emerge and design at the same time. Creating 'moments to signify' is needed to create commitment for those involved.[8] People need to share experience for ideas to become sustainable and materialize in the real world.

Special signifying moments offer unanticipated impact. In situations of trauma and tragedy the human mind accelerates. When bearing witness to moments of trauma, human beings dramatize to communicate impact.[9] In these traumatic 'imaginative' moments inner-time dominates perception. Stories of trauma may even include perceptions of experiences that never took place. However, they reveal an inner experience of impact that needs to be signified to be able to communicate. Signifying moments are necessary for meaning to emerge. Offering a shared experience and/or offering an intense personal experience, they are fundamental for cultures to sustain. Any next nature that includes human life will be faced with the human need to signify. Moments to share the process of signification can be designed or will emerge. In these moments human inner time interacts deeply with surrounding rhythms and shapes culture.

Duration of Engagement

One's short-lived presence on Facebook can be as authentic as a real-life land ownership spanning 80 years.[10] Where authenticity used to be a property of being in one place for long stretches of time, in today's world this notion is replaced by being engaged in an activity for specific durations of time. Duration of engagement qualifies participation, validates contributions and therefore deeply influences human lives. Consequentially, it is not enough to be just present any more. Individuals need to prove existence by constantly transacting.[7] The formulation of 'duration of engagement' stresses the fact that there is a beginning and an end to activity. From simple time designs to more complex situations in which time emerges, people have to adapt to beginnings and endings continuously, just as birth and death are fundamental to human existence.

For human beings the transformation between the start and end of engagement is crucial to their wellbeing because it generates 'empty time' in between. In empty time, whether one is bored or not, feelings, emotions and a different thinking surface and human presence emerges. When such empty time is not granted, as in the Global Service Delivery model in the outsourcing industry in India in which people are monitored 24 hours a day, human beings' wellbeing is seriously jeopardized.[11] To generate empty time, robust structures of time design are needed.[12] Only in moments of empty time can people experience the situation they are in and act on their wellbeing.

Duration of engagement is needed for authentic human participation to emerge. However, longer durations of engagement need to include empty time for human experience to surface and to offer people the opportunity to sustain the duration of engagement. When duration of engagement is not properly designed, including a start and end with empty time included, human beings lose wellbeing in significant ways. Next natures will have to accommodate human beings' need for both duration of engagement and empty time therein.

Communities of Practice

When accepting the proposition that recognizing spatio-temporal trajectories of other beings is fundamental to the ability to share concepts and develop language, any next nature that includes human presence will have to facilitate this recognition. In current nature, systems-time is especially challenging to the human mind. Its scale and speed can only be partially perceived and it does not seem to have a rhythm of its own. Human beings find solutions to integrate it anyway, but it is not a given that people will be endlessly capable of doing this. If next nature includes human presence it has to take into account that human beings integrate their own rhythm with the environment, synchronise performance of presence to be able to communicate and create moments to signify. Thus meaning emerges. Meaning in turn needs specific durations of engagement, with a beginning and an end, and has to include empty time to sustain human wellbeing and survival.

In the tension between emergence and design, human presence in past, current and next natures is shaped. The experience of time influences the experience of place, how we relate to each other and our scope of possible actions. Any next nature will also be defined by its time design in which integrating rhythm, synchronizing performance, moments to signify and duration of engagement will define how human beings will be able to create communities of practice in which concepts, language, social structures and cultures will emerge.

REFERENCES

1. KUHN, THOMAS S. 2000. THE ROAD SINCE STRUCTURE, PHILOSOPHICAL ESSAYS, 1970-1993, WITH AN AUTOBIOGRAPHICAL INTERVIEW. EDS. JAMES CONANT AND JOHN HAUGELAND. CHICAGO: THE UNIVERSITY OF CHICAGO PRESS.
2. WILSON, REBEKAH. 2008. WITNESSED PRESENCE AND SYSTEMS ENGINEERING, INTERVIEWS BY CAROLINE NEVEJAN CONDUCTED IN THE CONTEXT OF INTERDISCIPLINARY RESEARCH WITNESSED PRESENCE AND SYSTEMS ENGINEERING (FACULTY OF TECHNOLOGY, MANAGEMENT AND POLICY, TU DELFT & NL NET). HTTP://WWW.SYSTEMSDESIGN.TBM.TUDELFT.NL/WITNESS/INTERVIEWS/RW/INTERVIEW-RW.HTML (ACCESSED 21-06-2010)
3. OLIVER, KELLY. 2001. WITNESSING, BEYOND RECOGNITION. MINNEAPOLIS/LONDON: UNIVERSITY OF MINNESOTA PRESS.
4. BRAZIER, F. & VEER, G.VAN DER. 2009. "INTERACTIVE DISTRIBUTED AND NETWORKED AUTONOMOUS SYSTEMS: DELEGATION PARTICIPATION". WORKSHOP PAPER ACCEPTED BY THE WORKSHOP HUMAN INTERACTION WITH INTELLIGENT & NETWORKED SYSTEMS, ORGANIZED BY THE 2009 INTERNATIONAL CONFERENCE ON INTELLIGENT USER INTERFACES, SANIBEL ISLAND, FLORIDA. (HTTP://WWW.IIDS.ORG)
5. KUMAR, SIRISH. PERFORMANCE AND THE FUTURE OF BROADCAST MEDIA LAB, PERFORMING ARTS LABS, UK. HTTP://WWW.PALLABS.ORG/PORTFOLIO/TIMELINE/MAY_2001_PERFORMANCE_AND_THE_FUTURE_OF_BROADCAST_MEDIA_LAB/ (ACCESSED 21 JUNE 2010)
6. GILL, S.T., KAWAMORI M. KATAGIRI W. SHIMOGIMA A. 2000. "THE ROLE OF BODY MOVES IN DIALOGUE". INTERNATIONAL JOURNAL FOR LANGUAGE AND COMMUNICATION (RASK), VOLUME 12 PAGES 89-114.
7. ABRAHAM, SUNIL. 2008. WITNESSED PRESENCE AND SYSTEMS ENGINEERING, INTERVIEWS BY CAROLINE NEVEJAN CONDUCTED IN THE CONTEXT OF INTERDISCIPLINARY RESEARCH WITNESSED PRESENCE AND SYSTEMS ENGINEERING (FACULTY OF TECHNOLOGY, MANAGEMENT AND POLICY, TU DELFT & NL NET). HTTP://WWW.SYSTEMSDESIGN.TBM.TUDELFT.NL/WITNESS/INTERVIEWS/SA/INTERVIEW-SA.HTML (ACCESSED 21-06-2010)
8. SOLOMON, DEBRA. 2009. COLLABORATING IN A COMMUNITY: ARTWORK DEVELOPED IN THE CONTEXT OF INTERDISCIPLINARY RESEARCH WITNESSED PRESENCE AND SYSTEMS ENGINEERING (FACULTY TECHNOLOGY, MANAGEMENT AND POLICY, TU DELFT & THE NETHERLANDS FOUNDATION FOR VISUAL ARTS, DESIGN AND ARCHITECTURE). HTTP://WITNESS.BEING-HERE.NET/PAGE/2112/EN (ACCESSED 21-06-2010)
9. OPHUIS, RONALD. 2009. METHODS FOR PAINTING. ARTWORK DEVELOPED IN THE CONTEXT OF INTERDISCIPLINARY RESEARCH WITNESSED PRESENCE AND SYSTEMS ENGINEERING (FACULTY OF TECHNOLOGY, MANAGEMENT AND POLICY, TU DELFT & THE NETHERLANDS FOUNDATION FOR VISUAL ARTS, DESIGN AND ARCHITECTURE). HTTP://WITNESS.BEING-HERE.NET/PAGE/2110/EN (ACCESSED 21 JUNE 2010).
10. HAZRA, ABHISHEK. 2008. WITNESSED PRESENCE AND SYSTEMS ENGINEERING, INTERVIEWS BY CAROLINE NEVEJAN CONDUCTED IN THE CONTEXT OF INTERDISCIPLINARY RESEARCH WITNESSED PRESENCE AND SYSTEMS ENGINEERING (FACULTY TECHNOLOGY, MANAGEMENT AND POLICY, TU DELFT & NL NET). HTTP://WWW.SYSTEMSDESIGN.TBM.TUDELFT.NL/WITNESS/INTERVIEWS/AH/INTERVIEW-AH.HTML (ACCESSED 21-06-2010)
11. ILAVARASAN, P.VIGNESWARA. 2008. "SOFTWARE WORK IN INDIA: A LABOUR PROCESS VIEW". AN OUTPOST OF THE GLOBAL ECONOMY, WORK AND WORKERS IN INDIA'S INFORMATION TECHNOLOGY INDUSTRY, EDS. CAROL UPADHYA AND A.R.VASAVI. NEW DELHI: ROUTLEDGE.
12. FEIGL ZORO. 2009. MOVEMENT THROUGH TIME. ARTWORK DEVELOPED IN THE CONTEXT OF INTERDISCIPLINARY RESEARCH WITNESSED PRESENCE AND SYSTEMS ENGINEERING (FACULTY OF TECHNOLOGY, MANAGEMENT AND POLICY, TU DELFT & THE NETHERLANDS FOUNDATION FOR VISUAL ARTS, DESIGN AND ARCHITECTURE) HTTP://WITNESS.BEING-HERE.NET/PAGE/2111/EN (ACCESSED 21-06-2010)

JAR
ctricBinary-8.04.ja
r
MacS...
linden.rtf
radiostrean
MacS...
utils
spam
ad
5
ProcessV
app
MacS...
fusion_
beautifulc+
MacS...
MAPPE, DB SATELLITI
e CANALI....tClipping
Senza titolo.rtf
progettosei
MacSlo...
hash
b
symantec
MacS...
TGZ
testlib.tgz
alarm-0.3

MacS...
ce.w
attilio
fltk-1.1.7
clock
attilioold
threads++
MacS...
MacSlow's Cairo-Clock
urce
clock1
libevent-1.1a
qgit-1.5.2
soldi
MacS...
MacS...
PGP
index.html
musica
dancing-color.pdf

ROLEX

WE SERVE OUR SYSTEMS AS MUCH AS THEY SERVE US

System Animals

What animal is so naïve to come into this world as a naked and crying infant, completely vulnerable, helpless, and an easy prey for any predator? Newborn sheep or giraffes can walk within a few hours, but it takes humans years and years to learn to take care of themselves. Yet, despite our physical vulnerability, we've proven able not only to survive, but even to dominate the planet. How come?

Unlike other animals, which have specific organs, skills and reflexes that enable them to survive in their proper environment, humans have never lived in an environment for which we are specifically equipped. The human physique implies that there is no such thing as a 'purely' natural environment for us. We are system animals: technological beings by nature. Compared to other animals, humans are physically maladjusted, primitive and undeveloped. Yet, we compensate by being the most enthusiastic system builders of the entire animal kingdom. Although this trait is typically seen as advancement from our primate ancestors – an astronaut has some augmented attributes a monkey must live without – we rarely realize that our system craving not only empowers us, but at times also imprisons us.

From the dawn of humankind we have created systems. It is safe to say that we have co-evolved with them, like bees have co-evolved with flowers. And as in any symbiotic relation, traffic goes two ways. With every technological 'upgrade' we relinquish a piece of ourselves. A fur for a coat, hunting for farming, singing for writing, memory for websearch. Through our systems we domesticate our environment, yet every new system also causes a new situation, a next nature, which eventually also domesticates us. We serve our systems as much as they serve us. Hence the importance of having systems that maintain the life potencies of human beings. |KVM

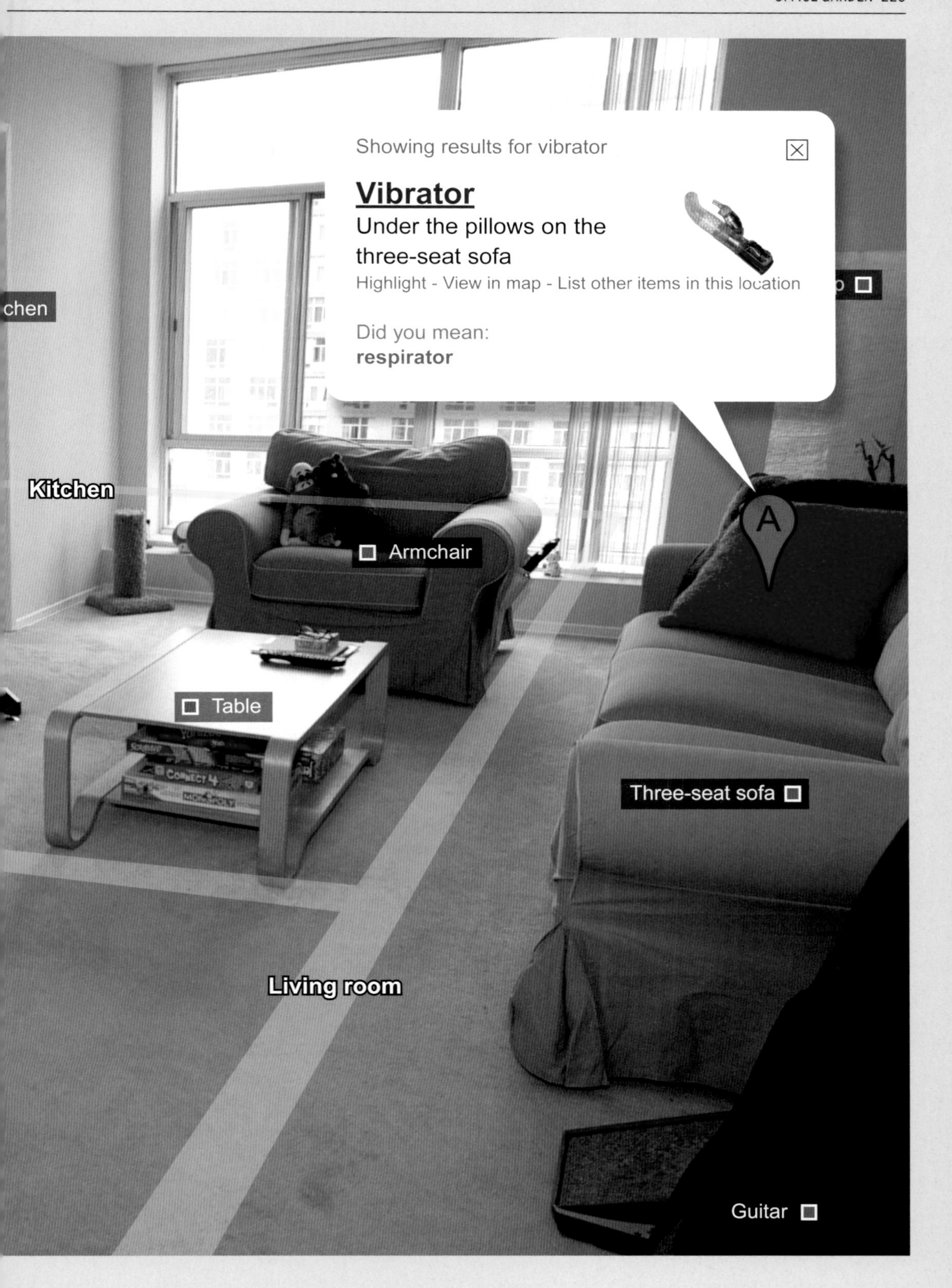
Showing results for vibrator
Vibrator
Under the pillows on the three-seat sofa
Highlight - View in map - List other items in this location
Did you mean:
respirator
chen
Kitchen
Armchair
Table
Three-seat sofa
Living room
Guitar
A

Clouding the Brain

By Ties van de Werff

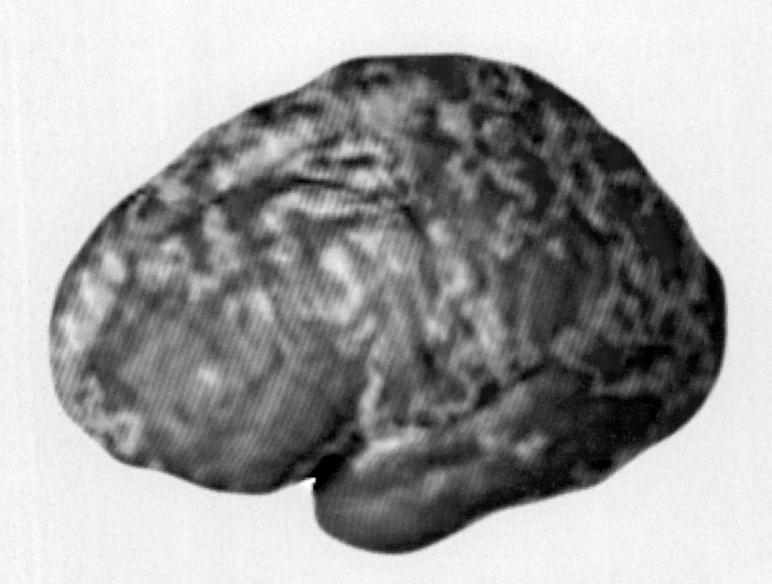

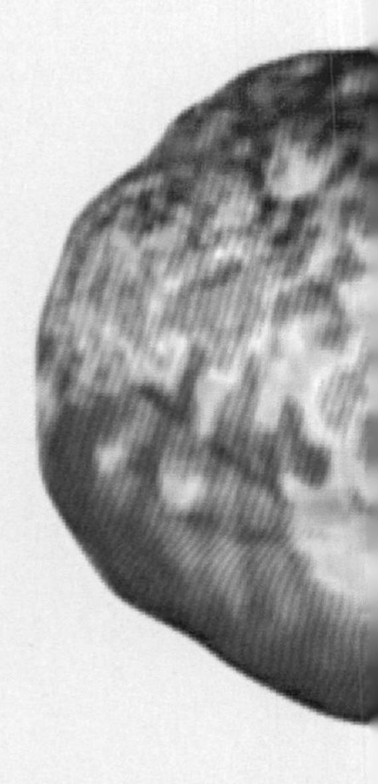

Man is a flexible species. We tend to adapt rapidly to new environments. For millions of office workers, the everyday outlook on the world starts with a Google search bar or Microsoft's rolling hills desktop. We have adjusted easily to this reality. But to what extent and how fast can these adaptations turn into new evolutionary traits? How does leading a largely virtual life change our way of thinking?

Some neuroscientists suggest that our brains change due to the extensive use of the Internet and other digital technologies like smartphones. Digital natives, those born in the Internet era, apparently have difficulties concentrating for longer periods of time. Studies of Stanford students show that multitaskers are more easily distracted and have less control over their attention.[1] Smart phones and the relentless connectivity to the cloud leads to a constant state of anxiety and even to new disorders such as 'vibranxiety', or phantom vibration syndrome. Many fear that our growing use of screen-based media could potentially decrease our ability for abstract reasoning, critical thinking and imagination. From an evolutionary perspective, however, we might get something in return.

Back in the day, as the story goes, we could remember whole bible stories. We could sing entire newspapers. Because the printed word was rare and expensive, we had to remember it all. That changed with the invention of book printing. Remembering became less important and instead, as philosopher Walter Ong and others claim, our brains could focus more on comparing and analyzing. Our analytical skills grew. Today, Google and other semantic search engines parse our questions and prioritize the possible answers in a split second. This outsourcing of our analytical thinking to the cloud opens up ways for our brains to develop new skills. Theorists and neuroscientists speculate that our future skills will consist of highly developed visual-spatial

intelligence. In the end, we might resemble Photoshop more than Google.

But evolution is a slow process. We will most likely not develop an extra visual cortex in this generation. So instead of looking for genetic alterations, many researchers turn to our changed intellectual culture to see whether the Internet changes the mind. There is a lot of natural variation to be found in the online world. Blogs, Twitter, Facebook and websites' comment sections display a huge variety of mind-numbing thoughts. Will the mother of all networks turn us into global netizens with narcissistic, promiscuous and exhibitionist character traits? Or will it give us many new ways to express ourselves? Instead of wasting our time passively watching TV like we did in the 20th century, we now spend our free time contributing, creating and participating on weblogs, Youtube and social media. Seen this way, the ecosystem of the Internet looks like a huge playground or a digital library of Alexandria rather than a socially oppressive global village square.

Nature changes along with us. But to what extent does the nature of man change too? The Internet and all other screen-based media we use, have the potential to change our way of thinking and our very brains – that lump of grey matter that defines our existence. While many fear we may become mere signal-processing units or living extensions of the cloud itself, others hope that we will develop new skills. So if you feel distracted, superficial or if you think your pants are vibrating all the time, feel comforted by the idea that at least as a species, humans are still evolving.

REFERENCES

1 OPHIR, E., NASS, C., WAGNER, A.D. (2009). 'COGNITIVE CONTROL IN MEDIA MULTITASKERS' PROCEEDINGS OF THE NATIONAL ACADEMY OF SCIENCES, 24 (2009).

Placebo Buttons

HAVING TROUBLE INVENTING AN OPTION? THERE'S A BUTTON FOR EVERY CHOICE. HOWEVER, THE DECLINE OF THE BUTTON MAY MEAN WE WILL HAVE TO FIND TV CHANNELS ON OUR OWN.

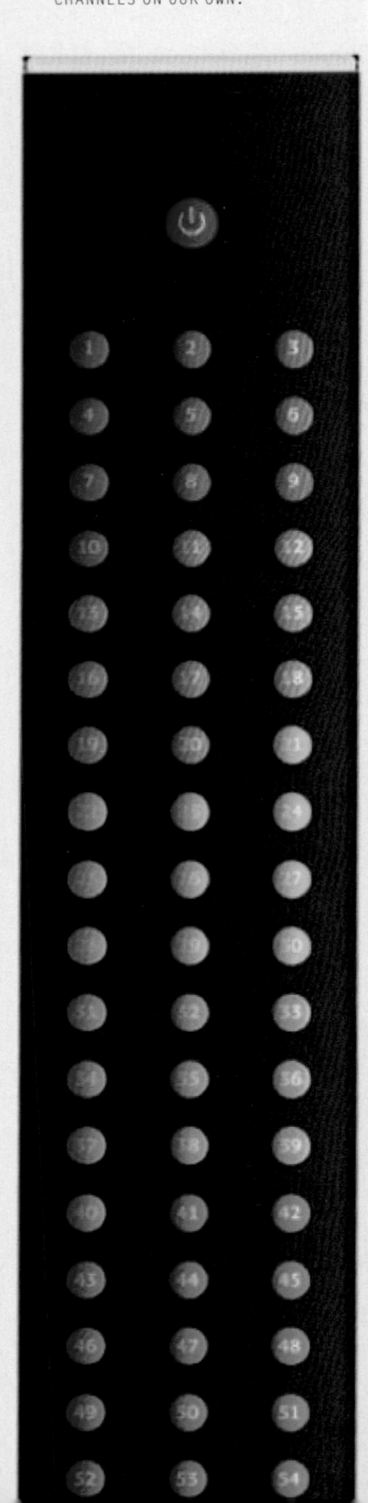

Buttons are everywhere: on phones, alarm clocks, keyboards, elevators, dishwashers and of course on the computer screen. Although buttons weren't around in old nature – assuming nipples aren't counted as buttons – these little symbols of control have become ubiquitous in everyday life. But for how long? As technology advances, buttons are replaced by sensors, gesture technology and autonomous systems.

It may well be that our grand-grand-children won't be pushing buttons like we do, yet as the current generation is so attuned to pushing buttons, they are increasingly applied as skeuomorphs which have no effect or function, except as decoration. These buttons are simply placebos. Examples of placebo buttons are unwired walk buttons at pedestrian crossings in New York City and door-close buttons in elevators, which have been replaced by sensors.

In some cases the button may have been functional, but may have failed or been disabled during installation or maintenance. Sometimes the button has been deliberately designed to do nothing except create the illusion of control in the mind of the user. Makes you wonder about that big red button in the White House... | KVM

WE'RE SO USED TO INSTANTANEOUS RESULTS, WE NEED SOMETHING TO EASE THE FRUSTRATION OF A SECONDS-LONG WAIT. FAKE BUTTONS GIVE US THE ILLUSION OF CONTROL.

Dominant Standards

Have you ever wondered why the letters on the keyboard are in the order they are? The origins of QWERTY go way back. This order was chosen to reduce the probability that mechanical typewriters' hammers would get entangled. Over time, computers replaced typewriters. Though various alternative keyboard layouts have been developed from a user-centered perspective, to enable faster and more comfortable typing, the QWERTY layout remains the standard today. Once a technology has become the norm, it seems to take on a certain aura of authenticity. To supplant it, an alternative must be significantly better.

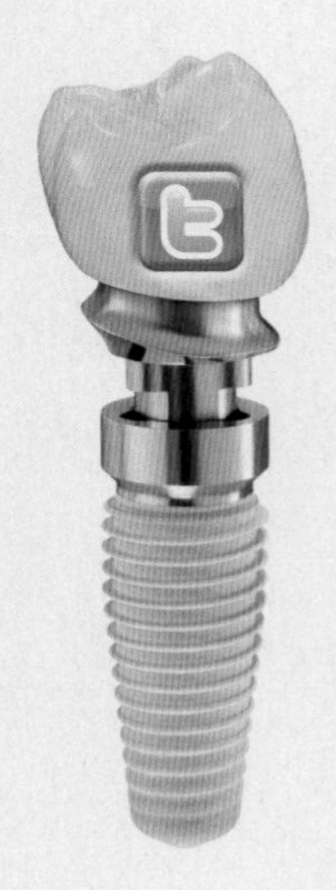

THE KEYBOARD IS A REMNANT OF THE TYPEWRITER, A TECHNOLOGY THAT BECAME OBSOLETE IN THE 1980S. VOICE-TO-TYPE TECHNOLOGY, NEURAL IMPLANTS, EVEN TWITTER TEETH MAY FINALLY SPELL OUT THE DEMISE OF THE QWERTY LAYOUT.

Racist Technology

The co-workers pictured here discovered that the face-tracking feature of their advanced Hewlett Packard webcam only recognized white faces. The device could not detect black faces. HP claimed that the webcam doesn't respond well to "insufficient foreground lighting." Too bad for those born with "insufficient foreground lighting." Techno-rhetoric euphemisms or blatant racism? Perhaps the hope that computing would override society's inequalities was a little naïve after all.

Domesticating Ourselves

By Allison Guy

Our bodies are maps of our ancestor's social lives. Humans evolved in complex, tight-knit social groups. Such intense sociability may have favored the very traits we favor in cows, cats and dogs. In both mind and body, human evolution mirrors the changes animals undergo during domestication.

'Neoteny,' the preservation of juvenile traits into sexual maturity, is a hallmark of domestication. Dogs are a classic example, hanging onto the floppy ears, wagging tail, and barking of a wolf puppy their entire lives. A Labrador looks like an immature wolf; a Chihuahua looks like a fetus. *Homo sapiens* appear child-like next to Neanderthals, and compared to chimpanzees, we're absolute babies. Humans are more fine-boned than our closest relatives, with the patchy body hair of youth. With weak jaws, bulging foreheads and wide-set eyes, the human skull is a dead ringer for an infant chimpanzee's. During the process of domestication, the animal brain shrinks by between 10 and 30%, depending on the species. For all our vaunted intelligence, humans are no different. Over the last 20,000 years, the average male brain has dwindled by 10%, a volume equivalent to a tennis ball. According to one theory, our minds have become more efficient compared to the bulky, energy-hungry brains of our ancestors. The gloomier interpretation argues that tight-knit, secure societies can protect the dopes from themselves. The same people that might have starved through lean times or been outsmarted by a lion are now buffered from misfortune by their clever peers, allowing the collective IQ to wither. Domestication goes deeper than physical changes. Our companion animals aren't just tame by training, but by genetics. The adrenal response, in charge of

the fight-or-flight reflex, is universally diminished. This could be thought of as neoteny in personality, a lifelong retention of the playful, social, and trusting nature of youth. In humans, retaining juvenile minds may have meant we kept the extraordinary mental plasticity of youth well past sexual maturity. Young chimpanzees find it easy to learn novel tasks; only humans find it just as easy in adulthood.

The theory that we're all adult-ified children is intriguing, but are humans really more tame than our ancestors? Certainly our willingness to work with strangers, our unusual altruism, and the somewhat orderly functioning of our societies speaks to a more domesticated creature. While humans can be astonishingly nasty to each other, constant violence is not rewarded when manageable group dynamics are key to survival. Future societies may engineer the most aggressive tendencies out of our genome, but why bother? We already medicate, incarcerate, or execute people on the extreme ends of antisocial behavior. It's possible that these traits may naturally continue to fall from favor in the gene pool.

Homo sapiens never really existed in a state of nature. We were too domestic for Eden. The social lives of ancient humans and near-humans made our minds and bodies modern. Human nature, at its core, is the result of millions of years of culture.

THE TRASH CAN METAPHOR.

FIRST CAME THE STEEL TRASHCAN. THEN CAME THE DIGITAL TRASHCAN, A DESKTOP ICON WHERE YOU CAN DRAG AND DUMP UNWANTED DOCUMENTS. IT MAKES INTUITIVE SENSE THAT DIGITAL GARBAGE SHOULD BE DELETED USING AN EASILY RECOGNIZABLE METAPHOR, RATHER THAN DISEMBODIED CODE OR AN ABSTRACT ICON. THE DESKTOP TRASHCAN HAS NOW BOOMERANGED BACK INTO REAL LIFE. CODECO.ORG CREATED A PAPER-CRAFT TRIBUTE TO THE FAMOUS PIXELATED ORIGINAL. THE DESIGN GROUP FRONT WENT ONE STEP FURTHER, OFFERING A PHYSICAL BIN THAT, LIKE THE DESKTOP VERSION, BULGES OUTWARDS WHEN IT'S FULL.

Boomeranged Metaphors

Metaphors are facilitators of change. At the start of the digital era, metaphors from everyday life were used – in what was then the new computer environment – to help us make sense of otherwise incomprehensible technology. Terms like 'digital highway' and the desktop metaphor with its windows, folders, buttons and trashcans made the computer world accessible to almost everyone.

Now though, the digital environment is understood and accepted almost everywhere. We see how established concepts from the digital realm are gradually seeping into our physical environment. These amusing 'boomeranged metaphors are everywhere: from the 'delete key' eraser to the 'pixel' oven gloves and the 'icon' watch.

However, it is likely that we will be confronted with more radical boomeranged metaphors in the near future. Are the advanced ranking systems found in Internet forums perhaps also applicable to the democratic voting process? Will the young players of SimCity turn out to be the urban planners of the future? Will we be able to select a new avatar for ourselves at the plastic surgeon's office?

Metaphors enable us to use familiar physical and social experiences to understand less familiar things. Meaning is produced by analogy. The boomeranged metaphors make us aware that once we spend enough time in a so-called virtual environment, it becomes so familiar to us that we start to refer to its objects and ideas in different contexts outside the virtual realm. As the so-called 'real' world is increasingly understood by analogy of the so-called 'virtual' world, we learn that the presumed separation between the two isn't really valid. Everything that enters our mind contributes to our experience of the world around us and through these mental concepts we understand and interact with our environment.

Undoubtedly the digital technologies that have emerged during our lifetime are well on their way to becoming just as familiar and common as light bulbs or cars, which at the time of their introduction were described metaphorically as electric candles and horseless carriages. And surely, in time, digital natives will be referring metaphorically to elements from the digital realm – their first nature – to familiarize themselves with the newly emerging phenomena of the future. | KVM

DO HUMANS DREAM OF ELECTRONIC SHEEP? THEY CERTAINLY DREAM OF LARA CROFT.

DOES THE PHYSICAL WORLD ECHO THE DIGITAL WORLD?

Live in Your World, Dream in Ours

A Playstation advertising campaign once invited users to 'Live in Your World, Play in Ours .' Looks like we'll have to add dreaming to that list. According to psychologist Jayne Gackenbach, committed video game players have more 'lucid dreams' than non-gamers, where the dreamer is able to control the setting and outcome of the dream. Manipulating one virtual, disembodied realm by day is practice for the brain to do the same thing at night. Digital environments retrain our sense of reality. Our home city echoes *SimCity,* and that hill in Sonoma will always exist on Microsoft Windows. | AG

WORLD OF WORLD OF WARCRAFT BY PARODY NEWS SITE THEONION.COM

Serious Games

World of World of Warcraft lets users play as another player playing the MMORPG (Massively Multiplayer Online Role-Playing Game) World of Warcraft. Meta-jokiness aside, video games have joined the slew of media that shapes perceptions of life. While the link between video games and real life violence is fiercely debated, it's hard not to think of remote-control warfare as an extension of video game culture.

STILLS FROM WIKILEAKS' *COLLATERAL MURDER*, THE ARCADE-STYLE SLAYING OF CITIZENS AND JOURNALISTS IN BAGHDAD

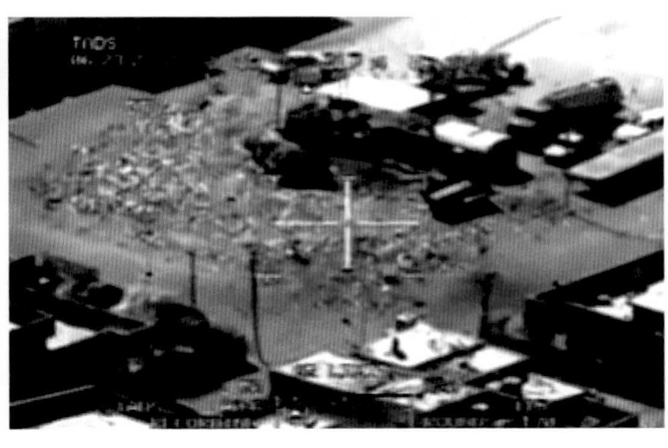

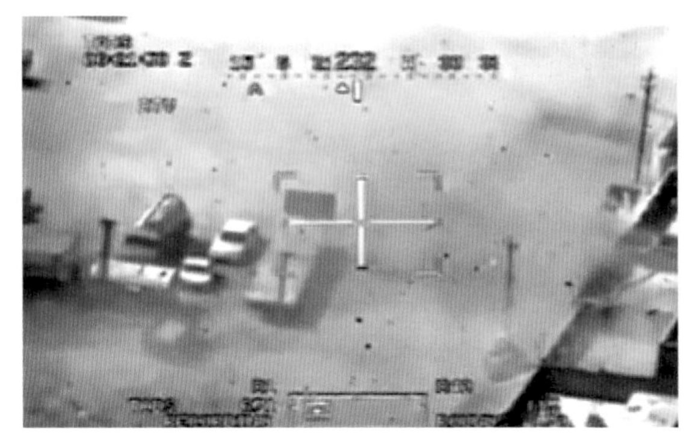

Jobs Become Games

Desktop Agriculture

In millions of offices and homes around the world, people are hard at work planting crops, feeding cattle and tilling their land. Welcome to Farmville, the digital rural world where the sun always shines, where beans take two days to grow, where pink cows produce strawberry milk, where farming is leisure.

Farmville has become a viral Internet trend since its launch as a Facebook application in 2007. According to Zynga, the company that brought FarmVille into the world, it has rapidly grown to over 70 million users – compare that to the one million traditional farmers active in the USA. Players sign up and get fields, infrastructure, and cash. Their task is to create bigger, better, and richer farms. The game starts off with a given piece of land and seeds that can be planted, harvested and sold for online coins. As you make money, you can buy products, from basics like pumpkin seeds and chicken to the truly superfluous, like elephants and hot-air balloons. Impatient players can use credit cards or a PayPal account to buy more assets, although purists tend to disapprove of the practice and constrain themselves to developing their farm through simple 'labor'.

Farmville has elements of the addictive Tamaotchi pet toy, but instead of feeding a little 'animal,' you're caring for a digital farm with insatiable livestock and crops that need regular clicking. Crops must be

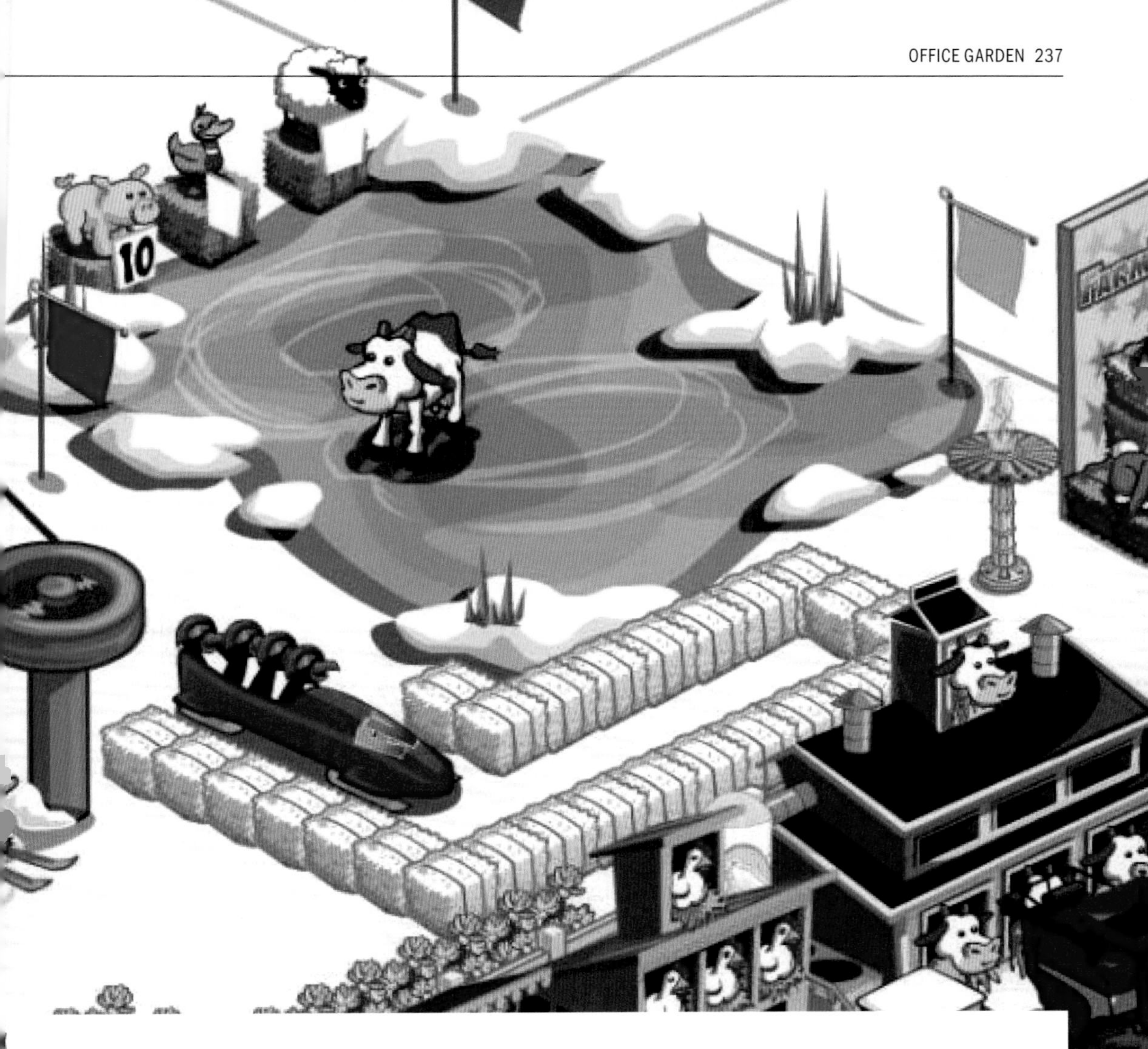

harvested in a timely fashion, cows must be milked, and social obligations – like exchanging gifts and fertilizing your neighbor's pumpkins – have to be met. Virtual farming provides an odd mash-up of social networking with rural nostalgia. In comparison to the often violent world of gaming, Farmville, with is emphasis on creation, cooperation and strategy, represents a form of virtual country calm that transports us somewhere else for a minute or an hour. In doing so, the game taps deep into the human psyche and the longing for an idyllic agrarian past that is long gone. For the larger part of human history food gathering and production was a daily obligation for almost everyone. Industrialization of our food production, however, has put nine out of 10 people on the planet in non-agricultural pursuits. This increased efficiency made room for other activities, yet we have also lost a few things: a clear connection between labor and income, direct contact with our surroundings through the caretaking of plants and animals and a life-rhythm in pace with the seasons and climate.

While the use of modern social networking infrastructure like Facebook to express old agrarian values may seem paradoxical, it makes sense once we realize that every superseded nature is destined to become a romantic escape within the new dominant nature. From that perspective Farmville is just the latest iteration of an ancient theme. | KVM

Games Become Jobs

Anshe Chung, the online avatar of Ailin Graef, is the Rockefeller of Second Life. From an initial investment of $9.95, Anshe became the first (real life) millionaire from the sale of virtual goods. Although the economy in Second Life has fizzled along with its hype, virtual economies are going strong in other massively multiplayer games. Although the physical parameters can be very different, the results look familiar: prostitution, real estate speculation, and outsourcing labor to sweatshops. | AG

ANSHE CHUNG ENJOYING HER VIRTUAL REALITY ESTATE IN SECOND LIFE

GOLD FARMERS TRYING TO MAKE A FEW YUAN IN NANJING, CHINA

BANNERS ON WEBSITES SUCH AS GOLDSOON.COM AND WOW7GOLD.COM

Gold Farming

Chinese workers are earning gold by slaying monsters for western customers. It sounds surreal, but it is a far from virtual reality for the so-called 'gold farmers,' who are working in 10-hour shifts to help players gain levels, and wealth, in online role-playing games like *World of Warcraft*. For thousands of Chinese workers, gold farming is a way of life. Workers earn between €90 – €140 a month which, given the long hours and night shifts, can amount to as little as 30 cents an hour. After completing a shift, they are given a basic meal of rice, meat and vegetables and fall into a bunk bed in a room shared with eight other gold farmers.

You can hire your own gold farming slave employee at wow7gold.com or goldsoon.com. According to an extensive report by Richard Heeks at Manchester University, a few hundred thousand Asian workers are employed in gold farming in a trade recently valued at about €800 million a year. With so many gamers now online, these industries are estimated to have a consumer base of five to 10 million, and numbers are expected to grow with increased Internet access. Recently, the Chinese government started taxing gold farmers. Games become jobs. And where there's a demand, China will supply it. | KVM

ANYTHING FOR A FEW LINDEN DOLLARS...

Virtual Prostitution

Despite its illegality in many countries, or perhaps because of it, prostitution has flourished in certain massively multiplayer online games. Second Life has its own strip clubs and brothels, staffed with avatars ready to accept real money for virtual sex. One avatar named Khannea charges 750 Lindens, about €2, for each half hour of pre-programmed sexual acts and sounds. According to another sex-worker, Palela, the going rate in Second Life is 2,000 to 3,000 Lindens for a full evening of entertainment, although clients often tip more. The most dedicated workers invest serious time and Second Life money in selecting physical characteristics and provocative clothing for their avatars. No matter how futuristic the platform, the world's oldest profession has proven enduringly popular. | AG

OFFICE REBELLION INSTRUCTION BY PACKARD JENNINGS

A
B
S.W.A.T.

WAG THE POD.

The iDog is a tiny tail that turns your smartphone into your best companion. Depending on who's calling the iDog wiggles its tail. It automatically learns from your calling behavior. You can even train it to automatically answer or decline calls from certain people.

LIKE WE DID TO OLD NATURE, WE MUST NOW CUL-TIVATE OUR SYSTEMS

'Information is the Oxygen of the Modern Age.'

Ronald Reagan

INFOBESITY BY ARNOUD VAN DE HEUVEL

Information Decoration

By Koert van Mensvoort

Houseplants with an opinion, paving stones that give directions and trains that blush before departing. What can we learn from old nature to solve our problems of limited human bandwidth and attention scarcity?

Picture this: it's 40,000 years ago, and you are an early *Homo sapiens*. You are standing on the savanna. Look around you. What do you see? No billboards, no traffic signs, no logos, no text. You might see grassland, a stand of trees, a bank of clouds in the distance. You are in a kind of vast, unspoilt nature reserve. Are you feeling wonderfully relaxed yet? Don't be mistaken. Unlike the woodland parks where you sometimes go walking on a Sunday, this is not a recreational environment. This is where you live. You must survive here, and the environment is full of information that helps you to do so. An animal you are pursuing has left tracks in the sand. Are the berries on that tree edible or poisonous? And that birdsong: does it mean there's going to be a storm and winter is on the way? Or are the silly birds just singing for their own enjoyment? You can't be sure, you have to interpret it all. And you are good at that. Good enough to survive in this environment.

Let's return to 2011. You are an average Western human being. You are looking at a spreadsheet or a Word document on a computer screen, trying to figure out what's going on. It is said that we live in an age of information. Although it is unclear what exactly this means, many of us suffer from wrist, back and neck pain. The copying, transformation and exchange of data is a daily job. Many have made this their profession. These people call themselves 'knowledge workers' (which is a fancy term for 'data pusher'; a bit like calling yourself a business developer when you're actually a salesperson). All of us have been born into a world full of abstract technologies and systems. We are forced to adapt to them in order to survive. Berries, grassland, birds and clouds have long since ceased to be the things we need to read in order to survive. Insofar as these elements still exist in our environment, they have taken on a recreational role. Instead, we live in a world of screens. We use these flat rectangular objects to inform ourselves of the state of our world. Screen workers are the assembly line workers of the 21st century. We use screens to check our e-mail, screens to monitor safety on the streets and screens to follow fashion. Our scientists use screens to explore the outer limits of the universe and to descend into the structures of our genes. A painful truth: many of us spend more time with screens than with our own friends and families.

Screens were originally found only in offices, but nowadays the screen virus has spread to fast-food joints, railway stations, public squares – more or less all public space is filled with them. This is done in the name of information, advertising, art and entertainment. According to governmental guidelines, a knowledge worker may spend a maximum of six hours a day working at a monitor. I don't know to what degree they took into account exposure to garish LED screens on the street when they formulated this rule. Personally, it gives me a nasty feeling when after a long day of knowledge work, on my way home or during a night out, I am once again forced to look at a screen hanging randomly in public space (screens featuring Windows error messages are particularly distressing). The really brutal thing about screens is that they seldom enter into a relationship with their environment. They are isolated, draining elements that do nothing but try to seize our undivided attention and turn our environment into a Swiss cheese of realities. Screens are even more obtrusive than, say, posters, which stand still and have a light intensity that is linked to their environment.

Human Bandwidth

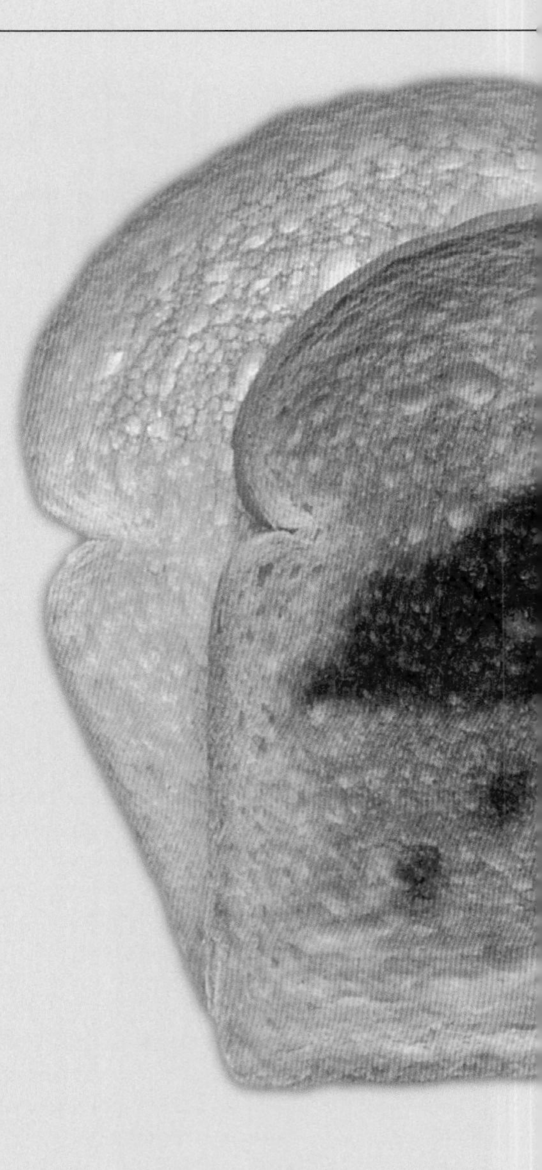

Perhaps you expect me to start arguing now for screen-free public space, but that is not what I am out to do. The merging of virtual and physical spaces is an inevitable development, and we should welcome it. After all, remaining seated in front of the computer, stiff from repetitive stress injuries, is no alternative. If we are charitable, we can look at the contemporary screen virus as a transitional phase – a growing pain, if you will, of the information age. Tiling our environment with screens is an extremely literal, and rather unimaginative, way to introduce virtuality into the physical world, simply piling it on where seamless integration was the original intention. It is said that we live in an information age; that is, we have a big problem with information management. Evidently, the manner in which we make information available is not sufficiently coupled with human perception, or more precisely, human bandwidth. This, too, can be explained as a growing pain. In the past, the reproduction of information was technologically complicated, so people were forced to adapt to the existing possibilities. Today, the duplication of data has become extraordinarily simple. It is high time we adjusted its presentation to suit human needs.

The million-dollar question is: how do we integrate all those indispensable information streams into our environment? Besides the fact that we can learn a lot from old nature, where information is present in a well-integrated way, I think we can learn from the decorative world. For centuries, people have been utilizing decorative patterns, indoors and out, with the aim of improving and giving an identity to the atmosphere around them. The primary goal is not information but aesthetics. What happens if we start looking at every pattern in our environment as a possible information carrier? Look around you, wherever you are. Try to recognize all of the forms and patterns in the space. The flowered wallpaper, the humming of the air-conditioning, the fish in the aquarium, a shadow on the wall. Do you realize how few of the patterns in our environment are being used as information carriers? Information overload? What information overload? The so-called information society has barely scratched the surface of our human bandwidth!

THE *POWER AWARE CORD* TRANSPORTS ELECTRICAL POWER WHILE SIMULTANEOUSLY VISUALIZING ENERGY USAGE THROUGH GLOWING PULSES, FLOW, AND INTENSITY OF LIGHT. DESIGN BY ANTON GUSTAFSSON AND MAGNUS GYLLENSWÄRD.

DESIGNER ROBIN SOUTHGATE DEVELOPED A TOASTER THAT TAKES METEOROLOGICAL INFORMATION FROM THE INTERNET AND THEN BROWNS YOUR BREAD WITH AN ICON OF THE WEATHER FORECAST THAT DAY.

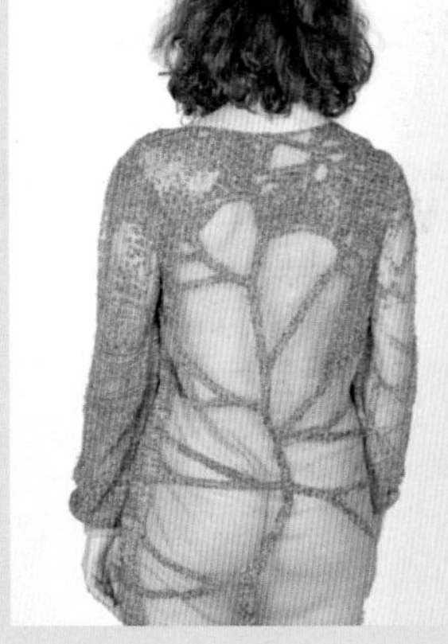

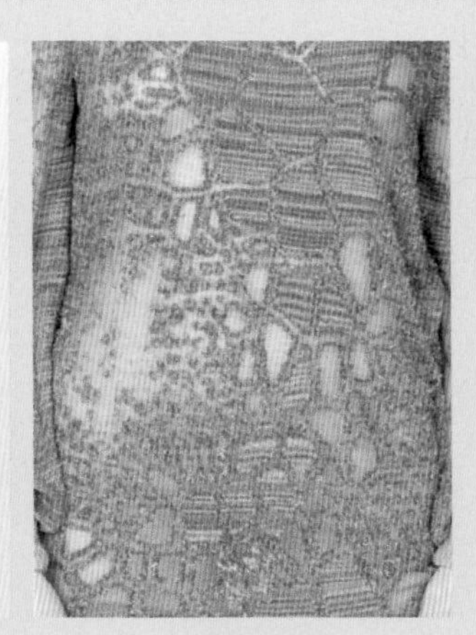

THE *DECAY* PROJECT EXPLORES HOW TRACES OF TIME AND USE CAN BE EMBEDDED IN TEXTILE. BY WEARING A CARBON FIBRE SUIT OVER A WHITE BLOUSE, DESIGNER MARIE ILSE BOURLANGES CAPTURED THE GESTURES OF THE BODY BENDING, STRETCHING, SCRATCHING AND RUBBING. THE IMPRINT ON THE BLOUSE WAS THEN TRANSLATED INTO A PATTERN OF LINES THAT EBB AND FLOWER ACROSS THE TEXTILE.

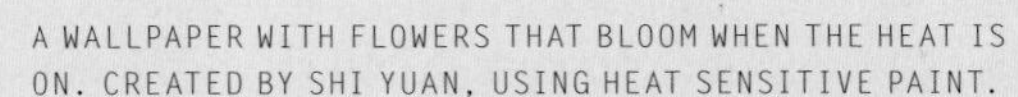

A WALLPAPER WITH FLOWERS THAT BLOOM WHEN THE HEAT IS ON. CREATED BY SHI YUAN, USING HEAT SENSITIVE PAINT.

THIS NECKLACE FOR WOMEN WITH BREAST CANCER SUBTLY COMMUNICATES HOW THEIR RECOVERY IS PROGRESSING: THE DEEPER THE COLOR OF THE LINK THE HEALTHER YOU ARE. DESIGN BY MARCO VAN BEERS.

The Magical Garden

I wish to argue for information decoration, which means seeking a balance between aesthetic and informational quality. Information designers are usually inclined to place the message at the centre of our field of attention (to make sure it comes across). Did no one ever tell them it was impolite to always come straight to the point? We humans have evolved precisely to attend to information at the edges of our field of attention, and when necessary transfer it to the center ourselves.

Of course, information decoration is not always appropriate. Some messages, such as fire alarms, are too urgent to work subtly into the wallpaper and must be brought to our attention unambiguously. Information decoration lends itself primarily to the kind of data we wish to have available at all times but to be able to ignore: the online status of my friends, the traffic update, the weather forecast, the number of unread messages in my inbox. I want to emphasize that information decoration is more than just making data look better; it requires a genuinely new information model. Traditional information theory usually advises against things like ambiguity and repetition. In information decoration, these factors play an influential role, because ambiguity and repetition are classic aesthetic means of achieving interesting images. The big advantage of information decoration is that if it's not informative, it's still decorative. That's more than you can say for most contemporary information carriers.

Is this the future of our environment as an information carrier – feeling as if you're being pounced on by a lion at every street corner? No thank you. 'Attention' is the scarcest resource in the information age. There is still sufficient space at the edges of our field of attention; let us utilize our human bandwidth sensibly. Our environment was previously made up of objects, now it consists of information. When architects design buildings, they will have to consider to what degree those buildings function as information carriers (if they neglect this, they run the risk of LED screens being attached to the buildings in due course). I want new wallpaper. I want new furniture. I want a houseplant that has something to say. Paving stones that show me the way. Trains that blush before they take off. When autumn comes, the street will be littered with flyers.

NIET STOREN
Dr Pepper
Heinz
DB
DIESEL
STAR WARS
Pedigree
B-36
WRONG WAY
BEGIN
END
MasterCard
BATTLESHIP
RAMP EXIT
DURACELL
JUNIOR G-MEN CLUB
nash
BP
Order Now!
DEPOSIT
12 MONTHS
WEST
THE GARDEN PROJECT
I-SPY SECRET CODES
PLAYBOY
PHILIPS
Google
PEPSI
end
VISA
SPITS
SAMSUNG
INTER
AL CAPONE
Chiquita

The Sound of the Blue Canary

By Berry Eggen

Blue is a beautiful color, but its sound is simply irresistible. It is the song of the unhappy and the depressed. It is a sound that touches people. It was also the sound of a little songbird, the *Serinus canaria domestica*, a sound that so moved me, I was led on a voyage of discovery into the world of birdsong. The *Serinus canaria domestica* is the man-made descendant of the Wild Canary, a finch originally from the Canary Islands, which nowadays exists in many different breeds. This essay deals with the cultivation of the song-bred canary and imagines how its story might lend inspiration to the sound design of electric cars.

Sounds 'exist in time and over space'.[1] You can hear a sound without having to face the source that produces it; you only have to be listening or recording at the right time. If you want to see an object, however, you have to be facing it. And, in most cases, you can re-view the object at different moments and for longer periods; visual objects therefore 'exist in space and over time.'

When you are a small bird living in dense foliage, leaves prevent effective visual communication. This makes sound an excellent alternative for warning or impressing your mates, or for marking out your territory. The volatile character of sound, however, makes its evolutionary development difficult to trace, whether it be birdsong or vocal communication in animals in general. We know from visual fossil inspection, for example, that there was a close relationship between dinosaurs and birds.[2] At the same time, though we have a sense of what dinosaurs looked like, we can only imagine their vocal expressions today. In *On the Origin of Species*, Charles Darwin explains how adapting to changing conditions in the natural environment results in survival for some living organisms and extinction for others. Biologists have discovered that this principle of 'natural selection' not only causes species to develop subspecies with very different characteristics that are determined by heredity, but also lies at the basis of the origin of new species.[3] The Domestic Canary is a subspecies of the Wild Canary and contains a wide variety of breeds that have not been scientifically classified. The origin of these (new) breeds is a result of 'human selection.' But what exactly does this 'human selection' principle entail? And can this principle inspire, or maybe even guide, the sound design of next nature? Before focusing on what comes 'next,' I will briefly review the current state of nature's infinite design space as cultivated by human breeders of the species *Canaria*.

Nowadays, three main groups of domesticated canaries can be distinguished: posture, color, and song canaries. The various breeds within these groups show a wide variety of different shapes (small, big, curved, bowed, curly-feathered, crested, and more) and colors (green, yellow, red, brown, white, orange, gray, and more, though no blue!). The song canary group comprises different breeds with clearly distinguishable songs. Unfortunately, the richness and uniqueness of these different songs cannot be captured in words, a 'sound' description would take pages. For now, I will introduce two of the most familiar breeds of this group: the Harz Roller (a.k.a. the German Roller) and the Waterslager (a.k.a. the Malinois).

The song of the Harz Roller canary was cultivated in the Harz Mountains in Germany, whereas the Waterslager originates in Belgium. The melancholy song of the Harz Roller is characterized by relatively slow, nostalgic, soft accents as compared to the jubilant song of the

Waterslager, which has a more animated rhythm with sound segments, called tours, that are more individually distinct.[4] Although these song-bred canaries sound very different from each other, the Wild Canary is their shared ancestor. What selection principles were involved in the breeding of these distinct songbirds?

...a Ferrari from the countryside will easily betray its origin by sounding completely different than an urban-raised Ferrari

To answer this question, we will assume the vantage points of the range of actors involved in the evolutionary process. The male bird is the main character – he's on lead vocals. He's the only one that sings; female canaries, and female birds in general, do not sing. And he had better sing well (!), to impress the female canary, create a bond, and bring about a successful mating. In our case, however, the act of singing clearly goes beyond the mating function: the male bird not only has to please the female bird, but the human breeder as well. Unbeknownst to the male bird, it is ultimately the breeder who decides for or against the composition of a possible breeding pair based on the song qualities of the male bird.

However, there is an important difference: the breeder's (human) selection criteria predominantly relate to the aesthetic qualities of birdsong, whereas the functional qualities of the song of the male bird seem to dominate natural selection principles. Female birds judge a male bird's physical fitness for reproduction on his vocal performance. Yet any person who has ever listened to the varied, beautifully nuanced, and apparently improvised phrases performed by a solitary songbird with no other birds in its direct vicinity might seriously wonder whether reproduction is the only intrinsic motivation for birds to sing.[5] The third principal actor is the female bird. She not only has to be susceptible to the male's singing courtship behavior, but she should also supply a good genetic blueprint for nesting behavior, as this is what determines the actual offspring produced in any generation.

In the case of the Harz Roller and Waterslager breeds, breeders' opinions about what made the perfect canary song differed sharply. The Harz Roller breeders preferred low, smooth, rolled sounds above shrill, noisy sounds, leading to the calm, melodic song of the Harz Roller as it is known today. For the breeders of the Waterslager canary, on the other hand, the song of the Nightingale was the model to emulate. This led to the interrupted, boiling and rolling water beats and metallic tone qualities that characterize today's Waterslager song. Already in the nineteenth century, breeders organized clubs to share knowledge and to hold song contests. Standards of song quality were first established within these clubs, and eventually led to worldwide standards describing the various song tours and their ideal qualities.

Scientists have recently discovered[6] that these canary breeds differ with respect to hearing sensitivity for high-frequency sounds. Waterslager canaries show impaired hearing in the frequency range in which their vocalizations contain the most energy. In other words, in order to contact a 'hard-of-hearing' female, a male Waterslager has to produce louder sounds. This finding demonstrates that the non-singing female birds have an equally important role in the evolutionary emergence of new song-bred canaries. At this point, the case of the *Serinus canaria domestica* has been introduced in sufficient detail to address the main question of this essay: how can the cultivation of traditional nature inspire next nature's sound design? For this purpose, and as a hypothetical example, I will consider a challenge currently faced by car manufacturers–sound design for electric cars.

Car manufacturers have known for quite some time that the sounds their cars produce need to be explicitly designed. While the functional quality of car sounds guarantees skilled and safe driving, their subjective qualities are crucial to the driver's overall experience, as well as the car company's brand image. Consider the subjective associations of a car door slamming or an accelerating car engine. A car that does not produce the right sounds has the same effect on the driver's experience as a silent movie played on a full-blown, state-of-the-art home theatre system.

By mapping the lead characters of the song-canary case directly onto the stakeholders involved in the sound design of future electric cars, some intriguing new interactions immediately pop out. The car (male bird) produces the sounds that will impress and seduce its future owner (female bird) into purchasing.

A future car, for example, could adapt its sound to its owner's driving style, or sonically radiate the driver's personality traits

The sound designer or car manufacturer (breeder) decides which car and corresponding sound set best matches a particular customer segment. This may sound like common practice, but songbirds and their 'designers' do things differently. First of all, their songs are dynamic and adapt gradually to the changing environment. Moreover, as we have seen, cultivated birdsong goes beyond the functional, and the aesthetics of expression are at the heart of its being. For future electric cars, this could mean that the basic 'brand specific' sound synthesis algorithms and the type of sounds they are able to generate will still be defined by the car manufacturer, but that individual cars may be able to learn sounds and adapt them to their own environments and driver preferences. In this scenario, a Ferrari will always sound like a Ferrari, but a Ferrari from the countryside will easily betray its origin by sounding completely different than an urban-raised Ferrari. More 'open' futuristic scenarios would allow any car to disguise itself as a Ferrari sound-alike,[7] or even audiomorph into a Batmobile destined to break the sound barrier.

Other adaptive schemes could breed 'cars with personalities.' A future car, for example, could adapt its sound to its owner's driving style, or sonically radiate the driver's personality traits. Such sonifications would enable drivers and their environments to become aware of behaviors which, if desirable, could boost self-esteem or, in the case of unwanted behaviors, could motivate behavioral change. And what about car-driver units synchronizing their sounds to those of other car-driver units, much like cicadas sometimes synchronize their songs, or as song-canaries have been trained to sing in pairs or in groups of four? Such emergent phenomena could create positive feelings of being connected and, at the same time, improve traffic flow.

Many more scenarios could be envisioned, but the most important challenge remains to create the right conditions for an ecosystem to emerge in which all stakeholders (car manufacturers, intelligent cars and car drivers) will be able to freely explore the opportunities offered by sound. As we have learned from the case of the song-bred canary, these explorations need to be determined by interactions between the various stakeholders. The conditions for interaction need to be defined properly in order for this kind of evolution to thrive–one in which brand-specific sound sets simultaneously reflect the personal preferences shaping the driver-car relationship. Only then will there be a chance that one day, at daybreak, I will be moved again, this time by the sad song of a lonely abandoned car, subtly standing out from the peaceful dawn chorus in my backyard.

REFERENCES

1 GAVER, W.W. (1989). THE SONIC FINDER: AN INTERFACE THAT USES AUDITORY ICONS. HUMAN-COMPUTER INTERACTION, 4 (1), 67-94, 1989.
2 RUBEN, J. (2010). PALEOBIOLOGY AND THE ORIGINS OF AVIAN FLIGHT. PROCEEDINGS OF THE NATIONAL ACADEMY OF SCIENCES, 107, 2733-2734, 2010. 2010; OR FOR A POPULAR SUMMARY SEE: HTTP://WWW.SCIENCEDAILY.COM/RELEASES/2010/02/100209183335.HTM
3 ORR, H.A. (2009). TESTING NATURAL SELECTION. SCIENTIFIC AMERICAN 300, 30-37, 2009.
4 WORLD CONFEDERATION OF ORNITHOLOGY: SONG STANDARD OF THE WATERSLAGER/MALINOIS CANARY. HTTP://WWW.WESTERNWATERSLAGER.COM/TEXT/ARTICLES/SONGSTANDARDS/COM/COMSTD.HTM RETRIEVED ON 02-09-2010.
5 ROTHENBERG, D. (2005). WHY BIRDS SING - ONE MAN'S QUEST TO SOLVE AN EVERYDAY MYSTERY. PENGUIN, ALLEN LANE, GREAT BRITAIN. 978-0-713-99829-6; ALSO SEE ACCOMPANYING WEBSITE: WWW.WHYBIRDSSING.COM RETRIEVED ON 02-09-2010.
6 OKANOYA, K., DOOLING, R.J. AND DOWNING, J.D. (1990). HEARING AND VOCALIZATIONS IN HYBRID WATERSLAGER-ROLLER CANARIES (SERINUS CANARIUS). HEARING RESEARCH, 46, 271-276.
7 HUKAR OZYASAR (2010). HOW TO MAKE MY CAR SOUND LIKE A FERRARI. HTTP://WWW.EHOW.COM/HOW_6576564_MAKE-CAR-SOUND-LIKE-FERRARI.HTML RETRIEVED 21-10-2010.

WallSmart®

ONE BUCKET. A THOUSAND COLORS.
NANO
PAINT

FOR ALL TOO LONG, PEOPLE HAD TO ADAPT THEMSELVES TO THE TECHNOLOGY THEY INVENTED. THIS SMS THUMB IS A PERFECT EXAMPLE OF THAT. HUMANE TECHNOLOGY IS A SERIES OF STRATEGIES THAT REVERSES THIS PROCESS - HUMANS FIRST!

Humane Technology

Too often, technology frustrates us. It constrains us to predefined directions, rather than encouraging discovery or unexpected outcomes. It aims to overcome, not augment, our hominid instincts and physiological processes.

Even worse, technology may knock us out of alignment with our personalities or fundamental values. 'Humane technology' has the opposite effect. It acts as a partner in improving the human condition, rather than a rival. It recognizes that we are biological organisms, evolved to live in an information-rich, three-dimensional environment. Neurologists have counted between nine and twenty-one unique human senses, yet conventional technologies engage with only the senses of sight, touch and hearing. Humane technology resonates with the whole set of body senses, and beyond that, with our social sense, our sense of freedom, and our sense of self. Humane technology, as a concept, can be tricky to pin down. What is humane in one circumstance is irritating or destructive in another. A cellphone may be more humane than a landline, permitting the talker to wander around, free to conduct business or call home from the far side of the globe. But cellphones may be inhumane for precisely this reason. A Blackberry can seem less like an indispensable fifth limb than a second mouth that just won't shut up. Technologies are never defined as entirely humane or entirely inhumane; we should not expect simple answers on how 'humane' a particular technology is.

Nevertheless, discussing its humaneness helps us understand how technology extends (or numbs) our human potential. The concept of humane technology may provide guidelines in deciding which technologies we allow to penetrate into our lives – and which ones we prefer to keep out. In the end, asking the question is of more importance than having the answer. |AG|KVM

Technology that feels natural, rather than estranging

Medicine can be hard to swallow, and a vaccine needle can make patients squirm. Is there a friendlier way to get what's good for us? Humane technology recognizes that humans are not one-size fits-all. What works like a charm for me might feel like a curse to you. Like a biological organism, humane technology is so adaptable you might not even notice that it's there. Just don't be surprised if your doctor prescribes medical-grade sushi.

THE TASTE OF OUR OWN MEDICINE? NOT SO GREAT.

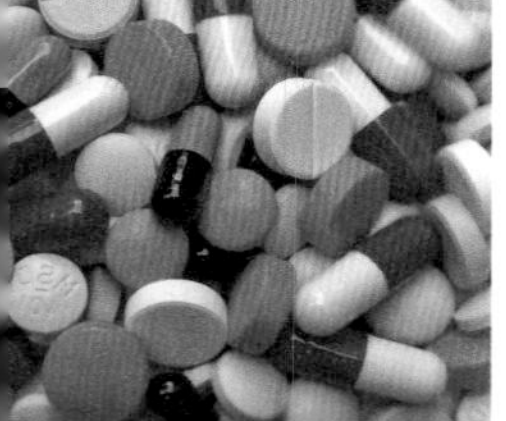

BETTER THAN CHICKEN SOUP? IN THE FUTURE, PHARMASUSHI MIGHT DELIVER MEDICINES VIA GENETICALLY ENGINEERED FISH.

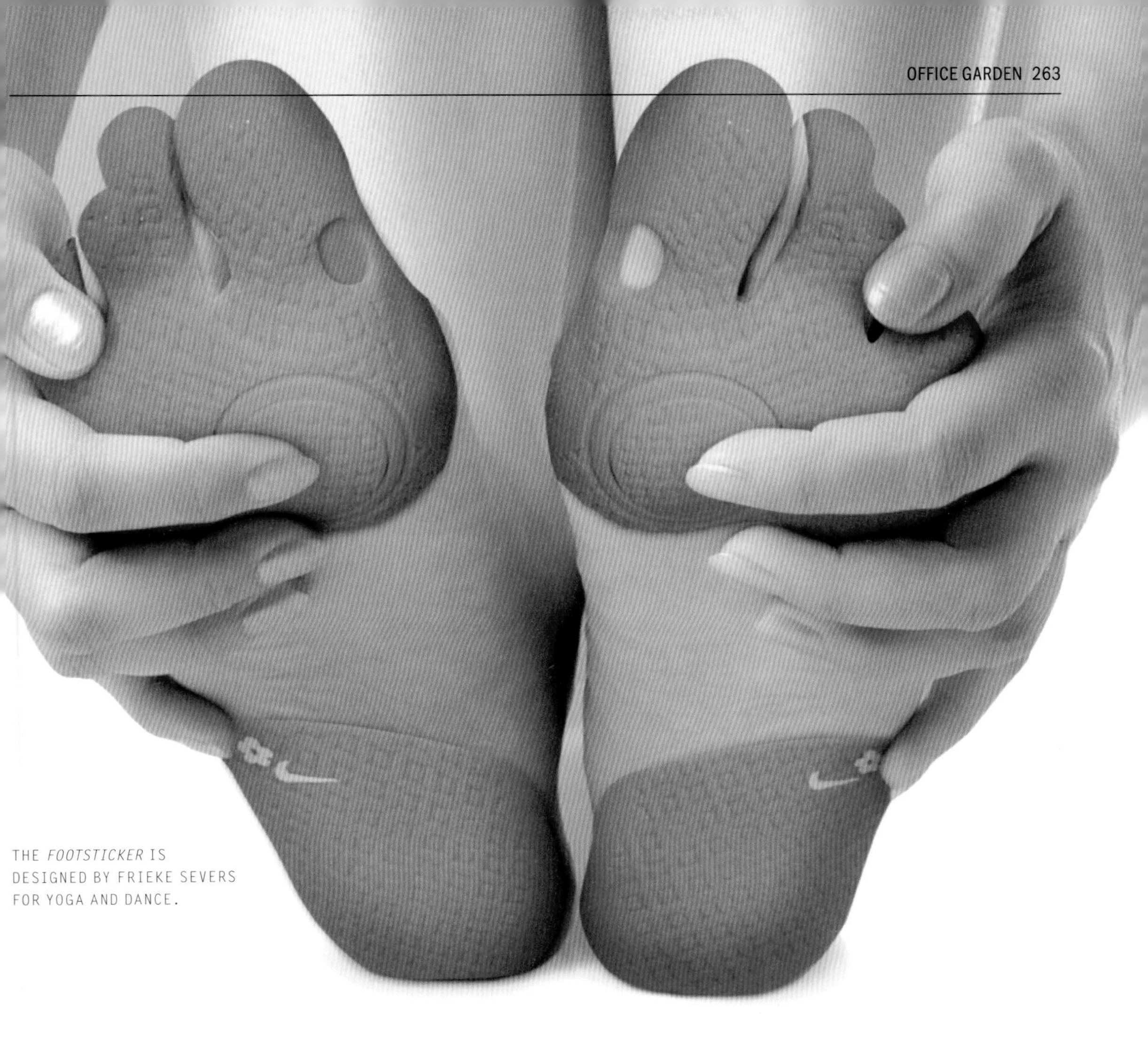

THE *FOOTSTICKER* IS DESIGNED BY FRIEKE SEVERS FOR YOGA AND DANCE.

Technology that revives human intuitions, in particular those we might have forgotten about

Technology may aim to overcome our hominid instincts and physiological processes, but isn't it better to augment them? Some recent shoe designs recognize that 'barefoot is best,' while protecting millions of years of evolution from more recent developments: broken glass and slippery floors. Humane technology will take us back to the tribe.

AIR CONDITIONING WORKS AGAINST OUR SWEAT GLANDS, NOT WITH THEM.

Technology that takes human needs as a cornerstone of its development

THE *HIPPO WATER ROLLER* WAS INVENTED IN 1991 BY TWO SOUTH AFRICANS; MR. PETTIE PETZER AND JOHAN JONKER. INITIALLY CALLED THE *AQUA ROLLER* THE NAME WAS CHANGED TO *HIPPO WATER ROLLER* IN 1993. THE ASSOCIATION WITH WATER, A ROUND BODY AND THICK SKIN COMPARES WELL TO THE FAMOUS AFRICAN HIPPOPOTAMUS, WHICH ALSO MAKES IT A NICE CASE OF BIOMIMICMARKETING (P.292)

FAST, CHEAP, AND EXPENDABLE. IS THIS THE WORKER OR THE PRODUCT?

Technology doesn't have to be expensive or electronic to be humane. Think of it as the Occam's Razor of humane technology: the simpler the solution, the better the outcome. The Hippo Water Roller makes it significantly easier for poor, rural communities to haul water back to the village. Rolling water, rather than carrying it, reduces stress on the body and frees up time for other tasks.

THE PHILIPS WAKE UP LIGHT AIMS TO NATURALIZE ALARM CLOCKS.

Technology that resonates with the human senses, rather than numbs them

WEARING A SCREEN ON YOUR HEAD DOESN'T MAKE IT MORE NATURAL.

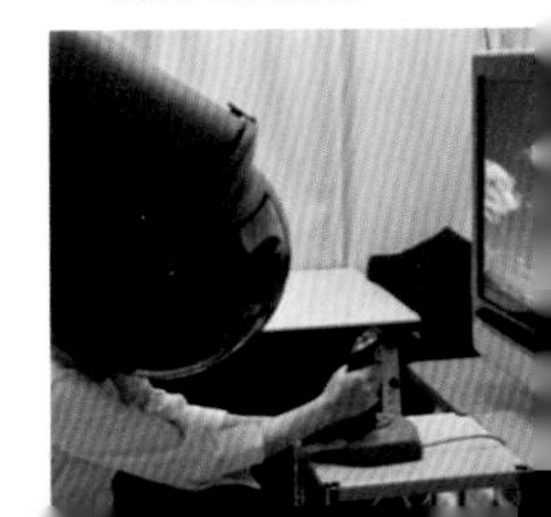

Depending on your job, you may spend most of your waking hours staring at a screen, and not tasting, touching, or smelling much of anything. Humane technology recognizes that humans are sensory organisms, made to live in nature. We prefer to be woken by the gradual glow of a sunrise-style lamp, rather than a blaring alarm in the dark.

Technology that doesn't outsource people, but empowers them

How healthy is it to have an escalator to the gym? Humane technology doesn't exist to replace the human mind or body. Rather, the human and the machine should have a mutually beneficial partnership. Cheetah Flex-Foot prosthetics seamlessly integrate with a user's leg to create an artificial foot. Users are at least as speedy as runners with flesh-and-blood feet, and may actually be faster thanks to the springy metal.

RUNNER OSCAR PISTORIOUS RUNS ON METAL LEGS THAT MIMIC THOSE OF A CHEETAH.

FOR HUMANE TECH, SOMETIMES LESS

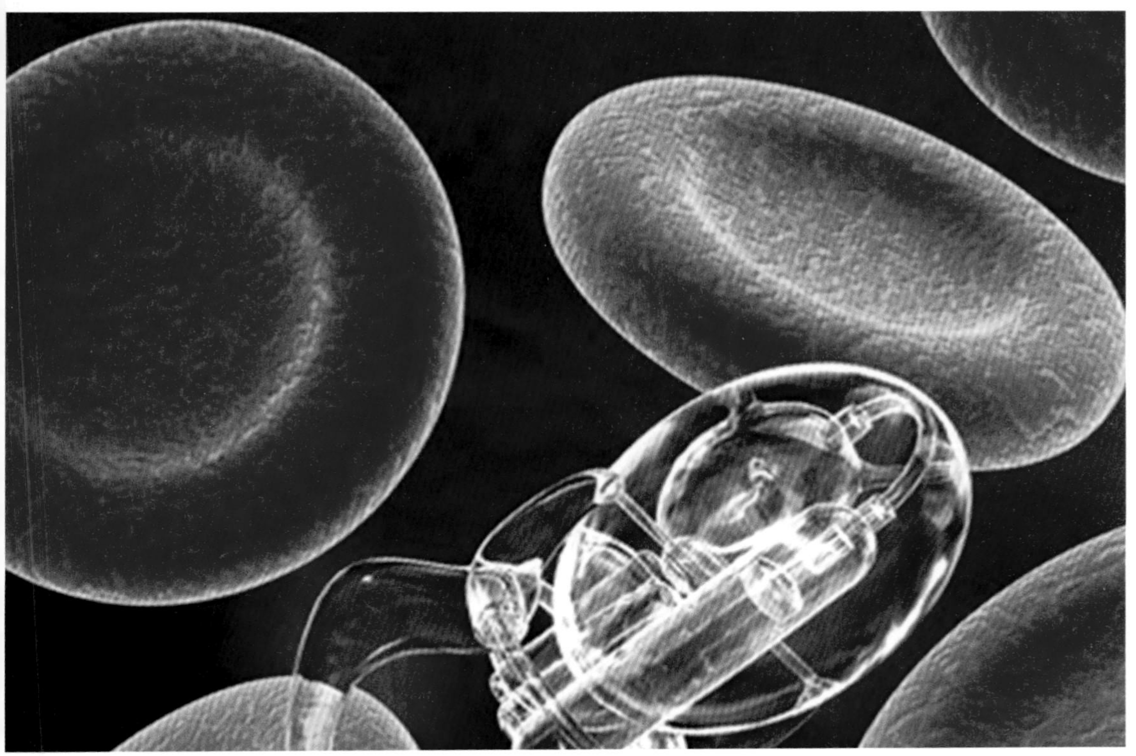

NANOBOT BLOOD CELLS MAY SOMEDAY REPAIR THE BODY FROM THE INSIDE OUT.

Technology that improves the human condition and realizes the dreams people have of themselves

KEEP THAT GUN ROBOT AWAY FROM MY HUMAN CONDITION.

We don't need better military technology. The atom bomb already showed that we've perfected the science of killing each other off. What we need is technology that adds to the very best in ourselves- our health, our minds, and our dreams for the future. Naturalist E.O. Wilson's concept of biophilia should not limited to humans. Our technology should love life as much as we do. | AG

12:21
BlackBerry
PREPARE YOUR CHILD FOR THE OFFICEGARD

NEW MEDIA.
NEW ENVIRONMENT
NEW THINKING.

TH THE BLACKBERRY MEDIA SKILLS TRAINING PROGRAM. AGE THREE AND UP.

#4 OFFICE GARDEN

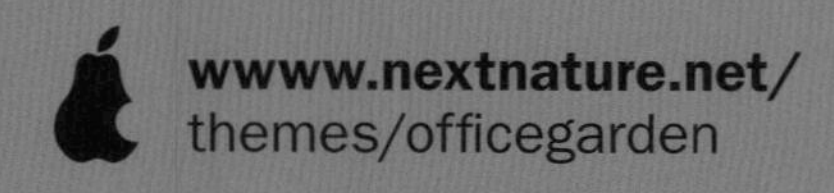

wwww.nextnature.net/
themes/officegarden

NEXT NATURE

ISBN 978-84-92861-53-8
9 788492 861538

WE ARE LIVING IN THE FUTURE, AND WE FIND IT PRETTY BORING.

WE WERE PROMISED FLYING CARS, BUT WE GOT TWINKIES INSTEAD. THE SUPERMARKET SHOULD ASTONISH US. THE FRUIT IS ALWAYS IN SEASON, THE MEAT IS ALWAYS FRESH, AND DESSERT IS SHELF-STABLE FOR YEARS. WE FIND THIS HI-TECH FUTURE BORING BECAUSE TO US THE SUPERMARKET IS AS NATURAL, AND EXPECTED, AS APPLES IN AUTUMN.

HAVE HUNTING AND GATHERING BECOME TOO EASY?

UNTIL HUMAN NATURE IS GENETICALLY ALTERED, WE WILL ALWAYS HAVE THE SAME TASTES FOR FATTY, SALTY, AND SWEET. ARGUABLY, IT IS PERVERSE THAT THE SUGAR IS PLACED IN A FIXED LOCATION IN THE SUPERMARKET AND WE KNOW EXACTLY WHERE THAT IS. MAYBE ONE DAY THE SUPERMARKET WILL REVERT TO THE SAVANNAH. WE WILL TOSS SPEARS AT FRIED CHICKENS, FORAGE FOR CANDY, AND PULL FISH FILLETS STRAIGHT FROM THE RIVER.

STONE AXES AND SPEARS HAVE BEEN REPLACED BY SHOPPING CARTS.

IMAGINE A CAVEMAN WALKS INTO A CONTEMPORARY SUPERMARKET. THE VAST MAJORITY OF THINGS IN THE SHOP WOULD NOT MAKE SENSE WHATSOEVER. COKE BOTTLES? CEREALS? POWDERED SOUP? BODY LOTION? WASHING POWDER? SHAVING CREAM? ALL MYSTERIES. OVERWHELMED BY THE HI-RES ENVIRONMENT OF SATURATED COLORS HE WOULD ASSUME HE WAS NO LONGER ON EARTH, BUT RATHER IN HEAVEN.

AND WE HAVE NO IDEA WHERE OUR PRODUCTS ARE GROWN.

AN OLD JOKE HAS IT THAT CHOCOLATE MILK COMES FROM CHOCOLATE COWS. BUT WHERE DOES IT REALLY COME FROM? FOR MANY CONSUMERS, THE FOOD CHAIN STOPS AT THE STORE WHERE THEY BUY THEIR PRODUCTS. THE DEPICTED CORPORATE SYMBOLISM OF THE DUTCH WAREHOUSE V&D CERTAINLY UNDERPINS THIS IDEA; PORK IS NATIVE TO THE MEAT AISLE, AND SHOES ARE GROWN ON TREES.

OUR BIODIVERSITY INCREASES EVERYDAY.

OUR ANCESTORS CONSUMED WHATEVER SPECIES THEIR BODIES COULD DIGEST. THERE ARE ABOUT 1,500 SPECIES OF PLANTS WITH EDIBLE LEAVES, YET IT'S LIKELY THAT IN THE PAST WEEK YOU'VE ONLY EATEN *LACTUCA SATIVA* OR *BRASSICA OLERACEA*. TODAY, MOST OF OUR CALORIES COME FROM A FEW TYPES OF GRAINS AND A HANDFUL OF ANIMALS. MODERN FOOD SCIENCE HAS MUTATED THESE INTO A LUSH ECOSYSTEM: THE AVERAGE NUMBER OF PRODUCTS IN AN AMERICAN SUPERMARKET HAS GONE FROM 15,000 IN 1980 TO 50,000 TODAY.

WE CONSUME ILLUSIONS.

WE USED TO BE ABLE TO IDENTIFY FOOD BY ITS COLOR, FEEL, AND SMELL. OUR MEALS NEEDED NO PACKAGING. NOW, EVEN BANANAS COME BRANDED AND WRAPPED IN CELLOPHANE. BUT THE IMAGES PRINTED ON THE BOX DON'T ALWAYS MATCH UP TO THE FOOD INSIDE. EVEN IF THE VISUAL EFFECT CAN BE FAKED, THE TASTE CANNOT. MANY OTHER MARKETING CLAIMS – WHOLESOME, ALL-NATURAL AND FARM-FRESH – ARE JUST AS ARTIFICIAL. IF IT'S TRUE THAT WE EAT WITH OUR EYES FIRST, THEN WHAT WE EAT ARE ADS.

THE SUPER-MARKET: OUR NEXT SAVANNA?

By Koert van Mensvoort

Our spears have been replaced with shopping carts, yet, we still have the urge to hunt and gather. Are modern supermarkets learning to mimic the savannah?

We are living in the future and we find it boring. The best place to gather evidence for this claim is the supermarket. Try and have a fresh look at the word: Supermarket. It is so futuristic word, yet we use it mindlessly. If only the supermarket wasn't such a mundane part of our life, we would realize how exceptional this environment really is. As a thought experiment, imagine we would put a caveman (the hunter-gatherer type that lived 40,000 years ago) in a time machine with the final destination: the supermarket around your corner. Surely our friend the cavemen would be astonished after opening the capsule; overwhelmed by all saturated colors he would likely assume he was no longer on Earth. The majority of objects would not make any sense to him. Body lotion? Shaving cream? Except for the fruit department – which is merely stunning by its huge variety – the majority of the food products would not even be recognized by him as food. Milk from cardboard packages? Coke bottles? Cereals? Powdered soup? All mysteries.

Possibly the meat department could bring our cavemen into appreciation of our 21st century consumer culture. As far as he would recognize the meat as meat – nowadays it has become so abstract we hardly recognize the animal in it anymore – he would find the preprocessed chopped pieces quite an improvement on his own painstaking food production efforts. Hunting and gathering is easy in the supermarket: Stone axes and spears are replaced by shopping carts, just run around and gather the stuff you need. Who knows after some further explorations, our caveman would conclude he was in heaven – or some caveman-like concept of an ideal world. Undoubtedly there would still be many surprises. He would not realize that it requires money to pay for all the treats, which has to be gathered elsewhere in some daytime job. How complex and inconvenient! Also, he would be clueless on where the foods come from and how they are produced. But then again, that's one mystery we share with the caveman: of the vast majority of the products we buy we haven't got the slightest idea where they're from.

One wonders how the caveman would look at us: oddly dressed, cart-pushing beings, with sneakers and haircuts. Certainly he would perceive us as peculiar people; possibly he would not even see us humans, but rather as post-human beings of some sort. Obviously he would be mistaken then, as we are humans just like him: genetically we are not different from cavemen, only our environment has changed. It is just that unlike our friend the caveman, we are living in the future. Humans evolved in an environment where food had to be gathered and hunted and in many respect the concept of the supermarket is just as strange for us as for a caveman. Looking at the recent developments in supermarket design it becomes apparent that retailers understand that the supermarket is our next savanna and that it should be designed as such. Moving from warehouses stacked with boxes, supermarkets are increasingly transformed into 'experiences' where the consumers can wonder and gather around to fulfill their needs, while at the same time establishing a relation with the environment. One of the leading chains in this development is Whole Foods Market, the 'natural and organic grocery' based in the U.S, Canada and U.K, where food is not merely presented in take away packages, but also freshly prepared and ready for consumption on the spot. All offerings are kept and presented in its purest state, avoiding artificial additives, sweeteners, colorings, and preservatives. Consumers can choose from six different types of chicken, all organically bred. Videos are shown on how the products are grown.

The supermarket might turn back into a forest someday

Shopping at Whole Foods reminds one of Roald Dahl's book *Charlie and the Chocolate Factory*, whose main attraction is the Chocolate Room in which everything is edible: the pavements, the bushes, even the grass. There even is a chocolate river, where the chocolate is mixed by a waterfall. Although the sugar and obesity levels are a lot higher in the Chocolate Room than at Whole Foods Market, the offerings at both places are organic and Willy Wonka even proclaims this as his unique selling point: 'There is no other factory in the world that mixes its chocolate by waterfall.' Can it get any more organic than that? Perhaps in the long run, supermarkets will become more like savannas again. A place to hunt, gather, farm and have a picnic with the family. If this ever happens, we would be full circle. Not so much back, but rather forward to nature.

The
More
Natural
Water

WE ARE THE ONLY ANIMAL CAPABLE OF CONSUMING IMAGES

The Packing Replaces the Product

REAL LEMON

THE ICONIC JIF LEMON JUICE BOTTLE, DESIGNED BY EX-WWII FIGHTER PILOT STANLEY WAGNER, COULD BE CONSIDERED THE MOTHER OF MODERN-DAY BIOMIMIC PACKAGING. MILLIONS HAVE BEEN SOLD, FIRST UNDER THE BRAND NAME "REALEMON" AND LATER UNDER THE NAME "REALEM".

Portable, prepackaged, and single-serving, fruits may be the ultimate convenience food. That doesn't mean they can't be improved. If peeling an orange rind is a hassle, peeling back a tinfoil lid is always satisfying. Hungry consumers routinely pass up the real deal for processed products that ape the appearance of the original. Food designed to look natural may be more appealing than natural food itself, and it makes sense. Nature designs with no audience in mind, but humans design nature for each other. There's only so many taste combinations that the human tongue can sense, but the human eye can distinguish hundreds if not thousands of brands. Most apple juice, for instance, tastes approximately the same. Packaging tells the confused consumer whether to experience that juice as wholesome or as fun, as age-appropriate, or where and how to drink it. In the agribusiness cornucopia, it no longer matters that the product is not necessarily as advertised. The ad is the ingredient; the package is the product. | AG

FRUIT FOR ONE DAY

DOCTORS RECOMMEND TWO SERVINGS OF FRUIT A DAY. YOU MIGHT ACHIEVE THIS BY SLURPING AWAY A *QUICKFRUIT*, BUT HOW CONVENIENT WOULD IT BE IF ONE PRODUCT WOULD MIX THE TWO TOGETHER, WITHOUT THE PESKY NEED TO CHEW THROUGH AN ENTIRE APPLE. LIKE THE DUTCH FRUIT2DAY. OF COURSE, CONVENIENCE COMES AT A COST. THE PACKAGE OF JUICE AND 'FRUIT BITS' IS TWICE AS EXPENSIVE AS TWO PIECES OF ACTUAL FRUIT, NOT TO MENTION THAT THE CHERRY-GRAPE FLAVOR CONSISTS MOSTLY OF APPLE, BANANA, AND PEAR PUREE.

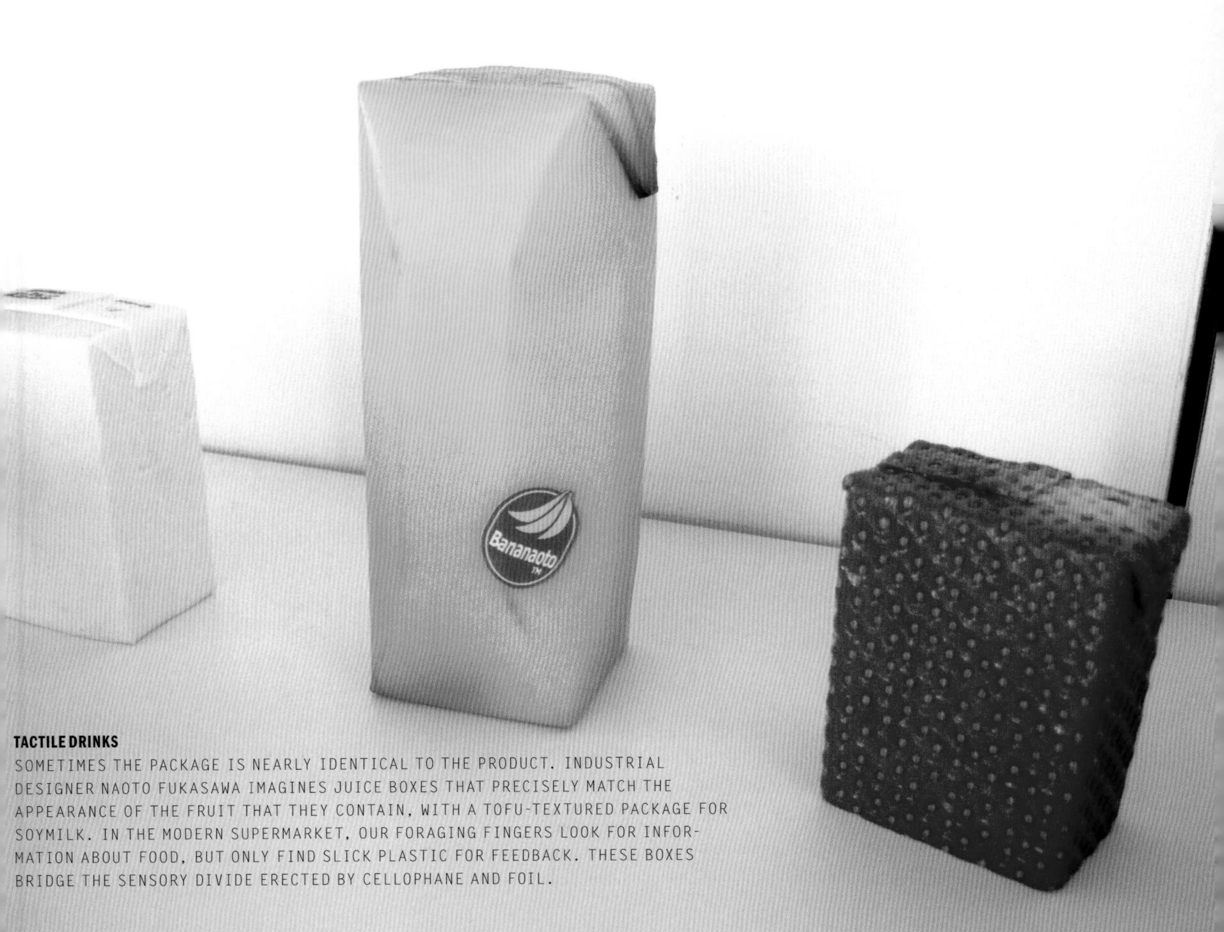

TACTILE DRINKS

SOMETIMES THE PACKAGE IS NEARLY IDENTICAL TO THE PRODUCT. INDUSTRIAL DESIGNER NAOTO FUKASAWA IMAGINES JUICE BOXES THAT PRECISELY MATCH THE APPEARANCE OF THE FRUIT THAT THEY CONTAIN, WITH A TOFU-TEXTURED PACKAGE FOR SOYMILK. IN THE MODERN SUPERMARKET, OUR FORAGING FINGERS LOOK FOR INFORMATION ABOUT FOOD, BUT ONLY FIND SLICK PLASTIC FOR FEEDBACK. THESE BOXES BRIDGE THE SENSORY DIVIDE ERECTED BY CELLOPHANE AND FOIL.

Healthy Eating Disorder

Healthy eating is, of course, healthy, but under some circumstances it can have the opposite effect. *Orthorexia nervosa* – healthy eating to the point of obsession – is a newcomer in the list of eating disorders that include anorexia (self-starvation) and bulimia, characterized by alternating periods of deprivation and bingeing.

Coined by American doctor Steven Bratman in 1997, the name comes from the Greek words for 'correct' and 'appetite.' Orthorexia is still a contentious diagnosis; the American Psychiatric Association and the World Health Organization don't recognize the disorder.

We are bombarded with hard-to-interpret information about diet. Eat carbohydrates for energy; never let carbs cross your lips! Alcohol gives you cancer; alcohol prevents cancer! Most of us respond to this overload of data about a healthy diet with the occasional (or constant) dose of willful ignorance. We feel more or less content to eat a cheeseburger, so long as it's followed up with a salad. Sufferers of orthorexia nervosa cannot digest information overload. They may spend as much time thinking about food as anorexics or bulimics, planning their meals days in advance and then taking no pleasure in the result. Sufferers can have an all-consuming desire to consume only organic produce, only whole grains or only raw foods. While the disorder is not dangerous in its initial stages, restrictive behaviors can snowball and lead to nutrient deficiencies or to emaciation. An orthorexic may begin by giving up all fish and meat, then all eggs and dairy products, then anything containing fat at all, even though certain fats are essential for good health. Each lapse is treated as a spiritual failing, and those who follow a less regimented diet are seen as morally inferior.

While anorexia is associated with the impossible beauty standards of popular culture, orthorexia appears to stem from the astonishing amount of information accessible with a single Google search. Thousands of studies and articles are there for the reading, but there is no easy way to prioritize it. While science agrees that the lifestyle of the information worker can lead to diabetes, obesity, and heart disease, it may be just as important to consider the psychological diseases that arise from hours on WebMD, or from absorbing the undifferentiated flow of news on blogs and television. It's been said that knowledge is power, but it may be more accurate to say that for information, the dose makes the poison. KVM|AG

Surrogate Sushi unmasked with DNA test

Two New York teenagers, Kate Stoeckle and Louisa Strauss, discovered that many of the city's sushi restaurants and seafood sellers were fibbing to customers about their fish. Cheap cuts of tilapia masqueraded as expensive white tuna, while seven 'red snapper' samples actually came from a variety of species, including the endangered Acadian redfish. While the classmates' sample size of 60 was not enough to indict the industry, it does provide a lesson to wary buyers. When all fish filets look about the same, it's easy for suppliers to switch species.

SUSHI DETECTIVES KATE STOECKLE AND LOUISIA STRAUSS

Perhaps the most exciting part of the story may be that certain methods used for DNA research, once restricted to academia and forensic labs, are rapidly being democratized. GPS, once a high-tech wonder, has been turned into an everyday gadget. Simple genetic fingerprinting devices may soon be available to anyone curious about their food's genome or, for that matter, their own. Our cellphones already help us navigate, surf the Internet, and take pictures. The newest models might also feature a handy DNA sequencer to help us decide if we should really purchase that dubious piece of toro. | AG

PRINGLES, DEFINITELY NOT POTATO CHIPS...

No Potatoes

Proctor & Gamble, the maker of Pringles, has successfully argued in a British court that their product is not a potato chip. The snack is also officially defined not to be a potato stick, wafer or even a potato product at all. Why is this a win for Proctor & Gamble? A ruling otherwise would result in a 17.5% value-added tax, from which most other foods are exempt. Proctor & Gamble's lawyers argued to the court that their product had a shape and color not found in nature, and that a typical Pringle is less than 50% potato by weight. We're not sure what makes up the other half of the mystery chip, but we are hoping it has nothing to do with the giant mustache of the man on the packaging.

BIOMIMIC MARKETING

Selling nature with five commercial strategies

Green electricity, organic shampoo, Jaguar convertibles, Alligator gardening tools, Red Bull, bio beef, Camel cigarettes and Puma sneakers. Once you develop an eye for it, it is quite astonishing to see how many products and brands – through their name or logo – refer to 'nature.' Called biomimic marketing, this technique uses images of nature to market a product.

Nature is a terrific marketing tool and corporations know this. Somehow, natural references provide us with a familiar feeling of recognition and trust. Biomimic marketing is applied in the most peculiar, unexpected ways. For instance, when having to choose between 18 different types of condoms, we are intuitively drawn towards the one with the word 'natural' on the packaging, thereby omitting the contradictory fact that condom-use in itself can hardly be called natural. But who cares? Biomimic marketing is not about nature as much as it is about marketing. Its goal is to promote the positive qualities of products in the minds of consumers. Nature – with its aura of authenticity, harmony and beauty – is among the best vehicles to achieve this. Here are five of the most popular biomimic marketing

SHARK
GA

BIOMIMIC MARKETING STRATEGY #1

Natural Aesthetics

ARTIFICIAL ISLANDS, DUBAI

The products in the first category have no link, or only a weak one, with the natural imagery being used. Lacoste is not about crocodiles, Linux is not about penguins, Bacardi is not about bats and Apple is certainly not a fruit company. The natural reference is mainly employed for aesthetic reasons. The placement of a cell phone antenna mast in a natural area can be sold to the public much more easily once the mast is dressed as a pine tree. Artificially created islands in Dubai are more attractive and have a higher market value when constructed in the shape of a palm tree. The marketers' reasoning is as simple as it is effective: we are familiar with natural phenomena and these resonate positively with us. How convenient for products to piggyback on the existing perception.

BIOMIMIC MARKETING STRATEGY #2

Nature As Metaphor

THE PUMA LOGO ON A VINTAGE PUMA SNEAKER

The border between the first and the second category is fluid. With the products in this category, the link is more content oriented. For instance, by calling a sneaker Puma, its makers not only apply the natural reference aesthetically, but also hope to transfer some of the positive qualities of this elegant, sporty animal onto their product. The relation is not meant to be taken literally – Puma sneakers are made from plastics, not from pumas – but rather as a metaphor. Other examples are the eagle symbol of the USA, the Twitter bird, the almost-extinct panda as the logo of the World Wildlife Fund (WWF), and the bunnies of *Playboy* magazine, which has nothing to do with rabbits, and everything to do with sex. Now just imagine for a moment that the U.S. government used a turtle as its mascot, or that Puma used an elephant, Twitter a dog, the WWF a dodo, Jaguar a cow and *Playboy* a spider as its logo.

DUREX CONDOMS: BRANDING THE NATURAL FEELING WITH A VISUAL PLAY OF TWO BODIES BECOMING A LANDSCAPE

BIOMIMIC MARKETING STRATEGY #3

A Natural Feeling

A third category of biomimic marketing is the group of products that sell themselves by emphasizing the fact that they provide the consumer with a 'natural feeling': bras so light and soft that it feels as if you're naked, hair-dye products that transform brunettes into natural blondes. Natural condoms also fall into this category. What this natural feeling means is usually not defined in much detail, but surely it must be positive. It's marketing, after all. Interestingly, the natural reference not only markets the product. As a side effect, it also enforces our notion of nature as something wonderful, harmonic, calm, and soothing.

BIOMIMIC MARKETING STRATEGY #4

Eco Friendliness

THE 'GREEN HUMMER', A CYNICAL OXYMORON OR CLEVER MARKETING STRATEGY?

The products in the fourth category don't necessarily provide the consumer with a natural feeling. Their biomimic marketing revolves around the claim to be 'friendly towards nature.' Typically, the claim is made in comparison with other 'less friendly' products from competing brands. Take, for instance, the Toyota ECO sports utility vehicle, which is less polluting than an average SUV, or the Honda Diesel GTI, which is advertised with singing flowers, birds and rabbits that welcome the engine because it's a little less polluting than the stinky bad diesel engines from the competition. Green electricity is another obvious example. Although electricity itself is not associated with nature, the label 'green' is added because a certain percentage of the electricity is created from renewable resources.

IT TAKES NATURE TO KILL NATURE

BIOMIMIC MARKETING STRATEGY #5

Naturally Made

Think eco-fruit, biological meat and organic shampoo. Contrary to regular products, naturally made products are marketed with the claim of being produced in a more wholesome way, on a smaller scale, with less pollutants and more sustainable production techniques. A naturally made product often uses a combination of other biomimic-marketing strategies. Its packaging is often brownish and made of natural materials, draped with images of the all-natural substances used in the product itself. Its 'authenticity' provides its buyers with a more healthful feeling and implies being friendlier to nature. It may be no surprise that naturally made products have gained a significant market share. What started at small farmers' markets with idealistic motives has become a global industry with rules and regulations for when a product may call itself bio, eco or organic.

THE BIGGEST MARKETING SCHEME OF ALL

Marketing Nature

URE®

Besides the countless products that reference nature in order to be liked and bought by consumers, there is another hidden, largely unconscious, yet even bigger marketing mechanism going on. Perhaps even the shrewdest marketers are not really aware that biomimic-marketed products not only promote themselves, but simultaneously also promote a one-dimensional, romanticized notion of nature. Along with the promotion of products, nature itself is being characterized as a sensible, harmonic, soothing, authentic, healthy, honest and beautiful force in life. The darker, more negative side of nature is consistently omitted by the biomimic-marketers. You can't sell products with diseases, death, hurricanes, floods, or the other extremely crude, unpredictable qualities nature has to offer. Nature itself is the most successfully marketed product of our time. |KVM

Comeback of the Ugly Fruit

Perhaps in the long run, historians will consider this the official end of modernity as we know it: The comeback of the curved cucumber, bent banana, and questionable carrot, at least in the European Union.

The EU recently scratched dozens of 'marketing standards' that placed strict regulations on the appearance of fruits and vegetables sold in stores. These laws mandated everything from the precise color of eggplants to the ramrod-straightness of cucumbers. The stipulation that "slightly crooked cucumbers may have a maximum height of the arc of 20 mm per 10 cm of length of the cucumber" sounds like a word problem from elementary school. Funnier still was the suggestive rule that bananas should be at least 14 centimeters in length and 'free of abnormal curvature.' In a rare circumstance where business and activist interests intersected, both Friends of the Earth and retailers like Sainsbury's lobbied against the legislation – the laws were responsible for the waste of one-fifth of all produce in Europe. Opponents claimed scrapping the laws would lower prices and be better for the environment to boot. The standards will remain for a significant chunk of the market, including lettuce, tomatoes and apples, although they'll be exempt if they're labeled 'intended for processing.'

More interesting might be the impetus for the ban on natural-looking produce in the first place. It makes sense to impose limits on pesticide levels or the precise definition of 'rotten,' but the physical appearance of a tomato has little effect on its taste or vitamin levels. In fact, as many farmers who grow heirloom varieties will attest, the standardization of produce has leeched out taste with it. The end to the dominance of cosmetically perfect produce may suggest that society is moving away from the assumption that visual perfection is a sign of actual quality. At least for food, appearance isn't everything. |KVM|AG

ORGANIC VEGETABLES AT A FARMER'S MARKET

BLACK SHEEP: A STRAIGHT BANANA AND A CURVED CUCUMBER

Mom, Why Are Carrots Orange?

The image above is not a collection of hyper-carrots genetically manipulated to mimic the colors of the rainbow...

This is the natural spectrum of colors carrots used to have, and in some regions of the world you can still find them in white, yellow, red and purple. Yet in most countries, the commercial varieties of carrot are almost always bright orange. Although the carrot has been a human food for centuries, the first concrete evidence for the iconic orange carrot first comes from early 17th century paintings.

It turns out that they may be this color for entirely political reasons. As legend has it, Dutch farmers in the 16th century cultivated orange carrots as a tribute to William of Orange – who led the struggle for Dutch independence – and the color stuck. Thousands of years of multicolored carrot history was wiped out in a generation.

The genetic origins of this carotene-rich variety are murky, although it most likely derived from selectively bred yellow carrots, or from crossbreeding with wild stock. The Long Orange Dutch carrot, first described in 1721, is the forebear of the orange Horn carrot varieties. The Horn Carrot derives from the Dutch town of Hoorn, where presumably it was developed. Coincidentally, the word 'carrot' comes from the Greek word that also means 'horn,' on account of the carrot's horn-like shape. Ultimately, all commercial types of carrots descend from these Dutch varieties: Hypernature *avant* la *lettre*. |KVM|AG

Modernist Food

For natural foods, form usually evolved before human function. Modernist food dispenses with evolved shapes for ones more suitable for slicing, stacking, and packing. The simple geometries of processed foods may be more food-like than actual food. In Japan, farmers grow watermelons in boxes with the same dimensions as grocery store shelves, creating square fruits suited for shipping with no wasted space. The finer ones make geat gifts, too. Despite its fruit-of-the-future shape, the watermelon's normal color and taste is unaltered. For modernist plants, the form is novel but the origin is clear. 'Biocubist' fruits and vegetables should read as themselves, but modernist meat should not read as an animal. Like any good modernist design, meat is stripped from its context, ground up, and abstracted. Animals have lives, deaths, eyes and organs, but fish sticks have no history except for a recent past in the freezer aisle.

Euclidean meat exists somewhere between the convenient and uncanny. SPAM is made from the shoulder muscles of *S. scrofa*, but the taste is pure pop-culture pork, as uniform as its tins. It could be that we can't bear to think of the animal in our meat, or it could be that rectangles are too seductive – clean, simple, and perfectible. | AG

YES. $300 FOR A WATERMELON. AND IT IS A CUBE. AND IT COSTS $300.

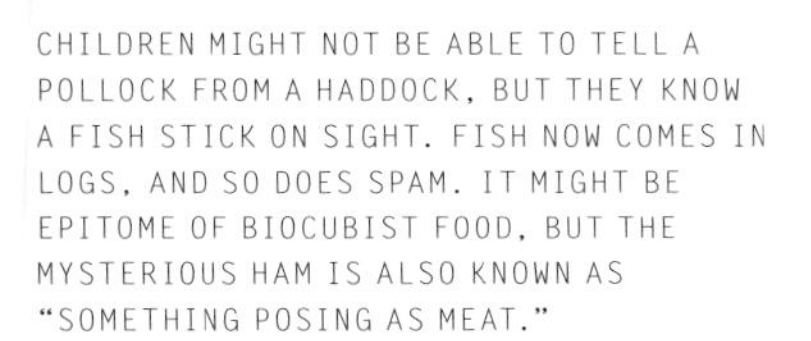

CHILDREN MIGHT NOT BE ABLE TO TELL A POLLOCK FROM A HADDOCK, BUT THEY KNOW A FISH STICK ON SIGHT. FISH NOW COMES IN LOGS, AND SO DOES SPAM. IT MIGHT BE EPITOME OF BIOCUBIST FOOD, BUT THE MYSTERIOUS HAM IS ALSO KNOWN AS "SOMETHING POSING AS MEAT."

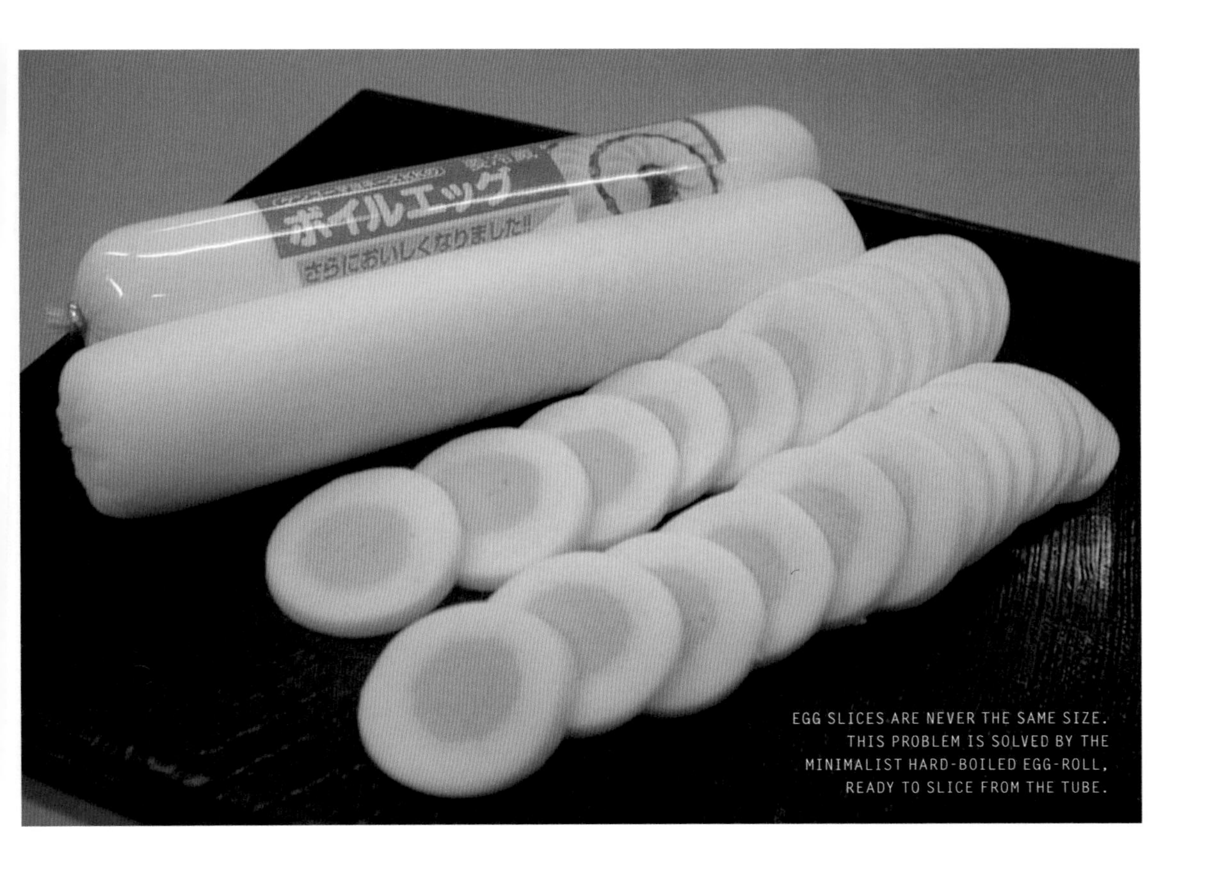

EGG SLICES ARE NEVER THE SAME SIZE. THIS PROBLEM IS SOLVED BY THE MINIMALIST HARD-BOILED EGG-ROLL, READY TO SLICE FROM THE TUBE.

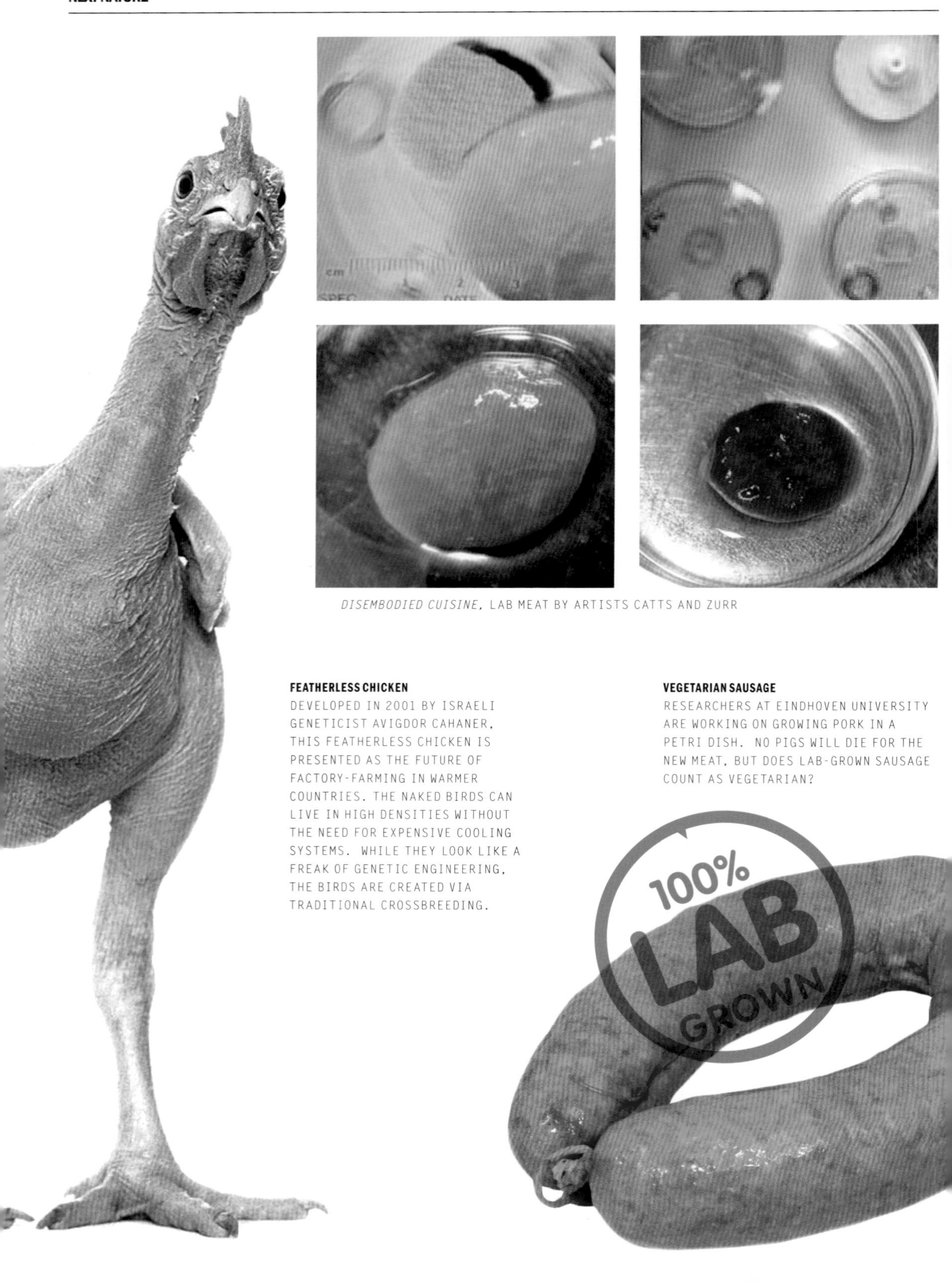

DISEMBODIED CUISINE, LAB MEAT BY ARTISTS CATTS AND ZURR

FEATHERLESS CHICKEN

DEVELOPED IN 2001 BY ISRAELI GENETICIST AVIGDOR CAHANER, THIS FEATHERLESS CHICKEN IS PRESENTED AS THE FUTURE OF FACTORY-FARMING IN WARMER COUNTRIES. THE NAKED BIRDS CAN LIVE IN HIGH DENSITIES WITHOUT THE NEED FOR EXPENSIVE COOLING SYSTEMS. WHILE THEY LOOK LIKE A FREAK OF GENETIC ENGINEERING, THE BIRDS ARE CREATED VIA TRADITIONAL CROSSBREEDING.

VEGETARIAN SAUSAGE

RESEARCHERS AT EINDHOVEN UNIVERSITY ARE WORKING ON GROWING PORK IN A PETRI DISH. NO PIGS WILL DIE FOR THE NEW MEAT, BUT DOES LAB-GROWN SAUSAGE COUNT AS VEGETARIAN?

DRESSING THE MEAT OF TOMORROW (2006) BY JAMES KING

Meat the Expectations

A friend once remarked he only eats meat if he "can not recognize the animal in it." This is a disturbing remark, but such 'consumer's preference' may also bring opportunities.

Why not disengage the animal from the meat? According to researchers, edible meat can be grown in a lab on an industrial scale. Winston Churchill – a carnivore to the core – predicted in 1936 that "we shall escape the absurdity of growing a whole chicken in order to eat the breast or wing, by growing these parts separately under a suitable medium." Today, growing meat in the lab still seems science fiction, but reality is not far behind. James King's *Dressing the Meat of Tomorrow* treats offal as biomedical cuisine. The artist proposes searching farmyards and feedlots for the most aesthetically pleasing swine, cattle and poultry. The animals would then be scanned head to hoof with an MRI machine, recording a precise portrait of their internal organs. The loveliest cross-sections would later be recreated as in-vitro meat, at once disassociated from the animal and intimately representative of it.

For *Disembodied Cuisine*, artists Catts and Zurr dined on a frog 'steak' grown in a bioreactor with living cells. Both works examine the shape of meat to come, proving that tasting like chicken will no longer mean looking like one, or even being one. |KVM|AG

Food Tech in Overdrive

Japan might soon be selling retro-futuristic steaks. In 2008, researchers there successfully cloned the legendary ox Yasufuku-go, forefather of 30% of Japanese black cattle, the breed famous for producing Kobe beef. In the bull's productive lifetime from 1980 to 1993, his sperm was used to bring 40,000 calves into the world. Kinki University (in Higashiosaka, Osaka) first cloned oxen in 2007 by using frozen testicular nuclei from bulls and injecting them into unfertilized egg cells from cows. Although the Yasufuku-go's testicles had been frozen for 13 years, the researchers found enough useable tissue to produce three viable calves. Unfortunately, the ghost of Yasufuku-go will not be able to haunt a new generation of beef cattle, as Japan currently forbids cloned animals from entering the food supply. Instead, scientists conducted the experiment to prove that prize livestock could be resurrected from long-frozen tissue. Dinosaurs might not be next on the menu, but maybe we can hold out hope for delicious, hypernatural mammoth burgers. | AG

THE FAMOUS HIDA-GYU BEEF, ADVERTIZED HERE IN A SUPERMARKET BROCHURE. THE SUPER LOW PRICE OF 17,980 YEN (A MERE 230 DOLLAR) PER KILO FOR THIS PREMIUM QUALITY MEAT IN THE A5 CATEGORY IS POSSIBLE BECAUSE THE SHOP BOUGHT THE WHOLE COW, NOT JUST A PART. *ITADAKIMASU!*

特上セット
一頭買いだから出来る
この価格!!
A５等級飛騨牛
特上カルビ・特上ロース
¥17,980/
合計
1kg

Don't like Chicken?

Why not try a new recipe for dinner tonight? First you take the meat of hundreds of birds; chickens 19% and turkeys 17% by weight. Second, you blend them in a large tank with water, corn, wheat, oil, fat, chicken collagen, salt, soy, aroma, modified cornstarch, milk parts, lemon juice, maltodextrin, dispersant (E450, E451, E452), rising means: E500, taste amplifier: E621, antioxidant: E341. Next, sculpt your mix into a shape that remotely resembles a chicken's leg – like the one folks know from the comics! Finally, place your meat-water-corn-mix on a bone-shaped piece of wood. Bread it and fry it for a crispy bite, then stick it in the freezer to enjoy later. Tasty!

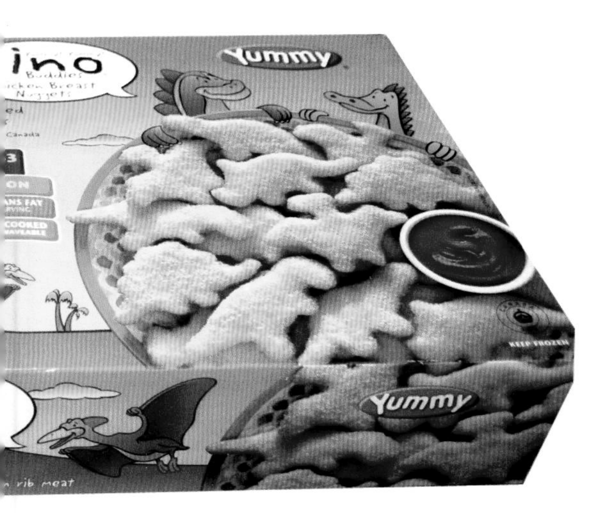

CHICKEN NUGGETS AREN'T MADE FROM WHOLE CHICKEN PARTS, BUT FROM A PALE PINK PASTE OF 'MECHANICALLY SEPARATED MEAT.' SINCE THE CHICKEN GOO HAS NO REAL STRUCTURE, IT CAN BE MADE INTO ANY SHAPE, INCLUDING CIRCLES, STARS, AND DINO BUDDIES.

Try Dinosaur!

We all know that chickens stem from dinosaurs. So should we see these nuggets as an innovative product to teach kids about the evolutionary relation between dinosaurs and birds, or simply as a cleverly marketed fantasy food? Judge for yourself.

Finland **McDouble Bacon**

U.S.A **McLobster**

The Netherlands **McKroket**

U.K. **McFish&Bacon**

Germany **Nürnburger**

Slovenia **McT**

U.S.A. **Angus Deluxe**

Austria **McVeggie**

U.S.A. **McRib**

Greece **Shrimpburger**

Egypt **Jalapeno Sandwich**

Egypt **Chicken Fillet**

U.S.A. **Angus Mushroom&Swiss**

U.S.A **Double Quarterpounder**

McAmerica
Since 1955

Brazil **Cheddar McMelt**

McAfrica
Terra Incognita

South Africa **Crispy Chicken Deluxe**

Argentina **McFiesta**

Peru **McPollo Italiano**

South Africa **Grilled Chicken Foldover**

Russia **Royal Cheeseburger**

akburger

Slovakia **McCountry**

Japan **Salt & Lemon Chicken Burger**

Turkey **Çıtır Dürüm**

Japan **Moonburger**

McArabia

100% Halal

rkey **McTurco**

Arabia **McRoyale**

Japan **Hot**

Arabia **Megamac**

China

Thigh Fillet Burger

Arabia **McFish**

India **Salad Sandwich**

India **Chicken Mcgrill**

bia **McArabia**

India **McGrill**

Arabia **Big Mac Chicken**

India **McAloo Tikki**

Philippines **McSpaghetti**

Malaysia **Bubur Ayam McD**

Indonesia **McRice**

McWorld

Experience the local specialties

Australia **Bacon BBQ Sauce**

Australia **McVeggie**

Australia **Grand Angus**

McAustralia

Second McBeefsource

Fast-food chain McDonald's is often seen as an exemplary example of the globalization processes that flatten the world and make things look, feel and taste the same everywhere. Why travel when cities have the same food, coffee and fashion chains? Increasingly, however, McDonald's offers local specialties. Have a Shrimp Burger in Greece, Teriyaki McBurger in Japan, McKroket in the Netherlands or a Nürnburger in Germany. The dishes show traces of traditional regional cuisines, allowing McBackpackers to get a taste of the world while keeping a safe level of comfort and recognition. Unfortunately, without a franchise, some places (most of Africa, Mongolia, Cuba and North Korea) won't be able to cater to fast food epicures.

YOU SHOULD TRY OUR TOMATO SOUP

THE SUPER-
MARKET IS
THE BEST
EVIDENCE
WE ARE
LIVING IN
THE FUTURE

I Want My Organic Coke

By Koert Van Mensvoort

Whoever you are, whatever you do, wherever you may be: You can't beat the real thing. It refreshes and delivers real satisfaction in every glass. It was only after America's favorite choice had been consumed for decades that a serious drawback arose: the real thing makes you *really* fat.

In 1982 the Coca-Cola Company addressed the negative aspect of its drink with the introduction of a sugar-free version of their ice-cold sunshine. Diet Coke was the first new coke brand since 1886 and signified a revolution – not only for wannabe thin women – but for the Coca-Cola brand as a whole. Numerous Coke by-products followed in the slipstream of the Diet Coke success: Cherry Coke, Caffeine-Free Coke, Classic Coke, Vanilla Coke, Coca-Cola Orange, Coke Zero, Coke Black Cherry Vanilla and many more. Today these spinoffs have saturated the market to a point that we can safely say nobody drinks the real thing anymore. End of story? Not really, evolution continues. The latest addition to the Coca-Cola product range could mean a new revolution in fast food. Diet Coke Plus not only reduces calories or adds flavor, it actually improves your health. Diet Coke Plus is fortified with vitamins and minerals. One can reportedly provides 25% of the daily required intake of niacin and vitamins B6 and B12, and 15% for zinc and magnesium, so if you drink four cans a day, you've got it all covered. Diet Coke Plus: better than the real thing!

The general trend for fast food is to remarket itself as health food. My guess is soon we will be eating burgers, fries, shakes, ice cream and chips that are engineered to make us healthy. Although this may sound like a hedonistic utopia, it's also uncanny. After all, humans have evolved with a powerful ability to tell

COKE PLUS, COKE ZERO, DIET COKE, COKE LIGHT PLUS: NOBODY DRINKS THE REAL THING ANYMORE

whether food is good or bad. When we were roaming the savanna 40,000 years ago, you could easily taste that those bitter berries were probably poisonous. But as food science progresses the ability to taste whether food is healthy becomes obsolete, or at least numbed. Nowadays we need long lists of nutrition facts on food packages in order to make sense of what we eat. Some of the important information isn't even on the package, because scientists simply don't have sufficient knowledge (yet). Aspartame, the sweetening ingredient in Diet Coke, has already been blamed by some scientists for possibly causing serious illnesses (cancer, brain tumors, brain lesions, and lymphoma) when consumed in large quantities. Though many studies are available that say aspartame is perfectly safe, the controversy has gained enough momentum for Coca-Cola to release 'Diet Coke Sweetened with Splenda', a Coke that instead of aspartame contains sucralose and acesulfame potassium, whatever that is. Thus, while engineering might make food healthier, it also makes food more abstract. Abstraction is not something people generally appreciate, whether it is in language, music, painting, or food. No surprise that people who can afford it move away from engineered food, packed with abstract chemicals and meta-substances, towards so-called organic food, which can be more or less classified as 'food produced in the way your grandparents prepared it.'

Historically, organic farms have been relatively small family run farms - which is why 'retro-food' was once only available in small stores or farmers' markets. However, since the early 1990s organic food has grown by about 20% a year, far ahead of the rest of the food industry. With the market share of organic food outpacing much of the food industry, many big corporations have moved into the market of retro-food production.

While engineering might make food healthier, it also makes food more abstract

GO FOR A ÜBER HEALTHY LIFESTYLE WITH THE *ZERO FAT HAMBURGER*

Now let's get back to Diet Coke Plus. I am not sure whether the people of the Coca-Cola Company realize it, but with Diet Coke Plus they are returning to their roots. In the late 19th century when Coca-Cola was invented by John Pemberton, he intended it as a patent medicine - similar to the cough syrups we still know today. Few people can remember early marketing slogans like "For headache and exhaustion, drink Coca-Cola" (1900) and "Coca-Cola is a delightful, palatable, healthful beverage" (1904) that clearly indicate Coca-Cola was originally marketed as a health drink. It was only after the salesmen realized that people didn't want to drink a 'medicine' on a daily basis that they decided to let go of this original image in order to increase consumption and market share.

So for over a hundred years, while we thought we were drinking the real thing, we were actually consuming a spin-off version of the original health drink - a fake! Only with this historical information does one realize that the new Diet Coke Plus health drink might just be the most original Coke produced in our lifetime. As I see it, the Coca-Cola Company is moving in an interesting direction with Diet Coke Plus, but they still have to push further in order to truly return to their origins. In their upcoming Coke launch they should drop all the chemically engineered additives and produce a truly authentic old-fashioned soft drink. After Diet Coke, Cherry Coke, Classic Coke, Coke Zero, Diet Coke Plus and Coke Splenda, I want my Organic Coke! Because you can't beat the original thing.

Organic
Coke®
FAIR-TRADE & NATURALLY PRODUCED HEALTH DRINK

THE STORY BEHIND OUR FOOD

By Maartje Somers

Every time we eat a piece of food, we take a bite out of the world. All these small bites tell a dozen stories. A carton of eggs presents the story of contented hens, a bottle of olive oil the tale of Italian grandmothers. Yet these pastoral scenes barely hide the realities of a food system that leaves one billion people starving and another billion overweight. Moving beyond food-based fictions, how should we react to the truth?

It happened in a trendy restaurant. A breadbasket and a small bowl of olives had just been brought to the table. Our hands reached out to take some, when the waitress stopped us. "Wait," she interrupted, "I have to explain the bread." Explain the bread? Yes, that one variety of bread had been baked with hard durum wheat from a village just south of Tuscany, the other one came from a bakery slightly north of Amsterdam. The olives were kalamata olives, imported from Thessaloniki, and olivas violadas (olives 'raped' by an almond) from Basque Country in Spain. It took the waitress about five minutes to finish her lecture. Then, finally we could dig in.

All our food comes with a story to tell, and usually it is the story we want to hear. In the supermarket the story is about the price of the food, in a restaurant it is about the taste and the origin.

These days all our food comes with a story to tell. Usually it is the story we want to hear. In the supermarket the story is about the price of food, in a restaurant or delicatessen it is about taste and origin. Very often stories about food focus on authenticity. That is the way food would like to be – authentic and natural – like in the old days when people harvested their own crops. And this is exactly what we want to believe. The jam in my fridge has 'a natural taste' and the milk is 'pure and honest.' Eat colour, it says on the posters in the street, displaying juicy red peppers. And these shiny vegetables almost jump from the page in the cookbooks by Jamie Oliver and Nigella Lawson, bestsellers the world over. But at the same time we are buying more and more ready-made meals.

The stories about food mask reality. It is quite obvious that our food is not natural and hasn't been so for a very long time. Deep down we all know this. And of course we know that the happy little lambs and fluffy little chicks we find in the supermarket around Easter time are in stark contrast to the reality of factory farming. Of course we know that Bertolli olive oil is not hand-pressed by those Italian grandmothers in the commercial, but that it is made in a factory owned by the multinational corporation Unilever. Very little is natural in modern food. More likely our food is a miracle of chemistry, logistics and technique. In 2007 artist Christien Meindertsma published her book PIG 05049, a thorough research of all the products made from one single pig with that specific number. The results are quite amazing. Pork chops and sausages, obviously. But also winegums and muffins and bread? Work gloves, porcelain, beer, wine, whipped cream, paper, fish food and bullets?

One might say that we have been alienated from our food. We don't have a clue what is inside those bags and packages we buy, but at the same time we feed our imagination with an idealized image of how our food is grown and produced. That is the reason why there are so many stories to tell. The distance between the pastoral story on the surface and the industrial reality behind the scenes is greater than ever. How did this happen? It is because of the success of our global food system. It all started after the Second World War. 'No more hunger,' was the motto. Western governments focused on an increase in production, on cost-reduction and efficiency. With the invention and use of fertilizers things changed really fast. In the period between 1947 and 1979 the production levels of global agriculture doubled. The wave of liberalization in the 1980s and the accompanying improvements in infrastructure, technique and logistics, were followed by a steady rise in the number of trade transactions. The sky was the limit: apples from New Zealand, meat from Brazil, shrimps from the Pacific.

Food and the landscape

All over the world the landscape changed. In line with the postwar process of upscaling, small streams were straightened, wooded banks were torn down and fields were combined in land consolidation transactions. Rye and wheat were replaced with corn, corn and corn. Farmers turned from mixed farming to either agriculture or cattle breeding. Their livestock disappeared inside the stables. Intensive farming took the upper hand, due to the low price of concentrate food, and subsequently soy. These changes in agriculture contributed to the schism between the city and the countryside, one of the most defining aspects of our modern world. In former times the cultivation of food and the slaughtering of cattle all took place inside and around the city. But with the increased separation of these functions, first after the industrial revolution and next after the Second World War, the city became a world of stone; food disappeared from the view of the city dweller and was transported to the countryside. In our day and age the only green areas in the urban landscape are the city parks, intended for leisure exclusively. The few bramble bushes I could find in my home town were recently removed and replaced with 'urban green.'

Food security

I must repeat: the global food system is a great success. According to the Czech demographer Vaclav Smil, two-fifth of the current world population would be dead if we didn't have fertilizers. The biggest boom in world population happened in the 50s, 60s and 70s of the last century. There is no shortage of food; there is more than enough to feed everybody. The world produces some three thousand calories per head each day. In the western world the term food security has gone out of fashion. Food is always available, in great abundance and it is amazingly cheap – we spend very little of our income on food. For people in the western word the choice is staggering. This hit me in the face once again when I overheard a toddler in a supermarket saying: 'Daddy, where can I find the carpaccio?' The Netherlands is one of the world leaders in intensive farming. After Brazil we are the leading exporter of food, especially meat, dairy products and vegetables.

The agribusiness makes up 10 percent of Holland's domestic product. But there is a dark side to our food system. Over the past ten years this has become more and more obvious. For example, the system is bad for our health. The huge concentrations of livestock are the cause of animal diseases that can also be dangerous to humans: mad cow disease, the swine flu, Q fever. Additionally, the complex, global mixture of our food carries a great risk. The poisonous powdered milk that killed babies in China was also found in lollipops in Europe. We are getting fatter and fatter. In Europe one in four people is overweight. In the United States diabetes has almost turned into a lifestyle. In emerging economies like China and Latin America, eating habits are starting to resemble those in the western world, resulting in a rapid rise of obesity. Fresh fruit and vegetables are expensive – bad food, made from cheap bulk ingredients like glucose (from corn), soy and palm oil and disguised in ever-changing colourful packages, is cheap. All over the world this creates a kind of food apartheid system: rich people who eat fresh and healthy food versus poor people who are simultaneously too fat and underfed because of a lack of nutritional value in the food they eat. In some countries we have this little kid that asks for the carpaccio, whereas other children don't even know what a carrot looks like, because all the food they know comes from a jar.

We have this little kid that asks for the carpaccio, whereas other children don't even know what a carrot looks like, because all the food they know comes from a jar.

From an ecological point of view our current agricultural industry is equally ill advised. Our food chain is completely dependent on our energy supply: oil is the raw material of fertilizers, insecticides and weed killers, and oil is the fuel needed to transport all this food all over the world. And in addition there is the problem of water consumption –the production of one kilo of beef requires approximately 15,000 to 20,000 litres of water.

Intensive meat production has a disastrous effect on the environment. Livestock, responsible for 18% of CO_2 emission, uses up a disproportionate amount of our food and water supplies. Enormous amounts of antibiotics and chemicals find their way into the environment, not to speak of manure surplus and acidification. The cultivation of soy for chicken and cattle feed destroys the tropical rain forest. Shrimp farms are a threat to the mangrove forests that form a natural protection against flooding. Very soon our oceans will be empty of fish. From an economic point of view monoculture, the large-scale cultivation of a single variety of tomatoes or apples, might be highly efficient, but from an ecological perspective it is not a smart move. Virtually all bananas the world over belong to one specific variety, the Cavendish, which is threatened by a fungal disease. At the same time bananas form the basic source of nutrition for the African continent.

The system is weakened by a huge number of economic problems. Distribution is the major problem, because more than a billion people are still starving. It is rather ironic that an equal number of people are suffering from severe obesity. For a brief moment in time the obese people even outnumbered the starving ones, but ever since the food crisis of 2008 and the financial crisis of 2009 the number of hungry people has been growing at a steady pace. All over the world the people at the bottom of the food chain – the farmers – are experiencing a slew of problems. Farmers are bled dry until they finally give up and move to the city. Around the year 2030 more than half the world population will be living in big cities. It is a fact that it is virtually impossible for a farmer to earn a decent living anymore. With the support of government subsidies and protection mechanisms, the giant corporations of agribusiness produce an abundance of food the small farmer will never be able to compete with. Should there be even more liberalization of agriculture to give the farmers in the developing countries a chance? The opposition will say that an honest food system will not survive the fierce competition on the world food market, where three companies run the seed market and four players have a monopoly in the buying and selling of grain and oil seed crops, where the almighty supermarket conglomerates push the prices down through the food chain and where China is buying up farmland in Africa.

On the contrary, we need more protection, but of a different kind, in order to safeguard the 'food sovereignty' of the poor countries. Advocates of liberalization point out that 'an unholy liaison' of romanticists and nostalgists makes the development of a competitive African horticultural business virtually impossible, that this is the very reason there still is hunger in the world. For as long as it has been around, this large, anonymous food system has brought forth its very own countermovement in the western world, from the health food stores of the 1970s to the current organic food stores selling pesticide-free food. Next came Fair Trade, a protest against economic injustice aimed at giving the farmer at the bottom of the food chain a fairer share. Some ten years ago we saw the rise of the slow-food movement, which opposed the uniformity of industrial food and the accompanying lack of taste and diversity. In the Netherlands a smart marketing expert of the ecological foundation Biologica came up with a 'adopt a chicken' campaign, followed by 'adopt an apple tree,' in order to reconnect the city dweller with his food and with the countryside. More recently a renewed interest in locally or regionally grown food was imported from the United States, where 'locavorism' is a separate movement. Local food equals sustainability; it has not been transported over long distances, it is in season, and it favours small-scale farmers.

Each time you eat, you put a piece of the world into your mouth. From the moment man stopped being a nomad, the stuff we put in our mouths has shaped the landscape.

The great thing about sustainable food is that it is usually high quality. The American author Barbara Kingsolver gave us the following delicious observation: 'Food is a rare moral arena in which the ethical choice is often the pleasurable choice.' In the meantime a growing number of people have become aware of this fact, judging by the popularity of farmers markets and the increasing overlap between delicatessen and health food stores.Each time you eat, you put a piece of the world into your mouth. From the moment man stopped being a nomad, the stuff we put in our mouths has shaped the landscape. When you look around you, the incongruity and the evils of the food system are clearly visible. The romantic picture postcard image of the traditional mixed farm is first and foremost in our mind, thanks to the commercials and the pictures on packaging and labels. But in reality were are stuck with this green industrial landscape of agribusiness (cornfields, closed stables, empty pastures) and a recreational landscape that is only intended for cycling trips. The landscape reflects the separation of functions and the monoculture of modern food production. In the American state of Ohio one single farmer can manage hundreds of acres of cornfields. In the urban landscape there are so-called food deserts, poor areas where apples or cucumbers are not available, where the only food in the stores is processed food. It is an example of tragic irony. Our food system that first came into being in times of food shortage now causes new types of scarcity that are looming on the horizon: fuel shortage lack of biodiversity, shortage of water and nature. Judging by the food crisis, the debate on climate change, the food scandal in China and the successive outbreaks of animal disease, we have hit a wall.

In the year 2050 the world population is expected to reach nine billion, with the large majority of people living in cities. How will we be able to feed all these people with oil running out and the effects of climate change increasing? The debate between the technocrats and the countermovement is in full swing. Until recently, the positions were clear. The technocrats proclaim that we will never be able to feed the world with 'nice' sustainable food. What the world really needs is even more intensive farming, no holds barred – including genetic modification. It is completely wrong to want to halt economic growth by advocating a kind of self-sufficient utopia. The other camp, however, argues that much is yet to be gained by turning production lines into production circles, by saving water and by clever imitation of natural processes. Walmart, the American supermarket giant with aisles and aisles stocked with anonymous food, embodies the former point of view. The nostalgic pick-your-own farm, where you can stroll along the strawberry patches in spring, personifies the latter proposition.

One of the major problems of our current food system – driven as it is by growth, expansion and world trade – is that is totally contrary to the circular, local character of nature itself. Should the production of food for the world population stay closer to nature, and what can we do to reach this goal? The most interesting solutions probably lie somewhere in between Waltmart and the pick-your-own farm. In an attempt to bring the technocratic and alternative opinions closer together, the American food writer Michael Pollan points out that progress does not necessarily equal technology, and that a back-to-nature attitude is not always driven by nostalgia.

One of the major problems of our current food system – driven as it is by growth, expansion and world trade - is that is totally contrary to the circular, local character of nature itself

One of those modern solutions with a retro feel is urban farming. Both the urban planners of the western world and the development experts of the world food organization FAO are experimenting with this concept. It will not be possible to transport all the food we need to the megacities of the future, so we will have to grow it inside the city limits. Food chains must be turned into food cycles, by returning to mixed farming and by recycling waste and water. Nature can be imitated by means of variable grazing, crop rotation, water purification and the re-use of surplus heat. We don't know if food grown this way will be sufficient to feed the world population of the future. But it seems foolish to focus on one system exclusively, however large-scale it may be. People are gradually becoming more aware of this. All of a sudden, the supermarkets that until recently were locked in a dead-end price war, are in a competition over sustainability. There is a heated debate over meat in the newspapers. Before the food crisis of 2008 you were considered slightly deranged when you started a discussion among non-peers about factory farming and unlimited meat consumption. It would seem stranger still that the topic would fill the editorial pages of the newspaper. Concerns about climate change and the outbreak of yet another animal disease have caused a shift in public awareness. Nevertheless, this doesn't mean that we are eating less meat.

Food shouldn't have to be explained to you.

One might ask what the ideal 21st century production landscape should look like. First and foremost this landscape must be able to narrow the existing gap between pastoral and large-scale, between nostalgia and industry, between pick-your-own farms and Walmart. This landscape should be able to blend city and countryside, or at least bring the two entities closer together. This landscape should be transparent, showing us where the food is coming from and how much work was involved, but it should also produce more than just 'nice' food. It should be as energy efficient and as environmentally friendly as possible. Preferably this landscape should match the urban world of the 21st century, where origin is no longer a matter of uniformity, where almost every human and organism at one time or another have been transplanted and uprooted before growing back together again. There is another gap to bridge, the one between global and local.

In a landscape like this, everyone should have his or her feet firmly planted in the mud. In a landscape like this, city dwellers will inhabit their own pantries again. Food shouldn't have to be explained to you – because you would know where it came from, and why.

THIS ESSAY WAS ORIGINALLY WRITTEN FOR 'PARK SUPERMARKT', A PROJECT REALIZED BY VAN BERGEN KOLPA ARCHITECTEN, WITHIN THE FOODPRINT MANIFESTATION BY STROOM, THE HAGUE.

THE NANO SUPERMARKET
OCTOBER 2010, THE NEXT NATURE STUDIO AND THE TECHNICAL UNIVERSITY IN EINDHOVEN (THE NETHERLANDS) JOINED FORCES TO SET UP A PROJECT ABOUT NANO TECHNOLOGY, WHICH RESULTED IN THE NANO SUPERMARKET: A MOBILE EXHIBITION OF SPECULATIVE PRODUCTS WITH NANO TECHNOLOGY, CREATED BY ARTISTS, DESIGNERS AND ENGINEERS FROM SEVEN DIFFERENT COUNTRIES.

NANO SUPERMARKET

NANO inside

LATRO

ECO FRIENDLY LAMP POWERED BY ALGAE

The Latro lamp is attractive and environmentally friendly. The Latro uses living algae confined in a uniquely shaped glass container. All you need to do is hang the lamp in a sunny spot so that the algae can collect energy, blow in the bulb once a month to provide the algae with CO_2 and refill with water once every six months. Consumer awareness made easy!

PHARMA SUSHI

WITH GENETICALLY MODIFIED FISH EGGS

Vitamins, blood thinners, calcium pills. More and more people are taking a substantial amount of medication on a daily basis. The pharmaceutical sushi set brings your personal medication together inside one very tasty dish.

Edible Software

UPDATE VIA THE ESOPHAGUS

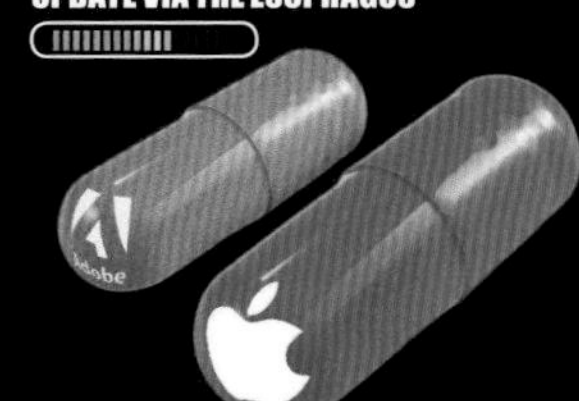

Liquid Glass

ULTRA THIN PROTECTIVE LAYER

A NEW SUPERMARKET IS COMING TO YOUR NEIGHBORHOOD: THE NANO SUPERMARKET MAKES THE IMPACT OF NANO TECHNOLOGY ON YOUR EVERYDAY LIFE TANGIBLE. OUR SHELVES ARE STOCKED WITH NANO PRODUCTS THAT COULD BECOME AVAILABLE ON THE MARKET BETWEEN NOW AND THE NEXT TEN YEARS.

Interactive wall paint, medicinal chocolates, a wine you can adapt to your own taste in the microwave at home, elastic stockings that pull themselves up around your leg, paper made from the cells of your own skin, an invisible security spray. Our products are innovative, useful and wonderful, but sometimes also strange, difficult or even frightening.

Maybe you have never heard of nanotechnology. It is about making things on an atomic or molecular scale; one nanometer is one billionth of a meter. Nanotechnology is an umbrella for a whole range of different technologies that take place on a scale between 1 and 100 nanometer. Companies and the government are currently investing millions of euros in nanotechnology, because it is generally thought that it will make our lives easier and better.

NANO REVOLUTION

Do you have products with nano technology at home? Yes, you do! Nanotechnology is already used in a variety of consumer products, from day creams to sports socks and from refrigerators to bread bins. Nanoparticles, for instance, provide extra UV-protection in cosmetics or improve the bounce of tennis balls. This is all very convenient, but at the same time there are a lot of things we don't know about nanotechnology; for example if nanoparticles are 100 % safe and, possibly more important: How will nanotechnology change our everyday life?

A wave of nanotechnology is flooding our society, but yet most people know very little about the impact of this new human achievement. Experts expect nanotechnology to have a huge influence on the way we live; comparable with historical technological developments like digitalization, the industrial revolution, electricity, the printing press, the introduction of the written word, or the beginnings of agriculture thousands of years ago. Nanotechnology radically changes our notion of what is natural. It can make the dreams people have about themselves come true, but it can also lead to products we possibly shouldn't even want to have.

PUBLIC DEBATE

The time is ripe for a public debate about nanotechnology, and this discussion is indeed already taking place. In newspapers, magazines, science cafés, in the theater and on television, and now also in the especially created NANO supermarket. Whereas most debates about nanotechnology address our fascination with making things on an incredibly small molecular scale or the safety of nanoparticles, the NANO supermarket tries to make the transformative influence of nanotechnology on mankind and society tangible and recognizable.

IN THE SUPER MARKET NEW HABITS, LIFE STYLES AND TECHNOLOGIES WILL BECOME COMMONPLACE

We have chosen the supermarket as our stage, instead of a traditional exhibition space, because a supermarket is both an everyday and a technological environment. Because of our frequent visits to the supermarket, we have come to view it as a normal or even boring place, but essentially it is a very futuristic phenomenon. Imagine what would happen if we put a randomly chosen person from the Middle Ages inside a contemporary supermarket: he or she would have very little understanding, or nothing at all, of the products on the shelves. Coke bottles? Body lotion? Energy bars? Milk cartons? Light products enriched with vitamins? Instant soup? Pregnancy tests? The supermarket is a place where new habits, lifestyles and technologies are made commonplace. Therefore it is the perfect location to explore our nano future.

IS EVERYTHING IN THE NANO SUPERMARKET 'REAL'?

The great majority of the products on our shelves are still in a theoretical phase. Young designers and technologists from six different countries submitted them, in response to our open call for nano product visions. A panel of leading professors of nanotechnology, designers, artists, philosophers and technical experts judged the submissions. The selection was made on the basis of originality, quality of design, technological feasibility, social implications and to what extent the product might promote discussion.

Only the most exclusive and exciting products are presented in the NANO supermarket. You shouldn't expect a complete survey, but a compact collection of personal product views. Strict attention is paid to the technological basis of our products. There are more than enough tall stories about nanotechnology going around. The products you find on our shelves all have a feasibility label. This gives an indication of the year the product might be technically realized and lists which nanotechnology is used in the product.

CONCEIVABLE FUTURE

Although we can look only a couple of years ahead and our products are based on a technology that is actually being developed today, we in the NANO supermarket are somewhat looking into a crystal ball. Will the bioelectric plant to charge your mobile phone really become available? Is Nano Lift the botox of the future? Will we all walk around with a Twitter implant? It should not be taken for granted that the products on our shelves will actually be realized. We should interpret them as explorations of a possible future and not an inevitable one. The products in the NANO supermarket haven't been designed to predict the future. They serve as something to hold onto in the here and now and can help us to determine what kind of nano future we really want.

Koert van Mensvoort
Managing director Nano Supermarket

VERY SPECULATIVE | IN DEVELOPMENT | REALISED

Paint your own electric circuits.
Because wires aren't *super*.
CONDUCTIVE BODY INK

WALLSMART

CHANGE YOUR INTERIOR WITH ONE CLICK OF A BUTTON

People love change. We often change the interior of our home to make it look fresh and contemporary again. Save yourself the trip to the hardware store and the trouble of painting. Use Wallsmart! With Wallsmart it only takes the click of a button to change the color of your walls. Paint the wall of your choice just once with Wallsmart paint and then launch the supplied Wallsmart application on your laptop or PC. All you need to do next is choose a color from the menu.

SELECTA DNA

INVISIBLE PROTECTION FOR YOUR POSSESSIONS

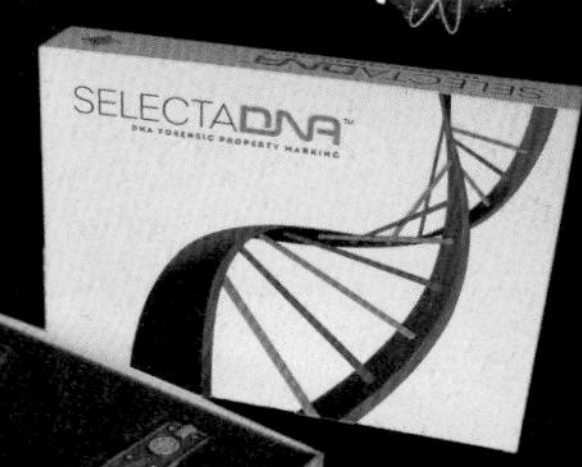

Protect your possessions with this clear marking spray. Stolen and found goods thus can be traced back to you as an owner. This spray increases the risk for an offender to get caught.

Various police forces actively look for traces of Selecta DNA. Criminals are aware of this too, so the spray considerably reduces your risk of becoming the victim of a burglary or robbery.

Self-cleaning Glass

DIRT-, GREASE- AND

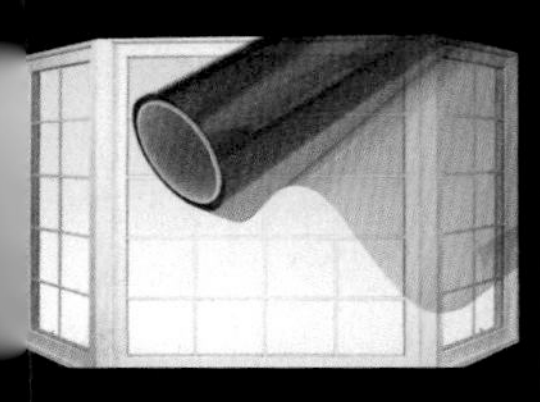

USB Stick 2TB

NOW DOUBLE THE STORAGE SPACE

2TB

BIO ELECTRICITY BONSAI

SUSTAINABLE ENERGY FROM PLANTS

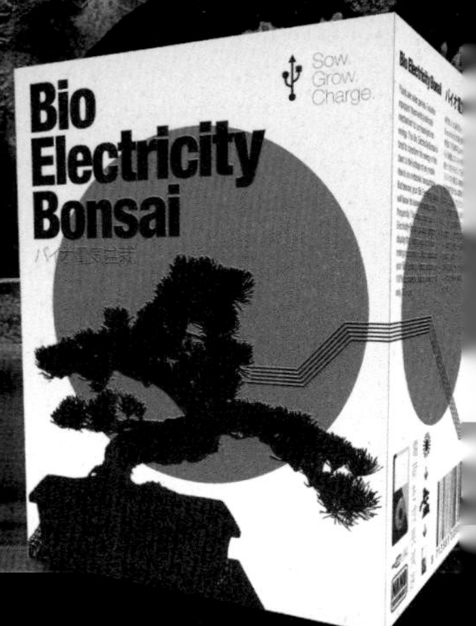

Plants are solar panels. Evolution has provided plants with a sophisticated mechanism to convert sunlight into energy. The bioelectric plant is designed to transform the energy of the plant into the appropriate voltage to charge a phone or mp3 player. 100 % sustainable. Please note: If you recharge your gadget too frequently this will cause damage to your plant. The bioelectric plant not only is a renewable energy source but also an intelligent display that makes you aware of your energy consumption.

Bullitproof Jacket

FASHIONABLE, V.I.P. AND SAFE!

Conductive Ink

LIQUID POWER CABLES

+/- Conductive Body Ink
paint your own electric circuits

Aware consumers choose Nano!

NANO

TRUE LOVE GOES TO THE BONE

BIOJEWELLERY
Your partner's bone tissue is cultivated and transformed into a beautiful and personal piece of jewellery. Make the symbol of your union physical and carry your loved one with you all the time!

FLU DOC

FLUORESCENT MICROBES REVEAL IF YOU HAVE THE FLU

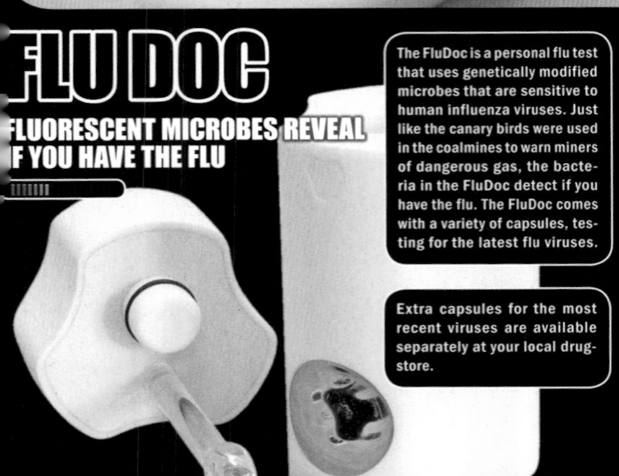

The FluDoc is a personal flu test that uses genetically modified microbes that are sensitive to human influenza viruses. Just like the canary birds were used in the coalmines to warn miners of dangerous gas, the bacteria in the FluDoc detect if you have the flu. The FluDoc comes with a variety of capsules, testing for the latest flu viruses.

Extra capsules for the most recent viruses are available separately at your local drugstore.

TWITTER IMPLANT

SHARE YOUR HEALTHY LIFE STYLE

WeightWatchers

Our health is worth a lot to us. We exercise, eat healthy food and try to live a balanced life. Your health-insurance company will help you! The Twitter implant monitors your body and alerts your doctor if something in your body isn't working properly. Therefore, medical interventions will occur at an earlier stage so you remain vital longer. The Twitter implant can also save you money on your monthly health-insurance bill. Your family doctor places the device free of charge. The Twitter implant can be supplied with various coaching programs including "Weight Watchers", "Club Sportive" and "Alcoholics Anonymous".

E CHROMI

LEARN TO READ YOUR STOOL

Our stool tells us a lot about our health. E.chromi is a product that helps us to better read our stool. E.chromi consists of a beverage and a corresponding color chart listing the most commonly known diseases in their own distinctive color.

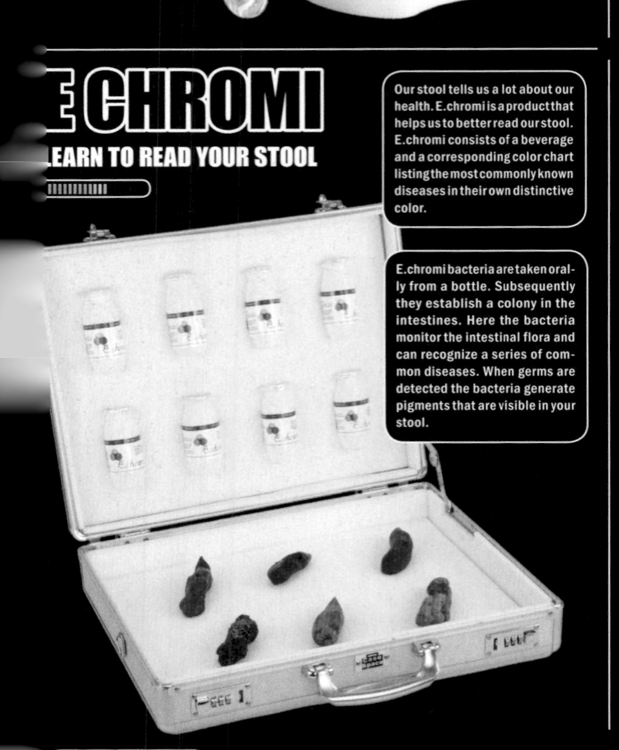

E.chromi bacteria are taken orally from a bottle. Subsequently they establish a colony in the intestines. Here the bacteria monitor the intestinal flora and can recognize a series of common diseases. When germs are detected the bacteria generate pigments that are visible in your stool.

Slim Fast

NOW AN EVEN FULLER TASTE

B12X Gumballs

CONCENTRATION IN A LITTLE BALL

Slim-O-Spread

IN THREE NEW FLAVORS

Home Test

CERVICAL CANCER

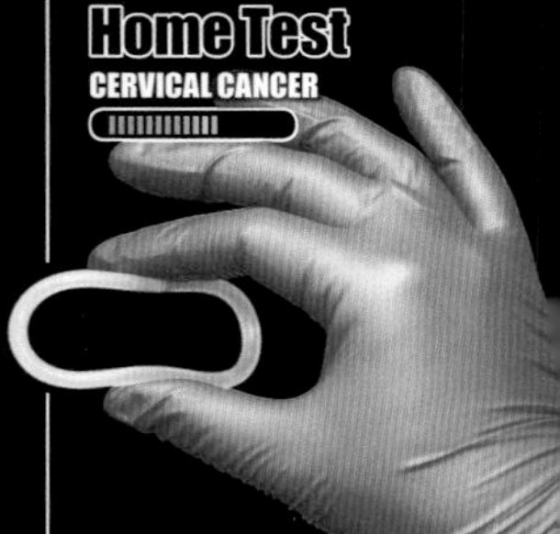

NANO Aware consumers choose Nano!

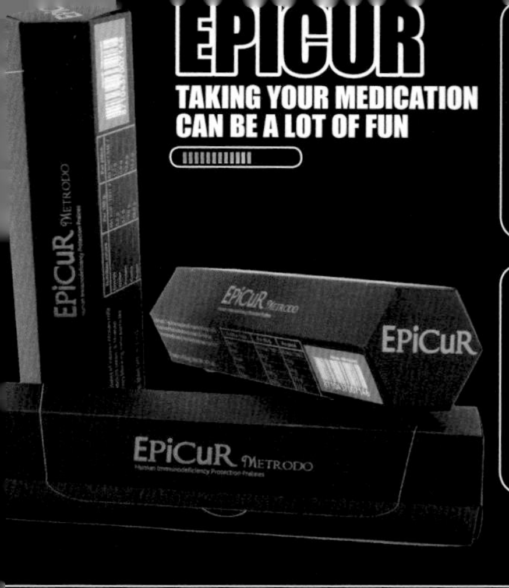

EPICUR

TAKING YOUR MEDICATION CAN BE A LOT OF FUN

Medication and (food) supplements have become an integral part of our lives. Whether it's blood thinners, painkillers or antidepressants: thousands of people swallow hundreds of pills per year. Even so, the administration of medication has remained relatively uncomfortable. But that is going to change. The Epicur Praline is an innovative new way of drug ingestion, especially developed for the hedonists among us. Epicur is a delicious chocolate with a medicinal filling that prevents HIV infection. The consumption of one chocolate capsule a day makes the consumer immune to HIV and prevents carriers of the virus to pass it on to others.

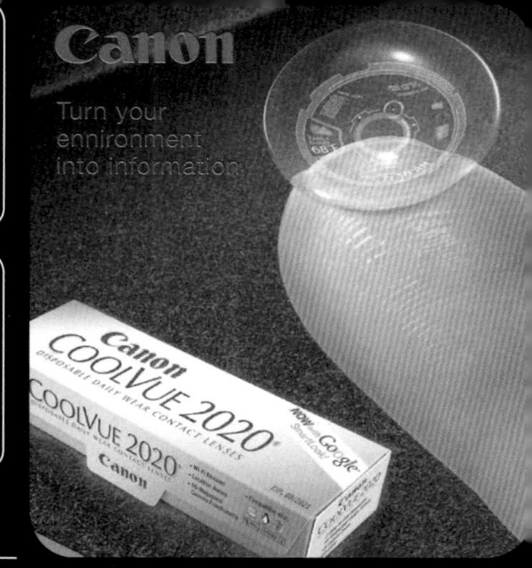

nno Band-Aid

PTS THE COLOR OF YOUR SKIN

nsaplast

Shaving Gel

SHINE AND SHAVE

Condom

FEATHERY SPIDER SILK

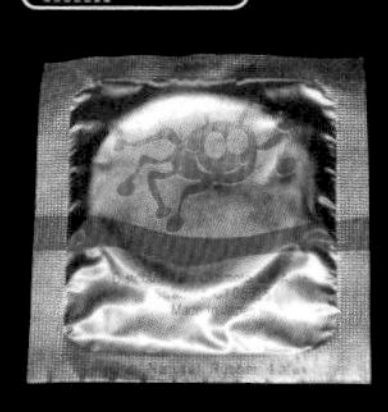

Clear Bear

SELF-CLEANING TEDDY

Thermo Cup

WITH INFRARED CONDUCTOR

rfect Sense

XING CANDY

Deodorant

WITH NANO SILVER

NECKLACE

SHARE YOUR RECOVERY WITH YOUR LOVED ONES

Breast cancer is a disease that comes with feelings of helplessness and loneliness. This beautiful crafted necklace allows you to have a feeling of control over your disease and to share your recovery with your loved ones. By gently pushing a link of the necklace onto your skin it withdraws a small blood sample without any feeling of discomfort. This link then changes color. The deeper the color of the link is, the healthier you are. In this way this unique necklace symbolizes your personal story, which you can share with everyone who is dear to you.

ndwich

ALL-NATURAL

Baby Curry

DISHES OF THE WORLD FOR TINY TOTS

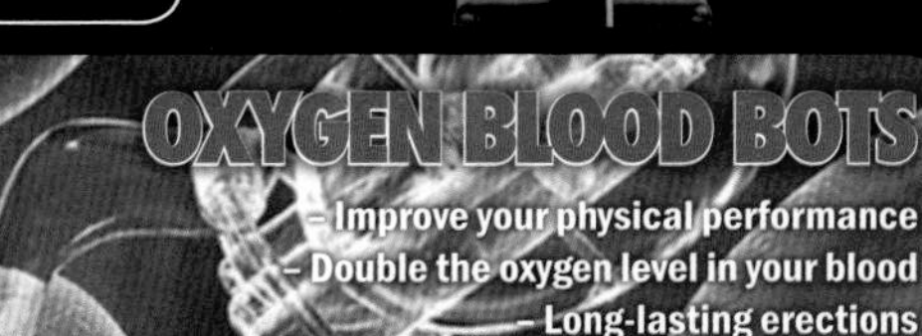

Aware consumers choose Nano!

NANO

SKINPAPER

THE ULTIMATE INTIMATE PAPER

Whether it is used to test cosmetics or to write a letter to your beloved, Skin Paper is the ultimate intimate paper. This innovative type of paper is grown with your own skin cells, which you can easily apply to the nano scaffolding: designed to provide a breeding ground for your skin tissue. Break through the boundaries of your body. Make a diary out of your skin.

NANOLIFT

HAVE THE FACE YOU ALWAYS WANTED

With Nano Lift it is now finally possible to not only manipulate your photo in Photoshop, but actually digitally improve your face. Similar to a Botox treatment the use of the Nano Lift system requires only a single visit to your cosmetic clinic. After the Nano Lift foundation is injected under your skin, you can use the included Nano Lift stick to shape your face for any occasion. Today you can enjoy the full lips of Angelina Jolie for a perfect date, and tomorrow you can show off a square jawline to score that great business deal at work. Nano Lift is painless, effective and can be used every day. It's the biggest innovation in beauty products since the invention of lipstick.

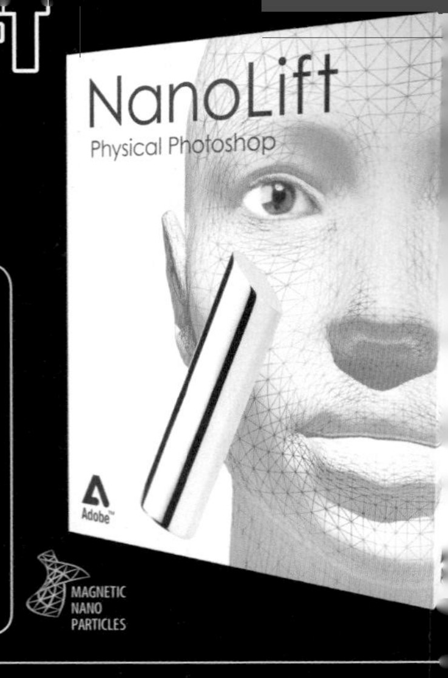

MAGNETIC NANO PARTICLES

NANOSOK

NEVER BEND DOWN AGAIN

Many elderly people in our society have difficulty bending over. Besides the inconvenience this condition brings, it also undermines their sense of independence. A common problem for this group is the daily routine of putting on socks and elastic stockings. The Nano Sock is the perfect solution!!

High-tech nanoparticles in the sock ensure that the sock pulls itself up bit by bit around your leg in a comfortable and safe way. The socks are on your feet in a minute, fitting like a glove. It is a perfect start to your day and you are no longer dependent on others. The Nano Sock is easy to use and is available in black, skin color, khaki and blue.

NANO DETECTOR

BECAUSE YOU WANT TO BE SURE

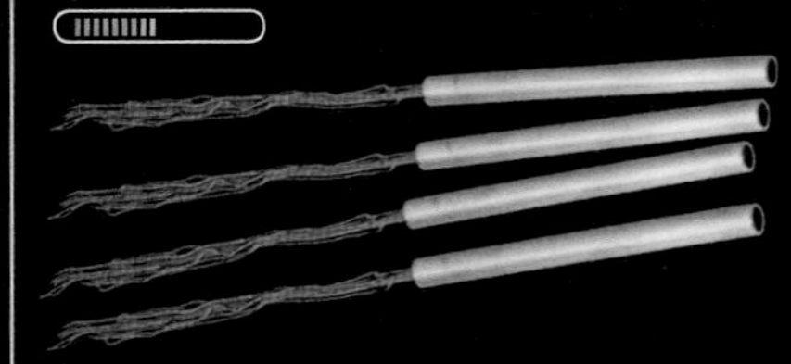

Our environment is increasingly c taminated with substances that invisible to the naked eye, but t can enter your body through y skin. The Nano Detector helps to identify such artificial nano ticles. This kit allows you to de unwanted nanoparticles on your before they enter your body. When small brush turns red there is a endogenous substance presen your skin. Contact certified agenc to remove these potentially harm substances from your body.

Organic Glam

MOOD RESPONSIVE MAKE-UP

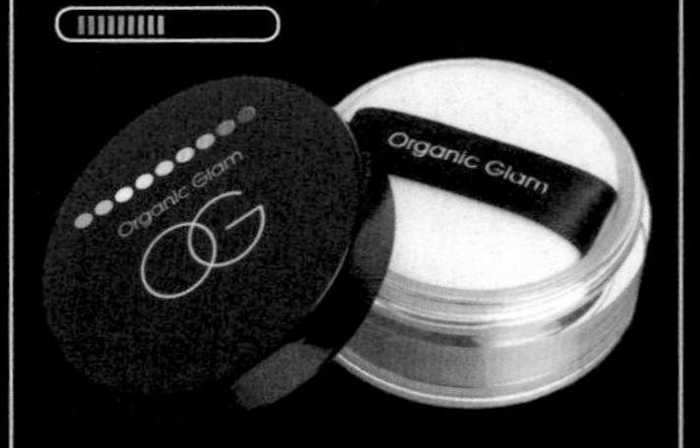

Color Matching Lens

MATCH THE COLOR OF YOUR EYES WITH YOUR OUTFIT

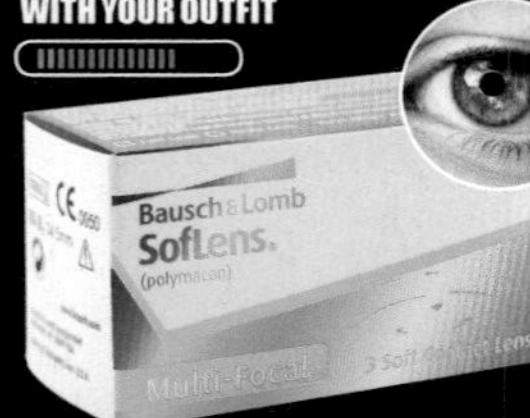

Scalable Heels

NO MORE SORE FEET

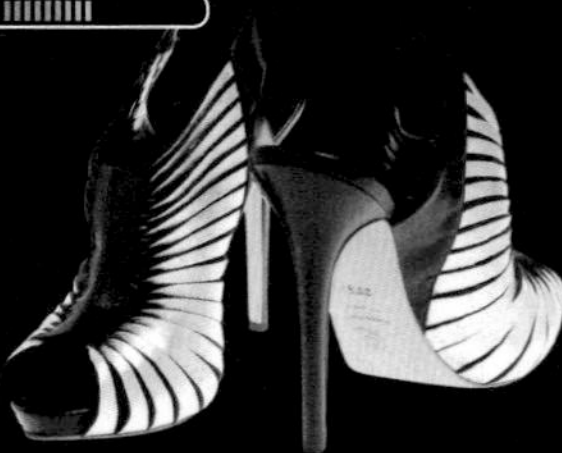

Lipstick

WITH 24-CARAT NANO GOLD PARTICLES

BONUS CHIP

- EXTRA DISCOUNT
- AUTOMATIC PAYMENT
- NO MORE WAITING IN LINE
- MATCH YOUR PURCHASES WITH YOUR FRIENDS

Aware consumers choose Nano!

NANO WINE

THE ENTIRE LIQUOR STORE IN ONE BOTTLE

mazing wine!
Chris Alblas

he endless variations provide
ou with the perfect wine
www.vinoo.eu

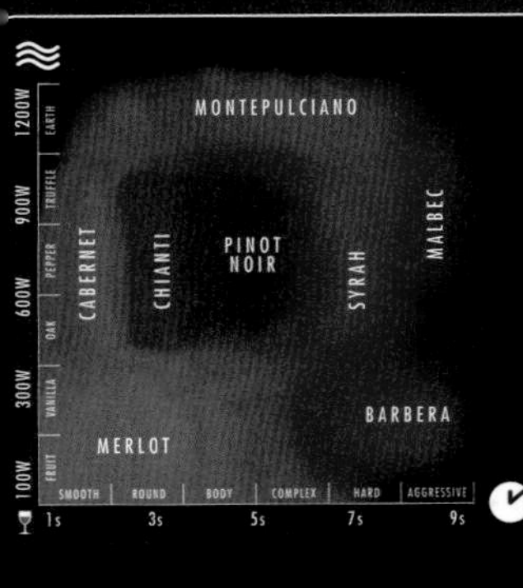

A delicious Montepulciano in only 6 seconds? This is now possible with the universal Nano wine. All you need is a microwave oven. In 5,64 seconds at 1000 watt you have a sublime Romanée-Conti. Or create a surprisingly young Mouton-Rothschild 1945 in only 2,34 seconds at 650 watt. The possibilities are endless. The wine contains millions of nano capsules which depending on your mood and taste preferences can be activated by microwaves. Inactivated nano capsules move unnoticed through the body, while the opened capsules alter the taste, smell and color of the wine.

NANO SUPERMARKET

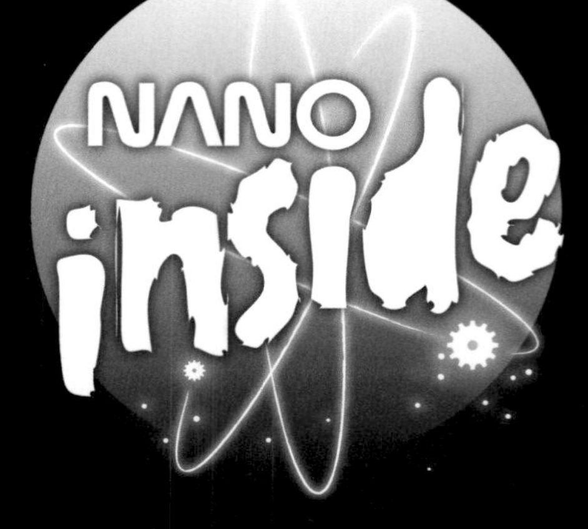

Everybody can be a chef
We bring the molecular gastronomy to your kitchen with the Philips Food Printer. Choose your basic ingredients and select aroma, color, structure, taste and nutritional value. Press 'print' and voilà; the 3D printed dish is on your table!

PHILIPS

ER ZIJN DAGELIJKS NIEUWE OPEN SOURCE RECEPTEN TE DOWNLOADEN OP WWW.DESIGN.PHILIPS.COM/PROBES/PROJECTS/FOOD

Steak 'Stegeman'

Stegeman SINDS 1858

- Cultivated in a laboratory, thus animal friendly
- Non-perishable, perfect for the BBQ
- Still juicy after 15 minutes on the grill

Placebo Bullets

HIGH IMPACT, NON-LETHAL

Non-Perishable Fruit

LANGER HOUDBAAR
DANKZIJ BESCHERMLAAG VAN VLOEIBAAR GLAS

COLOFON

PRODUCT DESIGN
Biojewellery: Tobie Kerridge, Nikki Stott, Ian Thompson / **Conductive Ink:** Matt Johnson e.a. / **E.Chromi:** Alexandra Daisy Ginsberg / **Epicur Pralines:** Stephan Hoes / **FluDoc:** Jan van der Asdonk / **Latro:** Mike Thompson / **Nano Detector:** Paul Frigout / **NanoLift:** Orestis Tsinalis / **Nano Sock:** Nicolas Nelson / **Nano Wijn:** Mensvoort, Grievink & Daas / **Nano World Map:** Niko Vegt / **Perfect Sense:** Katarina Brock / **Skin Paper:** Vanessa Harden, Tommaso Lanza / **The Necklace:** Marco van Beers / **Twitter Implant:** Hendrik-Jan Grievink / **WallSmart:** Jonas Enqvist

NANO SUPERMARKET TEAM
Creative Director: Koert van Mensvoort
Initiators: Marco Rozendaal, Koert van Mensvoort
Art Direction: Hendrik-Jan Grievink
Exhibition Design: Maze de Boer
Model Masters: Jan v.d. Asdonk, David Menting
Graphic Design: Hendrik-Jan Grievink, Ruben Daas, Michael Kluver
Commercial: Studio Smack
Organization: Marieke Kruithof, Lucas Asselbergs, Marco Rozendaal, Koert van Mensvoort
Tour Producer: Krista te Brake

JURY
Ronald van Tienhoven (Chair), Bas Haring, Bert Meijer, Dave Blank, Karin Spaink, Lucien Hanssen, René Janssen, Rinie van Est, Taco Stolk

PARTNERS & SPONSOR
The Nano Supermarket is a project of Technische Universiteit Eindhoven and the Next Nature Institute.

Realized with support of NanoPodium, TU/e, Stichting Doen and Mondriaan Foundation.

CONTACT
nanosupermarket@nextnature.net
www.nextnature.net/nano-supermarket

ARMY SURPLUS
NANO MATERIALS
X-tremist
• FEATHERLIGHT
• ULTRASTRONG
• INFLAMMABLE
• WATERPROOF

Running Jacket

COLOR SHOWS THE NUMBER OF CALORIES BURNT

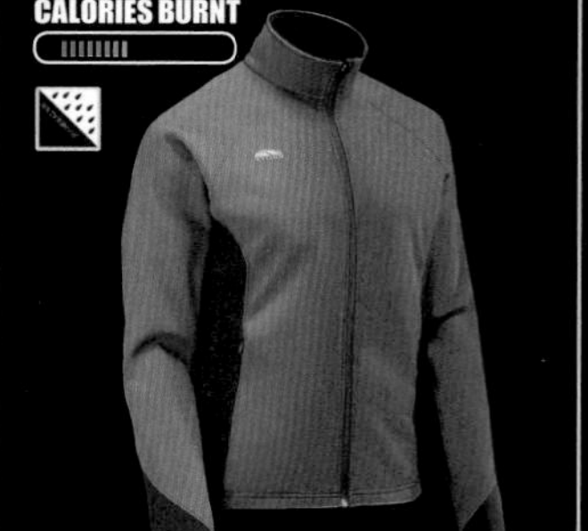

Toothbrush

WITH ANTIBACTERIAL NANO SILVER

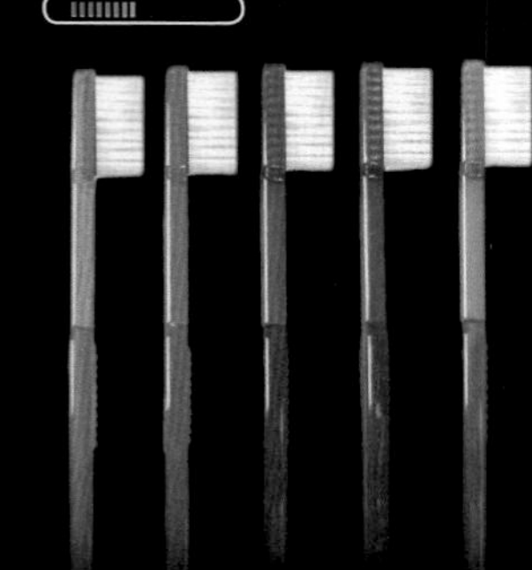

DE MOLECULAR SNACKBAR

★ SLIM SHAKE ★ WHITE SAUCE ★ DRY ICE FALAFEL
★ NANOSNACK ★ NANONAISE MILKSHAKE ★ LIQUID
★ HOT DOG FOAM ★ LIQUID EGG ROLL ★ HALAL COKE CHEESEBALLS
★ FROZEN

MOLECULAR SNACKS
TAKE AWAY AND DELIVERY
☎ 0900-NANOSNACK

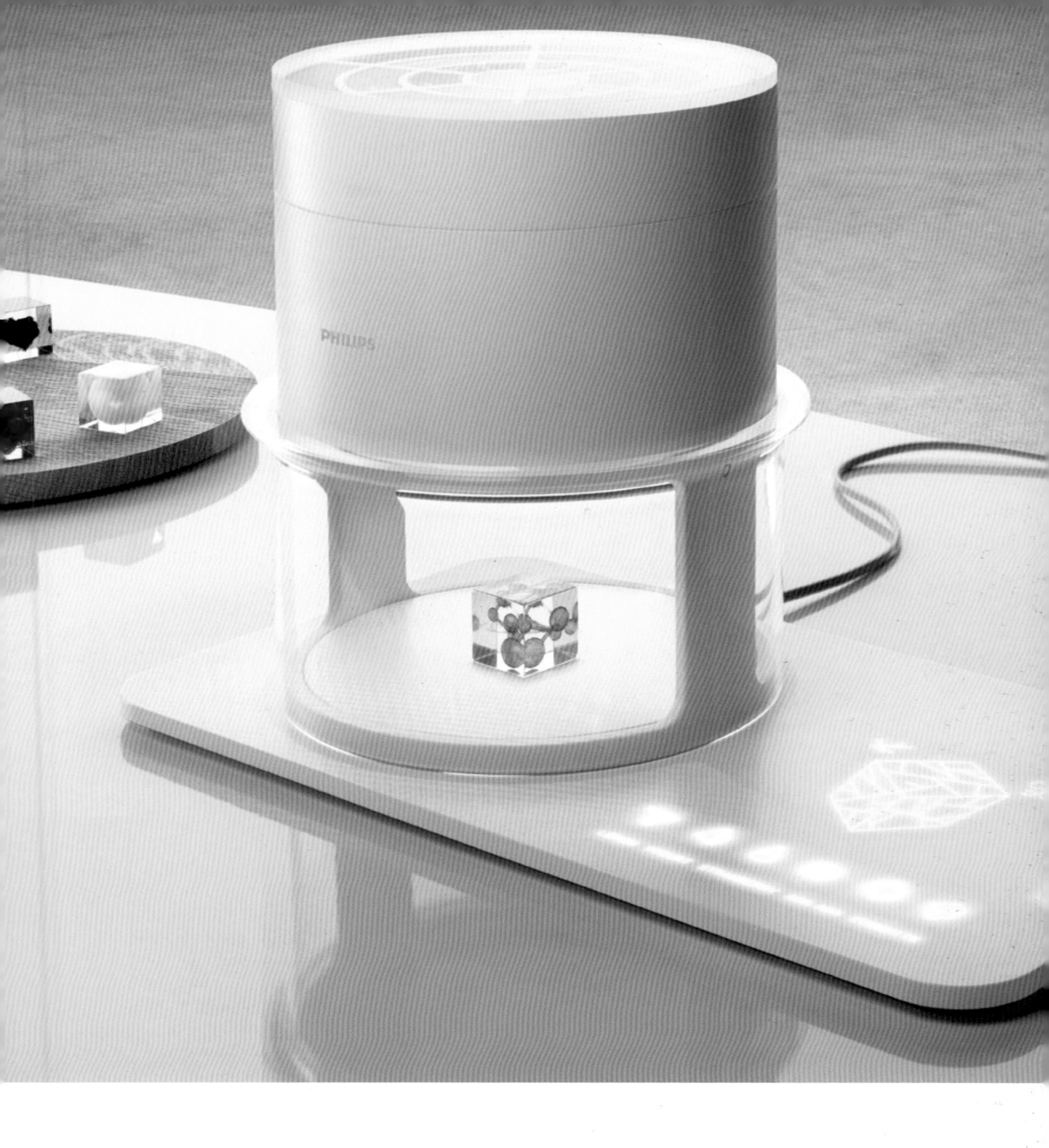
PHILIPS

#5 SUPERMARKET

wwww.nextnature.net/
themes/supermarket

NEXT NATURE

ANTHROPO MORPHOBIA

ISBN 978-84-92861-53-8
9 788492 861538

FROM CLOUDS TO CARS – PEOPLE RECOGNIZE THEMSELVES IN ALMOST EVERYTHING.

***PAREIDOLIA*, GREEK FOR 'FAULTY IMAGE,'** CAUSES US TO SEE SIGNIFICANCE IN RANDOM DATA. THIS PSYCHOLOGICAL PHENOMENON IS RESPONSIBLE FOR OUR TENDENCY TO PERCEIVE FACES WHERE THERE ARE NONE. OUR HIGHLY TUNED ABILITY TO PICK OUT FACES IS AN ADAPTATION THAT LET EARLY HUMANS INSTANTLY PICK OUT FRIEND FROM FOE. IT ALSO MEANS THAT IT CAN BE HARD NOT TO FEEL FRIENDLY TOWARDS A CAR WITH A FRIENDLY GRILLE.

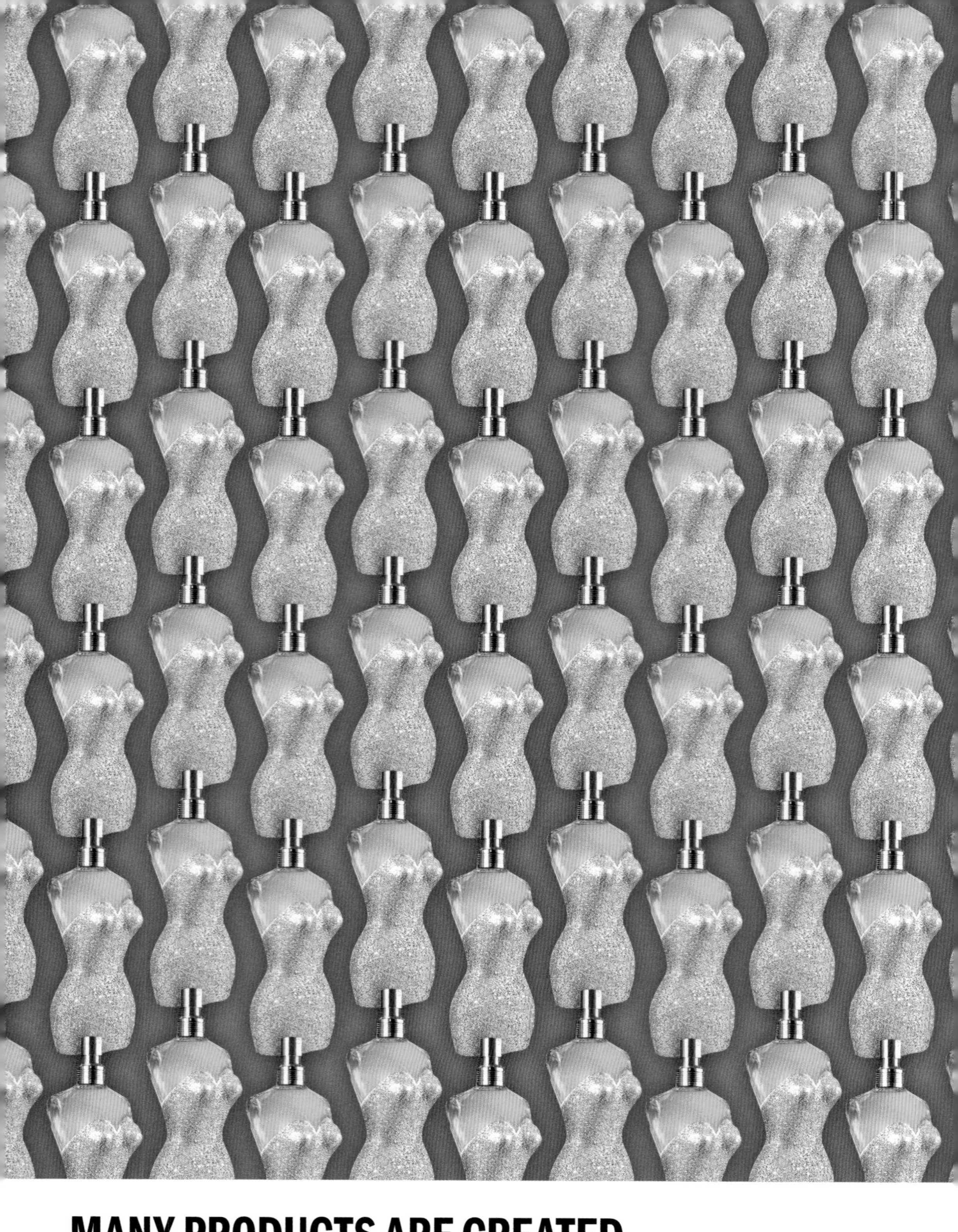

MANY PRODUCTS ARE CREATED IN THE IMAGE OF MAN.

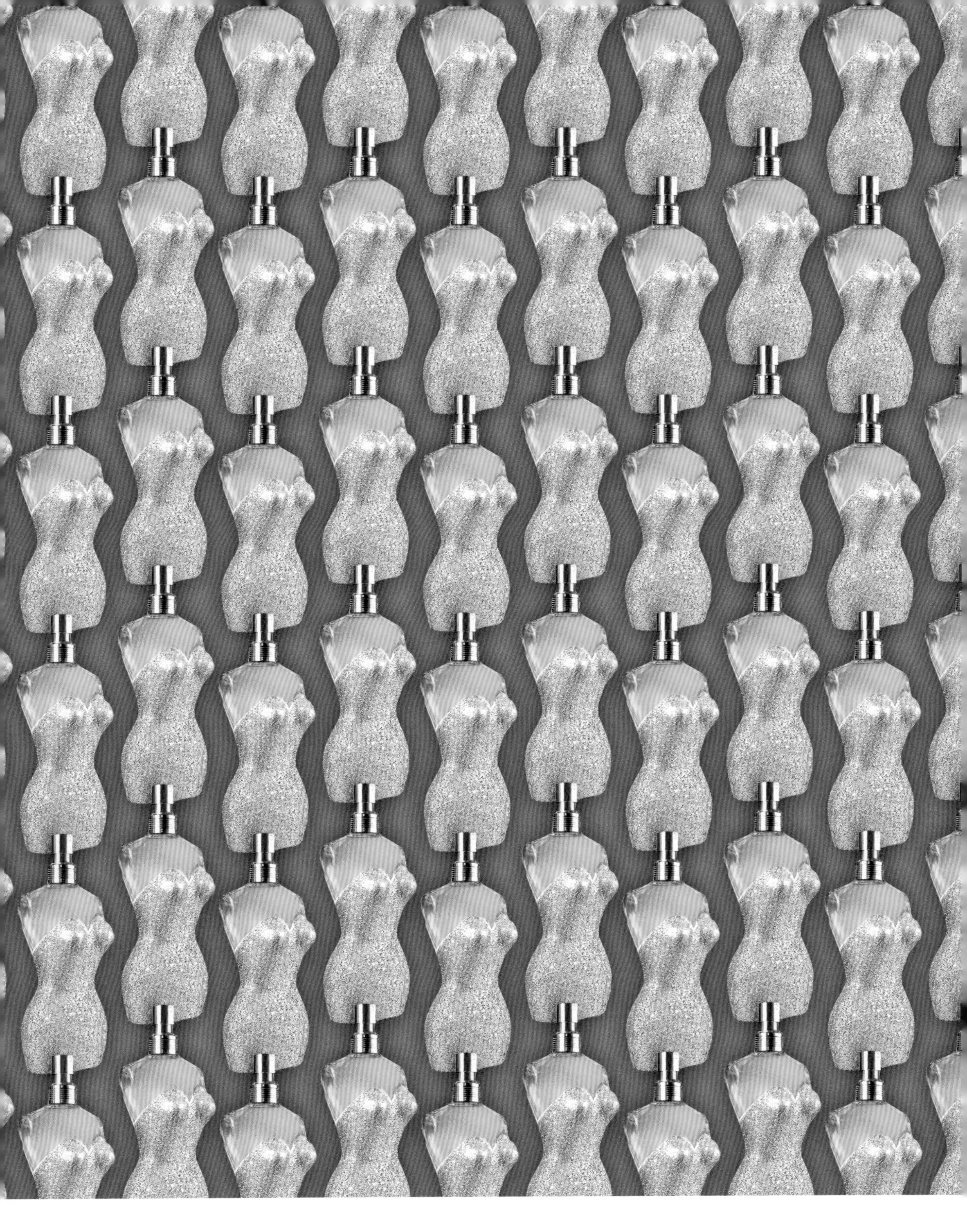

WHERE WE DON'T SEE ACCIDENTAL HUMANS, WE LIKE TO CREATE INTENTIONAL ONES. THE HUMAN FORM IS INNATELY APPEALING TO OTHER HUMANS. DESIGNER JEAN PAUL GAULTIER'S ICONIC PERFUME CLASSIQUE IS SHAPED LIKE A WOMAN'S TORSO. NEW MODELS ARE INTRODUCED EVERY YEAR IN DIFFERENT STYLES OF LINGERIE. ANTHROPOMORPHIC PRODUCTS ARE SEXIER, MORE PLEASING, OR SIMPLY MORE FAMILIAR THAN LESS-HUMAN GOODS.

BUT WHEN PRODUCTS BECOME PEOPLE...

THE FUNCTIONS AND MALFUNCTIONS OF COMPLEX MACHINES CAN SEEM LIKE THE ACTIONS OF A CONSCIOUS MIND. IS YOUR CAR PLOTTING ITS NEXT BREAKDOWN BEHIND YOUR BACK? DO YOU BERATE YOUR COMPUTER FOR LOSING DATA? "YOU SHOULD NOT ANTHROPOMORPHIZE COMPUTERS, THEY DON'T LIKE THAT" IS A CLASSIC JOKE AMONGST SOFTWARE DESIGNERS.

AND PEOPLE BECOME PRODUCTS...

WE AIM TO BE DESIRABLE COMMODITIES. WHEN NATURAL BEAUTY LAGS, THERE IS ALWAYS PHOTOSHOP TO IMPROVE THE PACKAGING. EVEN OUR DAILY HABITS - WHAT WE BUY AND WHO WE TALK TO - CAN BE BUNDLED UP AND SOLD TO ADVERTISERS. *HOMO ECONOMICUS* IS USED TO DEFINE HUMANS AS RATIONAL ECONOMIC ACTORS. THE TERM MAY BETTER BE RE-INTERPRETED TO DEFINE HUMANS AS ECONOMIC OBJECTS. THE IMAGE DEPICTS BEAUTY INDUSTRY ACTIVIST SUNNY BERGMAN, PORTRAYED BY MANON VAN DER ZWAAL.

THE DIFFERENCE BETWEEN PEOPLE AND PRODUCTS BECOMES VAGUE.

BROUGHT TO YOU BY THE DESIGNERS CHRISTOPHER BARRETT, EDWARD HEAL AND LUKE TAYLOR, THE SHRINK-WRAPPED MAN IS A TONGUE-IN-CHEEK COMMENTARY ON HUMAN COMMODIFICATION. OUR IDENTITIES ARE SHAPED AS MUCH BY OUR PURCHASES AS THEY ARE BY OUR PERSONALITIES AND IN-BORN TRAITS. PRODUCTS AND HUMANS MUTUALLY MANUFACTURE ONE ANOTHER. HOW OFTEN DO WE BUY OUR FRIENDS AND BEFRIEND OUR BELONGINGS?

THE TWI-LIGHT AREA BETWEEN PERSON AND PRODUCT

By Koert van Mensvoort

Are you familiar with the affliction? Anthropomorphobia is the fear of recognizing humans in non-human objects. The term is a hybrid of two Greek-derived words: 'anthropomorphic' means 'of human form' and 'phobia' means 'fear'. Its symptoms are irrational panic attacks, disdain, revulsion, and confusion about what it means to be human. Will it become public disease number one? Or can anthropomorphobia serve as a guiding principle in the evolution of humanity?

Luxury cars with blinking headlight eyes. Perfume bottles shaped like beautiful ladies. Grandma's face stretched smooth. Carefully selected designer babies. The Senseo coffeemaker shaped – subtly, but nonetheless – like a serving butler. And, of course, there are the robots, mowing grass, vacuuming living rooms, and even caring for elderly people with dementia.

Today more and more products are designed to exhibit anthropomorphic – that is, human – behaviour. At the same time, as a consequence of increasing technological capabilities, people are being more and more radically cultivated and turned into products. This essay will investigate the blurring of the boundary between people and products. My ultimate argument will be that we can use our relationship to anthropomorphobia as a guiding principle in our future evolution.

Anthropomorphism for Dummies

Before we take a closer look at the tension between people and products, here is a general introduction to anthropomorphism, that is, the human urge to recognise people in practically everything. Researchers distinguish various types of anthropomorphism.[1] The most obvious examples - cartoon characters, faces in clouds, teddy bears - fall into the category of 1) *structural anthropomorphism*, evoked by objects that show visible physical similarities to human beings. Alongside structural anthropomorphism, three other types are identified. 2) *Gestural anthropomorphism* has to do with movements or postures that suggest human action or expression. An example is provided by the living lamp in Pixar's short animated film, which does not look like a person but becomes human through its movements. 3) *Character anthropomorphism* relates to the exhibition of humanlike qualities or habits – think of a stubborn car that doesn't want to start. The last type, 4) *aware anthropomorphism*, has to do with the suggestion of a human capacity for thought and intent. Famous examples are provided by the HAL 9000 spaceship computer in the film *2001: A Space Odyssey* and the intelligent car KITT in the TV series *Knight Rider*.

Besides being aware that anthropomorphism can take different forms, we must keep in mind that it is a human characteristic, not a quality of the anthropomorphised object or creature per se: the fact that we recognise human traits in objects in no way means those objects are actually human, or even designed with the intention of seeming that way. Anthropomorphism is an extremely subjective business. Research has shown that how we experience anthropomorphism and to what degree, are extremely personal – what seems anthropomorphic to one person may not to another, or it may seem much less so.[2]

Blurring The Line Between People And Products

To understand anthropomorphobia – the fear of human characteristics in non-human objects – we must begin by studying the boundary between people and products. Our hypothesis will be that anthropomorphobia occurs when this boundary is transgressed. This can happen in two ways: 1) products or objects can exhibit human behaviour, and 2) people can act like products. We will explore both sides of this front line, beginning with the growing phenomenon of humanoid products.

Products as People

The question of whether and how anthropomorphism should be applied in product design has long been a matter of debate among researchers and product designers. "You shouldn't anthropomorphize computers, they don't like it" is a classic and frequently made joke among interaction designers; the punch line rests on the knowledge that people will always, to a greater or lesser degree, ascribe human attributes to products, whether or not they are not designed to exhibit them. Evidently it is simply human nature to project our own characteristics on just about everything.[3]

To understand anthropomorphobia we must begin by studying the boundary between people and products.

Some researchers argue that the deliberate evocation of anthropomorphism in product design must always be avoided because it generates unrealistic expectations, makes human-product interaction unnecessarily messy and complex, and stands in the way of the development of genuinely powerful tools.[4] Others argue that the failure of anthropomorphic products is simply a consequence of poor implementation and that anthropomorphism, if applied correctly, can offer an important advantage because it makes use of social models people already have access to.[5;6;7;8] A commonly used guiding principle among robot builders is the so-called uncanny valley theory,[9] which, briefly summarised, says people can deal fine with anthropomorphic products as long as they're obviously not fully fledged people –e.g., cartoon characters and robot dogs. However, when a humanoid robot looks too much like a person and we can still tell it's not one, an uncanny effect arises, causing strong feelings of revulsion – in other words, anthropomorphobia.[10]

The coffeemaker that says good morning and politely lets you know when it needs cleaning. A robot that scrubs the floor, and one that looks after the children

Most researchers agree that anthropomorphism can be advantageous as well as dangerous. On the one hand, it can encourage an empathic relationship between the user and the product. If the expectations raised are not met, however, disappointment and incomprehension can result. Personality, cultural background and specific context can also influence one's perception of the product, increasing the chance of further miscommunication.

Although no consensus exists on the application of anthropomorphism in product design and there is no generally accepted theory on the subject, technology cheerfully marches on. We are therefore seeing increasing numbers of advanced products that, whether or not as a direct consequence of artificial intelligence, show ever more anthropomorphic characteristics. The friendly soft drink machine; the coffeemaker that says good morning and politely lets you know when it needs cleaning. A robot that scrubs the floor, and one that looks after the children. Would you entrust your kids to a robot? Maybe you'd rather not. But why? Is it possible that you're suffering from a touch of anthropomorphobia? Consciously or unconsciously, many people feel uneasy when products act like people. Anthropomorphobia is evidently a deep-seated human response – but why? Looking at the phobia as it relates to products becoming people', broadly speaking, we can identify two possible causes:

1) Anthropomorphobia is a reaction to the inadequate quality of the anthropomorphic products we encounter.

2) People fundamentally dislike products acting like humans because it undermines our uniqueness as people: if an object can be human, then what am I good for?

Champions of anthropomorphic objects – such as the people who build humanoid robots – will subscribe to the first explanation, while opponents will feel more affinity for the second. What's difficult about the debate is that neither explanation is easy to prove or to disprove. Whenever an anthropomorphic product makes people uneasy, the advocates simply respond that they will develop a newer, cleverer version soon that will be accepted soon. Conversely, opponents will keep finding new reasons to reject anthropomorphic products. Take the chess computer – an instance of aware anthropomorphism, like HAL 9000. Thirty years ago, people thought that when a chess computer was able to beat a grandmaster, it would mean computers had achieved a human level of intelligence. But when world champion Garry Kasparov was finally vanquished in the 1990s by IBM's monster computer Deep Blue, the opponents calmly moved the goalposts, proposing that chess required merely a limited kind of intelligence and human intelligence as a whole entailed much more than that – emotional intelligence, bodily intelligence, creative intelligence, and so on. Never fear: computers couldn't touch human beings, even if they could beat us at chess! Then again, the nice thing about this game of leapfrog is that through our attempts to create humanoid products we continue to refine our definition of what a human being is – in copying ourselves, we come to know ourselves.

Where will it all end? We can only speculate. Researcher David Levy [11] predicts that marriage between robots and humans will be legal by the end of the 21st century. For people born in the 20th century, this sounds highly strange. And yet, if we think for a minute, we realize the idea of legal gay marriage might have sounded equally impossible and undesirable to our great-grandparents born in the 19th century. Boundaries are blurring; norms are shifting. Although I'm not personally interested in hopping into bed with a sophisticated sex doll, nor am I especially bothered if other people are. Robot sex has been a secret fantasy of both men and women for decades, and although I don't expect it will go mainstream anytime soon, I think we should allow each other our placebos. Actually, I'm more worried about something else: whether marrying a normal person will still be possible at the end of the 21st century. Because if we look at the increasing technologization of human beings and extrapolate into the future, it seems far from certain that normal people will still exist by then. This brings us to the second cause of anthropomorphobia.

People as Products

We have seen that more and more products in our everyday environment are being designed to act like people. As described earlier, the boundary between people and products is also being transgressed in the other direction: people are behaving as if they were products. I use the term 'product' in the sense of something that is functionally designed, manufactured, and carefully placed on the market.

Consciously or unconsciously, many people feel uneasy when products acts like people.

The contemporary social pressure on people to design and produce themselves is difficult to overestimate. Have you put together a personal marketing plan yet? If not, I wouldn't wait too long. Hairstyles, fashion, body corrections, smart drugs, Botox and Facebook profiles are just a few of the self-cultivating tools people use in the effort to design themselves – often in new, improved versions.

It is becoming less and less taboo to consider the body as a medium, something that must be shaped, upgraded and produced. Photoshopped models in lifestyle magazines show us how successful people are supposed to look. Performance-enhancing drugs help to make us just that little bit more alert than others. Some of our fellow human beings are even going so far in their self-cultivation that others are questioning whether they are still actually human – think, for example, of the uneasiness provoked by excessive plastic surgery.

It is becoming less and less a taboo to consider the body as a medium, something that must be shaped, upgraded and produced.

The ultimate example of the commodified human being is the so-called designer baby, whose genetic profile is selected or manipulated in advance in order to ensure the absence or presence of certain genetic traits. Designer babies are a rich subject for science fiction, but to an increasing degree they are also science fact. "Doctor, I'd like a child with blond hair, no Down's Syndrome and a minimal chance of Alzheimer's, please". An important criticism of the practice of creating designer babies concerns the fact that these (not-yet-born) people do not get to choose their own traits but are born as products, dependent on parents and doctors, who are themselves under various social pressures.

In general, the cultivation of people appears chiefly to be the consequence of social pressure, implicit or explicit. The young woman with breast implants is trying to measure up to visual culture's current beauty ideal. The Ritalin-popping ADHD child is calmed down so he or she can function within the artificial environment of the classroom. The ageing lady gets Botox injections in conformance with society's idealisation of young women. People cultivate themselves in all kinds of ways in an effort to become successful human beings within the norms of the societies they live in. What those norms are is heavily dependent on time and place.

Ever Met a Normal Human Being? What Did You Think of Them?

At the beginning of the 1990s, shortly after the fall of the Berlin Wall, I was in a European airport. The Cold War had just ended. Waiting to check in, I was standing between two queues for other flights, one of which was going to the United States – Los Angeles, I think – and the other to Bucharest, Romania. The striking difference between the people in the two queues made a powerful impression on me. In the queue for the United States stood a film crew and a Hollywood actor, who had been in picturesque Europe filming a romantic comedy whose name I have since forgotten. There were slightly too thin, stretched-tight yet elegantly dressed "Hello, how are you?" women and friendly yet superficially smiling teeth-whitened men like the ones I had seen in Gillette commercials. The whole thing made a sophisticated yet somewhat artificial Barbie-and-Ken-like impression. The contrast with the queue for the Eastern European flight was enormous. The latter was comprised of proud but bony people in grey fur coats with grown-out haircuts and too many suitcases – many times more authentic, but shabby verging on animal (I know that today Bucharest is a hip, fashionable city, but in 1990, just after the fall of the Wall, things were different).

As a Western European (deodorant, highlights, no Botox yet), I felt somewhere in between, with enough distance to reflect. Never before had I been so keenly aware of how relative our ideas about what a 'normal' human being is really are. Someone from the Middle Ages probably would have considered the Romanians' suitcases and fur coats unbelievably sophisticated. From the perspective of a cave-dweller, we would scarcely be recognisable as humans. I wouldn't be surprised if a caveperson experienced strong feelings of anthropomorphobia at the sight of the lines in the airport and presumed it was a landing zone for post-human aliens from a faraway planet.

Humans As Mutants

Throughout our history, to a greater or lesser degree, all of us human beings have been cultivated, domesticated, made into products. This need to cultivate people is probably as old as we are, as is opposition to it. It's tempting to think that, after evolving out of the primordial soup into mammals, then upright apes, and finally the intelligent animals we are today, we humans have reached the end of our development. Of course, this not the case. Evolution never ends. It will go on, and people will continue to change in the future. But that does not mean we will cease to be people, as is implied in terms like 'transhuman' and 'posthuman'.[12;13;14] It is more likely that our ideas about what a normal human being is will change along with us.

We should prevent people from becoming unable to recognise each other as human.

The idea that technology will determine our evolutionary future is by no means new. During its evolution over the past 200,000 years, *Homo sapiens* has distinguished itself from other, now extinct humanoids, such as *Homo habilis*, *Homo erectus*, *Homo ergaster* and the Neanderthal, by its inventive, intensive use of technology. This has afforded *Homo sapiens* an evolutionary advantage that has led us, rather than the stronger and more solidly built Neanderthal, to become the planet's dominant species. From this perspective, for technology to play a role in our evolutionary future would not be unnatural but in fact completely consistent with who we are. Since the dawn of our existence, human beings have been co-evolving with the technology they produce. Or, as Arnold Gehlen [15] put it, we are by nature technological creatures.

Because only one humanoid species walks the Earth today, it is difficult to imagine what kind of relationships, if any, different kinds of humans living in the past might have had with each other. Perhaps Neanderthals considered *Homo sapiens* feeble, unnatural, creepy nerds, wholly dependent on their technological toys. A similar feeling could overcome us when we encounter technologically 'improved' individuals of our own species. There is a good chance that we will see them in the first place as artificial individuals degraded to the status of products and that they will inspire violent feelings of anthropomorphobia. This, however, will not negate their existence or their potential evolutionary advantage.

Human Enhancement

If the promises around up-and-coming bio-, nano-, info-, and neuro-technologies are kept, we can look forward to seeing a rich assortment of mutated humans. There will be people with implanted RFID chips (there already are), people with fashionably rebuilt bodies (they, too, exist and are becoming the norm in some quarters), people with tissue-engineered heart valves (they exist), people with artificial blood cells that absorb twice as much oxygen (expected on the cycling circuit), test-tube babies (exist), people with tattooed electronic connections for neuro-implants (not yet the norm, although our depilated bodies are ready for them), natural-born soldiers created for secret military projects (rumour has it they exist), and, of course, clones – Mozarts to play music in holiday parks and Einsteins who will take your job (science fiction, for now, and perhaps not a great idea).

It is true that not everything that can happen has to, or will. But when something is technically possible in countless laboratories and clinics in the world (as many of these technologies are), a considerable number of people view them as useful, and drawing up enforceable legislation around them is practically impossible, then the question is not *whether* but *when and how* it will happen.[16] It would be naive to believe we will reach a consensus about the evolutionary future of humanity. We will not. The subject affects us too deeply, and the various positions are too closely linked to cultural traditions, philosophies of life, religion and politics. Some will see this situation as a monstrous thing, a terrible nadir, perhaps even the end of humanity. Others will say, "This is wonderful. We're at the apex of human ingenuity. This will improve the human condition." The truth probably lies somewhere in between. What is certain is that we are playing with fire, and that not only our future but also our descendants depends on it. But we must realize that playing with fire is simply something we do as people, part of what makes us human.

While the idea that technology should not influence human evolution constitutes a denial of human nature, it would fly in the face of human dignity to immediately make everything we can imagine reality. The crucial question is: how can we chart a course between rigidity and recklessness with respect to our own evolutionary future?

Anthropomorphobia as a Guideline

Let us return to the kernel of my argument. I believe the concept of anthropomorphobia can help us to find a balanced way of dealing with the issue of tinkering with people. There are two sides to anthropomorphobia that proponents as well as opponents of tinkering have to deal with. On the one hand, transhumanists, techno-utopians, humanoid builders, and fans of improving humanity need to realize that their visions and creations can elicit powerful emotional reactions and acute anthropomorphobia in many people. Not everyone is ready to accept being surrounded by humans with plastic faces, electrically controlled limbs and microchip implants – if only because they cannot afford these upgrades. Along with the improvements to the human condition assumed by proponents, we should realize that the uncritical application of people-enhancing technologies can cause profound alienation between individuals, which will lead overall to a worsening rather than an improvement of the human condition.

Understanding anthropomorphobia can guide us in our evolutionary future.

On the other hand, those who oppose all tinkering must realize anthropomorphobia is a phobia. It is a narrowing of consciousness that can easily be placed in the same list with xenophobia, racism and discrimination. Just as various evolutionary explanations can be proposed for anthropomorphobia as well as xenophobia, racism and discrimination, it is the business of civilisation to channel these feelings. Acceptance and respect for one's fellow human beings are at the root of a well-functioning society.

In conclusion, I would like to argue that understanding anthropomorphobia can guide us in our evolutionary future. I would like to propose a simple general maxim: prevent anthropomorphobia where possible. We should prevent people from having to live in a world where they are constantly confused about what it means to be human. We should prevent people from becoming unable to recognise each other as human.

The mere fact that an intelligent scientist can make a robot clerk to sell train tickets doesn't mean a robot is the best solution. A simple ticket machine that doesn't pretend to be anything more than what it is could work much better. An ageing movie star might realize she will alienate viewers if she does not call a halt to the unbridled plastic surgeries that are slowly but surely turning her into a life-sized Barbie – her audience will derive much more pleasure from seeing her get older and watching her beauty ripen. The 17-year-old boy who loses his legs in a tragic accident should think carefully before getting measured for purple octopus attachments, although that doesn't mean he should necessarily get the standard flesh-toned prostheses his overbearing mother would prefer. Awareness and discussion around anthropomorphobia can provide us with a framework for making decisions about the degree to which we wish to view the human being as a medium we can shape, reconstruct and improve – about which limits it is socially acceptable to transgress, and when.

I can already hear critics replying that although the maxim 'prevent anthropomorphobia' may sound good, anthropomorphobia is impossible to measure and therefore the maxim is useless. It is true that there is no 'anthromorphometric' for objectively measuring how anthropomorphic a specific phenomenon is and how uneasy it makes people. But I would argue that this is a good thing. Anthropomorphobia is a completely human-centred term, i.e., it is people who determine what makes them uncomfortable and what doesn't. Anthropomorphobia is therefore a dynamic and enduring term that can change with time, and with us. For we will change – that much is certain.

REFERENCES

1 DISALVO, C., GEMPERLE, F., AND FORLIZZI, J. IMITATING THE HUMAN FORM: FOUR KINDS OF ANTHROPOMORPHIC FORM. 2007.
2 GOOREN, D. ANTHROPOMORPHISM & NEUROTICISM: FEAR AND THE HUMAN FORM. EINDHOVEN: EINDHOVEN UNIVERSITY OF TECHNOLOGY, 2009.
3 REEVES, B. AND NASS, C. THE MEDIA EQUATION: HOW PEOPLE TREAT COMPUTERS, TELEVISION, AND NEW MEDIA LIKE REAL PEOPLE AND PLACES. CAMBRIDGE: CAMBRIDGE UNIVERSITY PRESS, 1996.
4 SHNEIDERMAN, B. 'ANTHROPOMORPHISM: FROM ELIZA TO TERMINATOR', PROCEEDINGS OF CHI '92, 1992, PP. 67-70.
5 HARRIS, R. AND LOEWEN, P. (ANTI-) ANTHROPOMORPHISM AND INTERFACE DESIGN. TORONTO: CANADIAN ASSOCIATION OF TEACHERS OF TECHNICAL WRITING, 2002.
6 MURANO, P. 'WHY ANTHROPOMORPHIC USER INTERFACE FEEDBACK CAN BE EFFECTIVE AND PREFERRED BY USERS'. ENTERPRISE INFORMATION SYSTEMS, VOL. VII, 2006, PP. 241-248.
7 DISALVO, C., GEMPERLE, F. FROM SEDUCTION TO FULFILLMENT: THE USE OF ANTHROPOMORPHIC FORM IN DESIGN. ENGINEERED SYSTEMS. 2003.
8 DUFFY, B.R. 'ANTHROPOMORPHISM AND THE SOCIAL ROBOT', ROBOTICS AND AUTONOMOUS SYSTEMS, VOL. 42, 2003, PP. 177-190.
9 MORI, M. 'THE UNCANNY VALLEY', ENERGY, VOL. 7, 1970, P. 33-35.
10 MACDORMAN, K., GREEN, R., HO, C. AND KOCH, C. 'TOO REAL FOR COMFORT? 'UNCANNY RESPONSES TO COMPUTER GENERATED FACES'. COMPUTERS IN HUMAN BEHAVIOR, 2009.
11 LEVY, D. LOVE AND SEX WITH ROBOTS: THE EVOLUTION OF HUMAN-ROBOT RELATIONSHIPS. HARPER, 2007.
12 ETTINGER, R. MAN INTO SUPERMAN. AVON, 1974.
13 WARWICK, K. I, CYBORG. UNIVERSITY OF ILLINOIS PRESS, 2004.
14 BOSTROM, N. 'IN DEFENCE OF POSTHUMAN DIGNITY'. BIOETHICS, VOL. 19, NO. 3, 2005, PP. 202-214.
15 GEHLEN, A. MAN: HIS NATURE AND PLACE IN THE WORLD. COLUMBIA UNIVERSITY PRESS, 1988.
16 STOCK, G. REDESIGNING HUMANS: OUR INEVITABLE GENETIC FUTURE. 2002.

AF NIKKOR EYE RETINA

HS AF

ϕ 22 - 24 mm

50 CPD

Nikon

Eye Retina

HIGH SPEED
FOCUS

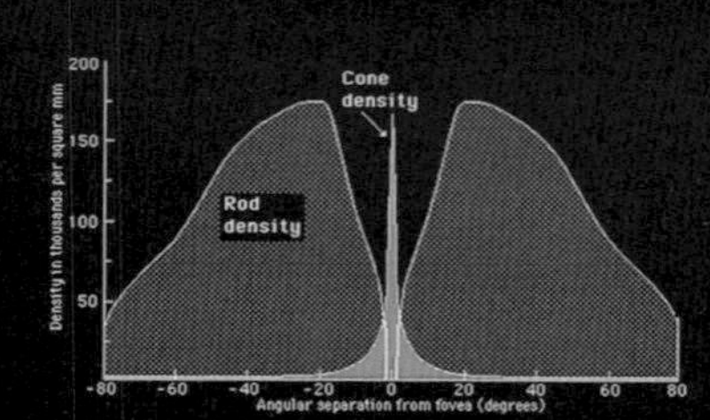

- 120 million sensitive rods
- 7 million cones for sublime bright colors
- High color vision & visual acuity

The new Nikon Eye Retina lens™ uses real human tissue for ultimate light sensitivity, sharpness and natural colors

PRODUCTS BECOME PEOPLE

The Life of Products

They are Born

The moment of birth occurs when the product emerges from its packaging. Out of the plastic comes an unsullied tabula rasa, ready to store our data, make our food, and play our movies. Sometimes the birth is quite literal. For the Tamagotchi toy, introduced in 1996, a pixelated egg hatches onscreen and quickly grows into a digital organism that demands care. An inattentive owner runs the risk of returning to a dead pet complete with angel wings, but there's no need to grieve. Reincarnation is only a button push away. Even more abstract births are commemorated as special moments for consumers. The Apple corporation, in particular, is known for the elegant design of its products' minimalistic 'womb.' Each component is lovingly swaddled, revealing itself in logical order. Owners proudly display boxes for years after purchase. Others post pics of their unwrapping ceremonies in online galleries that draw millions of views. Apple promises all the future good times of a child, with none of the hassle of dirty diapers. | AG

They Serve You

The bowed shape of the Senseo coffee maker is intended to echo the form of a butler, stooping to pour its employer's tea. Wildly popular in Europe, the Senseo is now on its way to America. The success of the Senseo shows the utility of injecting a social narrative into the human-machine interaction. We might not be able to afford a manservant, but we can entertain a fantasy of kindly appliances.

They Annoy You

Incessant beeps, flashes, and helpful alerts may be the sign of a healthy product, but that doesn't mean the human user won't wish for a neck to throttle.

Looks like you're committing suicide!
Office Assistant can help you write your suicide note. First, tell us how you plan to kill yourself.
Pills
Jump
Pastry
Tips
Options
Close

And Eventually, They Die

Who has not experienced a sense of grief, even betrayal, when the 'screen of death' appears on a beloved electronic device? Old age comes to humans and technology in the form of strange creaking and slow performance. Finally, we both crash. Biology has not designed humans for inconveniently long life; neither have corporations designed their products for an unprofitably long existence. But products, unlike people, do not break down into fertilizer after death. A global industry now copes with 50 million tons of the 'corpses' we throw out each year. Some recycling outfits are legal and environmentally responsible. Others ship our e-waste to developing nations. Unsafe practices expose workers to toxins and leech plasticizers and heavy metals into the soil and water.

Everyday Robots

From the perspective of next nature, robots represent a new kingdom of inorganic animals. Some mow our lawns, some mow down our enemies. While many robots never make it out of research institutions, others have been quietly laboring in the workforce for years. | AG

Thank The Maker!

C3PO in *Star Wars* (1977)

Housekeeper
When the Roomba vacuum seems too impersonal, accessories are available to dress it as a cat, maid, or Orca.

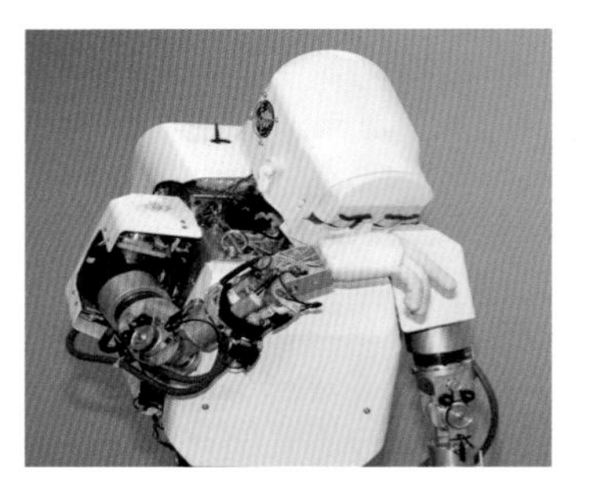

Friend
Like us, Kobian uses its whole body to express emotion. Unlike us, it is limited to seven basic feelings.

Substitute Teacher
Created by Professor Hiroshi Ishiguro, the initial research aim of this robot was to teach university classes.

Recycler
Part of an Italian research program, the DustCart collects trash, recycles and records data on pollution.

Cleaner
Robots don't just vacuum the floor – this one sucks up sediment from aquaculture tanks and reservoirs.

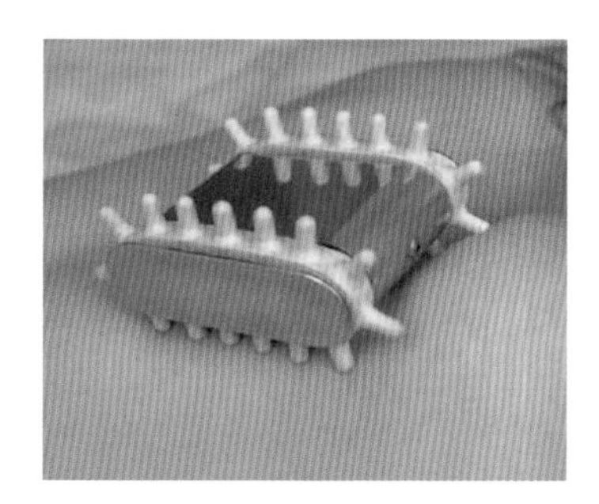

Masseuse
Tickle follows an unpredictable route over the body. Its rubberized treads create a tickling sensation on bare skin.

Gardener
An outdoor version of the Roomba, the Robomow uses a sensor to navigate while cutting the grass.

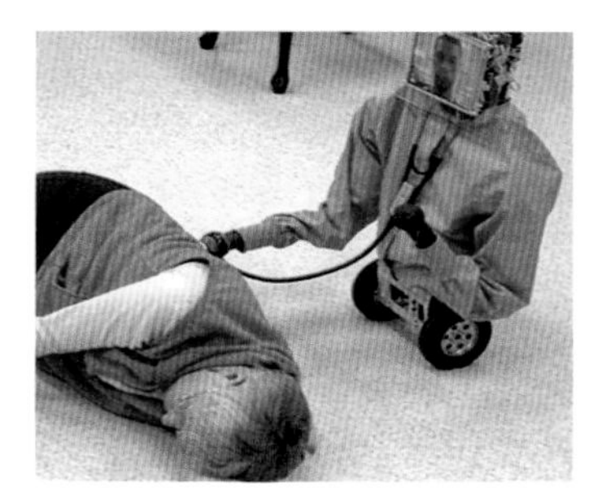

Medic
Robots may soon replace doctors for basic medical tasks. This robot can measure a patient's blood pressure.

Pet
Dream Cat Venus, courtesy of Sega Corporation, responds to its owner's affections with meows, purrs, and blinking.

Soldier
MAARS uses GPS to locate combat zones. Humans can switch its gun for a less-deadly mechanical arm.

Nanny
The PaPeRo can recognize children and play with them. The robot allows parents to remotely observe their offspring.

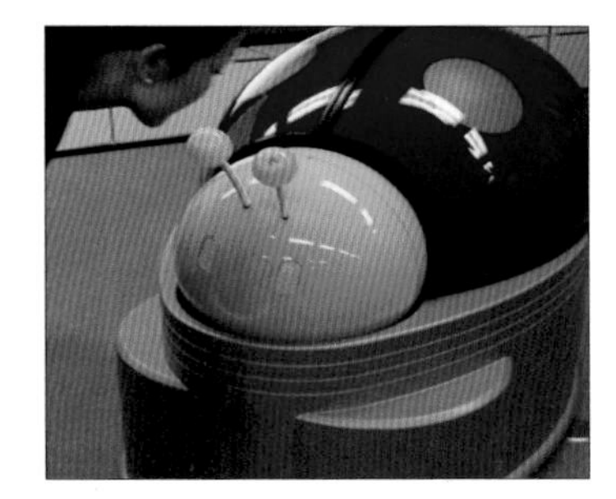

Bathroom Attendant
The LadyBird cleans toilets at highway rest stops. An Internet feed answers visitors' questions about traffic conditions.

Factory Worker
This industrial robot is a computer-controlled mechanical arm with six axes. It works on auto assembly lines.

Controller
ApriPoco learns to take over the function of appliances' remote controls by asking its owner questions.

Comforter
Paro, a robotic seal, is a therapeutic aid for children and the elderly. It offers the benefits of live-animal therapy.

People Versus Products *A Love Story*

MAN ON A LAWNMOWER BY DUANE HANSON

Man Arrested for Shooting Lawn Mower

A 56-year-old man from Wisconsin (USA) has been arrested after shooting his lawn mower in his garden because it would not start. The man was charged by police with disorderly conduct and possession of a sawn-off shotgun. Police officers said the man told them, “It’s my lawn mower and my yard, so I can shoot it if I want.” Police found the shotgun, a handgun and a stungun, as well as ammunition when they detained the man in the basement of his house. He could face a fine of up to $11,000 and a maximum prison sentence of six and a half years if convicted. Witnesses told police that the gunman appeared to have been drinking. The lawn mower was found sitting outside the house. A local retailer said that the gunman might now have difficulty getting his lawn mower repaired. “Anything not factory recommended would void the warranty,” said Dick Wagner, of Wagner’s Garden Mart in Milwaukee.

Man Charged for Dumping Silicone Girlfriend

Fifteen policemen rushed to the scene, after a couple discovered a "corpse" while out walking their dog in a mountain forest in Izu, central Japan. The officers discovered a human form wrapped in plastic and tightly bound around the neck, midriff and ankles, with hair protruding from one end. The body was left untouched and taken away for examination, and the crime scene duly secured by a police cordon. By mid-afternoon, the body was in the hands of police pathologists. But when they sliced open the wrapping, they were confronted not by a decomposing corpse, but by a life-sized sex doll.

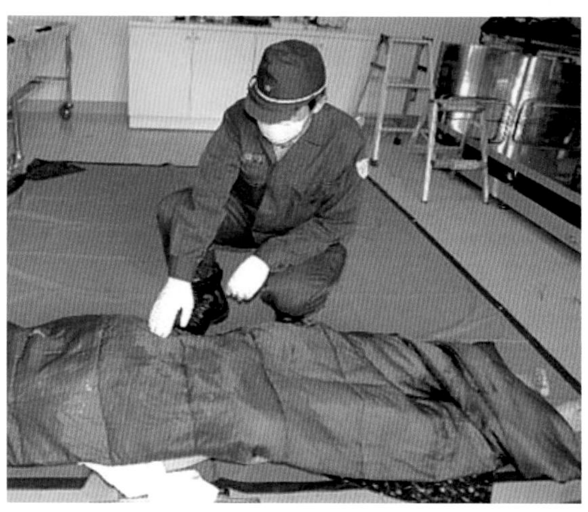

DUMPING YOUR SILICONE GIRLFRIEND IN A FOREST?

Two weeks earlier a 60-year-old unemployed resident of Izu had wrapped his 1.7-meter tall, 50-kilogram silicone girlfriend in a sleeping bag, driven to a remote wooded area, and dumped her. A nice, clean break-up, he thought. According to investigators, the man had lived with the sophisticated doll for several years after his wife passed away, but decided to part with her after making plans to move in with one of his children. "It seems he grew attached to the doll over the years," said the chief investigator. "He was confused about how to get rid of her. He thought it would be cruel to cut her up into pieces and throw her out with the trash, so he proceeded to dump her illegally." The man, who regrets his life-like doll was mistaken for a corpse, now faces fines for violating Japan's Waste Management Law.

Professor builds his own Android Twin

Although there are numerous researchers out there creating humanoid robots, none are as explicit about the very close relation between narcissism and anthropomorphism as Professor Ishiguro's android twin. Professor Hiroshi Ishiguro, working at ATR Intelligent Robotics and Communication Laboratories, decided that if he's going to be a roboticist he might as well create an "angry eyes" twin-brother version of himself, the Geminoid HI-1. The remarkable realism comes from silicone molds cast from Ishiguro's own body. Ishiguro is using his robot twin brother to teach his classes for him, and creep out students with life-like movements such as blinking, "breathing" and fidgeting.

The robot can be remotely controlled via a motion capture system that tracks Ishiguro's mouth movements and allows the robot to speak his voice – or that of an assistant if he's feeling particularly uninspired. The aim of the project is to experiment with the viability of tele-presence and find out if he can really command the attention of a classroom with a robot doppelgänger. Ishiguro thinks future business meetings will have androids and humans side by side at the table. Well, lets be positive, it sure would save a lot of business-class flights.

HIROSHI ISHIGURO WITH GEMINOID HI-1: WHO'S WHO?

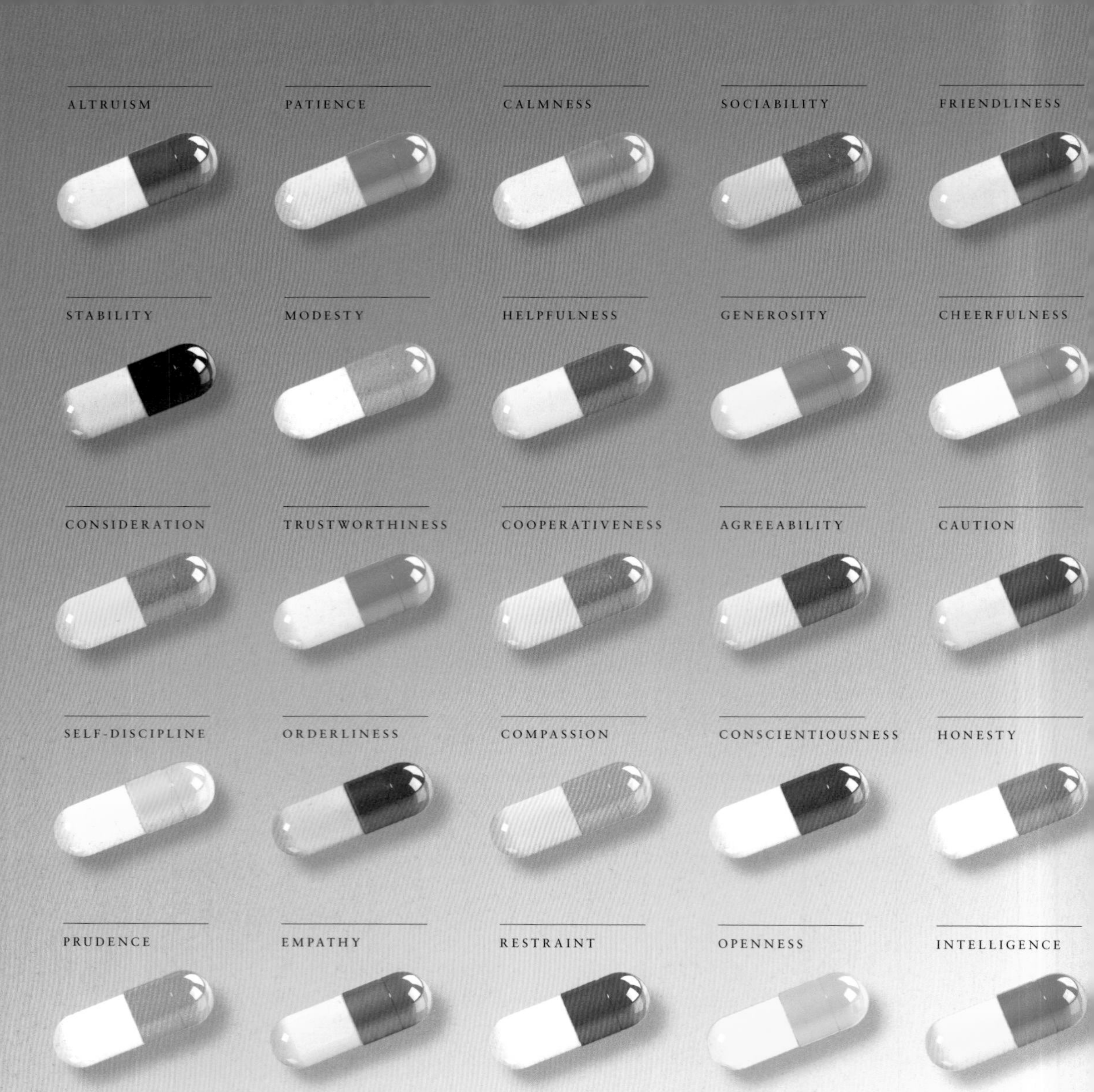
ALTRUISM
PATIENCE
CALMNESS
SOCIABILITY
FRIENDLINESS
STABILITY
MODESTY
HELPFULNESS
GENEROSITY
CHEERFULNESS
CONSIDERATION
TRUSTWORTHINESS
COOPERATIVENESS
AGREEABILITY
CAUTION
SELF-DISCIPLINE
ORDERLINESS
COMPASSION
CONSCIENTIOUSNESS
HONESTY
PRUDENCE
EMPATHY
RESTRAINT
OPENNESS
INTELLIGENCE

Be Normal.

SOCIETY
OF
MEDICATION

PEOPLE BECOME PRODUCTS

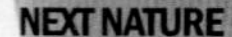

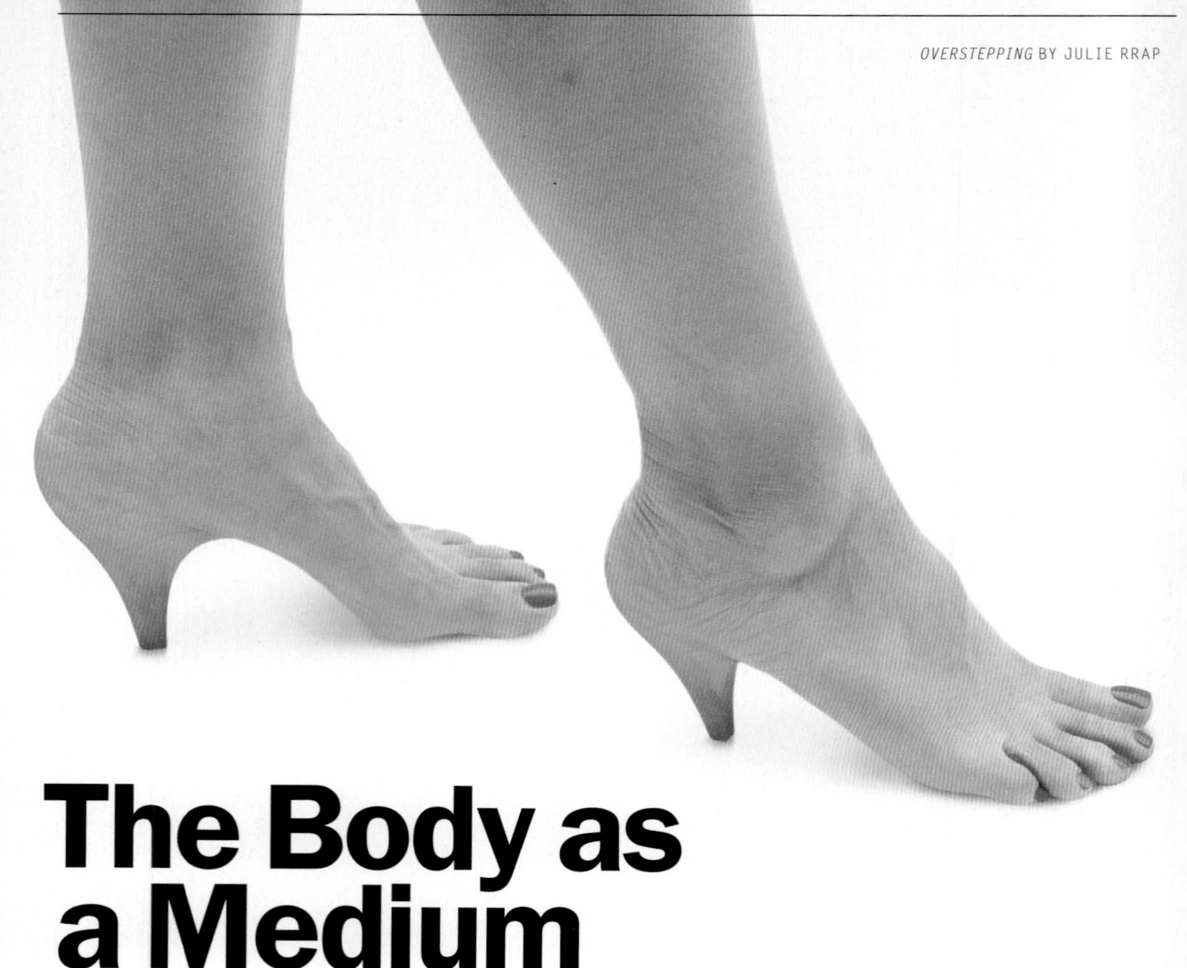

OVERSTEPPING BY JULIE RRAP

The Body as a Medium

Truly sophisticated technology makes itself invisible. It empowers its user by functioning as an extension of the body. But increasingly, we start to see the body itself as a technology that can be manipulated and upgraded.

With the increasing regulation of our lives by electronics and a drive towards the miniaturization and portability of electronics on and around the body it seems only logical to place electronic circuits on the surface and perhaps even inside the body. The Royal College of Art has developed conductive ink that is applied directly to the skin to bridge the gap between electronics and the body. The material allows users to create custom electronics and interact with technology through intuitive gesture. It also allows information to be sent to the surface of the skin from person to person or person to object. The formulation is carbon based, water-soluble, skin-safe and non-invasive. It may be applied in a number of ways including brushing on, stamping or spraying and has future potential for use with conventional printing processes on the body. Potential application areas may be: dance performances, music, fashion, security, military, audio/visual communication and medical devices.

From Necessity to Opportunity

Lost your wisdom teeth? Why not go for the upgrade? James Auger and Jimmy Loizeau envision a 'tooth phone' consisting of a tiny vibrator and a radio wave receiver that can be implanted into a tooth through regular dental surgery. Information can be received anywhere and at any time. Sound that comes into the tooth as a digital radio signal is conveyed to the

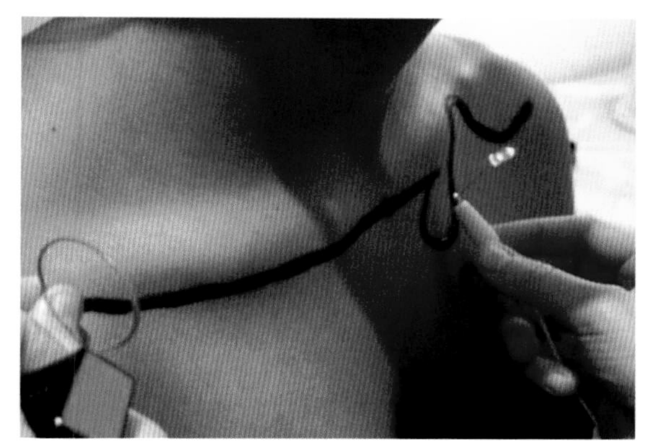

CONDUCTIVE SKIN BY PILDITCH, JOHNSON, LIZARDI, NELSON

TOOTH PHONE BY JAMES AUGER AND JIMMY LUIZEAU

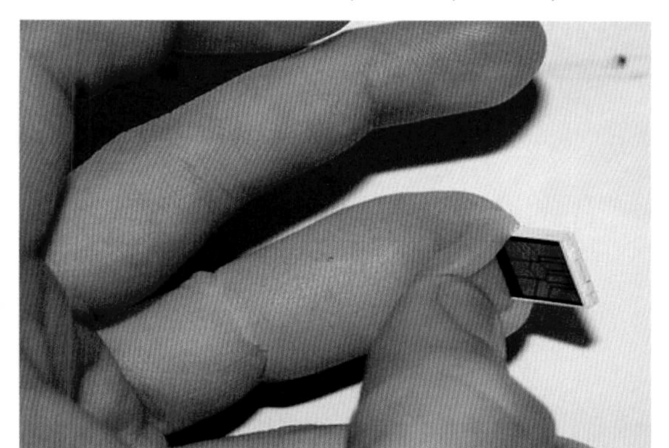

USB PROSTHETIC BY JARRY JALAVA

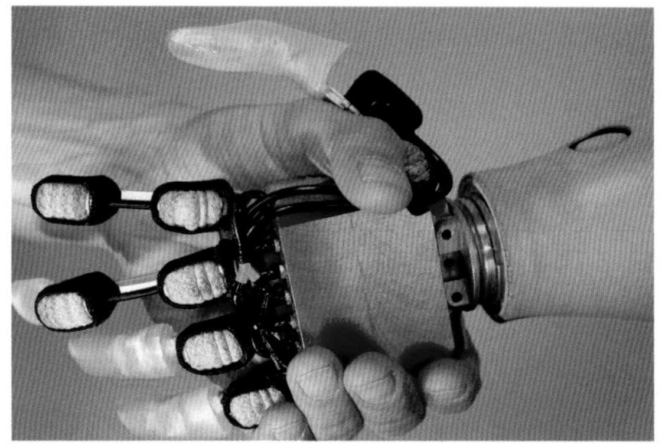

THE *FLUIDHAND* BY THE RESEARCH CENTER IN KARLSRUHE

listener without anyone noticing – using the inner ear bone as a resonance box. Until now the tooth phone is still a speculative design as no functioning prototype has been realized.

Beyond Biomimic Prosthetics

Medical necessity is often the father of innovation. After a motor accident Finish hacker Jerry Jalava decided not to use an off-the-shelf prosthesis and felt that a prosthesis with an integrated USB memory stick would suit him better. One wonders why the developers of artificial limbs usually stick to the somewhat boring and predictable simulations of the human limbs they are replacing. If you have lost your hand and you need a prosthesis, why not grasp the opportunity to go for an upgrade? Wouldn't it be liberating if disabled people could receive more creative (and advanced) limb replacements? Think for example of a prosthesis with an integrated GPS–compass, MP3–player, USB–stick, smartphone or a fashionable eight finger octopus design? Admittedly, non-prosthetic people might be slightly scared off – or even experience signs of anthrpomorphobia – at the sight of a prosthetic that doesn't try to look like the human limb it is replacing. But then again, a fake plastic hand that desperately tries to look like an original human hand but never really makes it, also causes uncanny feelings. So why not create something that's better, or at least different than the real thing? Can we move beyond skeuomorph prosthetic designs, in which the visual appearance of the original human limb is retained merely as an ornament of a lost functionality. Wouldn't it be much more empowering to transform the disabled into the enabled? | KVM

Tempted By the Upgrade

Throughout the evolution of many animal species, creatures capable of pinpointing and mating with strong and healthy mates gained a reproductive advantage over those with less discriminating tastes. Their offspring would be more likely to survive to adulthood and have offspring of their own, propagating genes for successful mate-selection through thousands of generations.

Over time this simple evolutionary rule has equipped humans with highly advanced sensibilities to detect and select suitable mates. Your mere existence proves you come from a long line of mate-picking experts. Evolution is not only responsible for our preoccupation with choosing worthy mates, but for our urge to present ourselves as successful partners. Think of the exuberant tail of the peacock, which serves no purpose other than to dazzle the peahen. Likewise, we humans have a wide range of visual features to consider when determining the quality of a potential mate, from facial symmetry, height, and body shape to the glossiness of the hair and tightness of the skin.

Technological Fitness Boosters

In contrast to the peacock, which depends on the luck of genes and nutrition for its impressive tail, mankind has developed artificial features that intervene with truthful fitness assessments. Cosmetics, clothes and surgery allow us to manufacture all sorts of physical properties that make us more attractive to our peers. Lipstick, hair dye, Botox, breast augmentations, steroids, anti-hair loss shampoo and solariums are just a few of the inventions intended to influence how potential mates perceive our fitness. These artificial fitness

THE PUSSYCAT DOLLS
CERTAINLY NOT THE LATEST LINE OF BARBIE DOLLS, ARE A FORMER LA BURLESQUE ACT, UPGRADED TO BE A POP MUSIC SENSATION AND THE NEW FACE OF FEMALE EMPOWERMENT

enhancers only appear to make you younger, healthier or stronger. In general, these improvements are superficial and short-term. A thirty-year-old face slapped on a sixty-year-old body does not magically instill fertility. Nonetheless, people will shell out fortunes for these augmentations. Why is that? Perhaps our instincts tell us that it doesn't really matter if cosmetic improvements don't actually make us younger, healthier or stronger. The illusion they evoke in the minds of our peers is already enough. On a rational level, while every man knows that the 99 cent lipgloss of that sexy young woman is just a decoy, his inner caveman is attracted just the same. It's hard to undo million-year-old instincts with simple logic. The fascinating thing about our technological fitness-boosters is that they have now become so sophisticated they can make us look more attractive than a normal person could possibly hope to be. With advances in cosmetic enhancements, beauty has turned into 'hyper-beauty,' a simulation of physical attractiveness more cutting-edge than anything found in nature. Our evolutionary ability to recognize fitness indicators is pushed to the limit. To a certain extent, perfection has become possible. Just open a fashion magazine and you'll understand. Evolution equipped us with a desire for the fitter specimens of our species, and both Hollywood and the gym-loving boy next door aim to make good use of that. Yet at a certain moment our senses are saturated, and hyper-beauty turns into a caricature.

The Uncanny Valley of The Pussycat Dolls

The Pussycat Dolls are a former burlesque troupe, upgraded to a pop music sensation that enjoyed mainstream popularity from 2005 to 2008. With all

THE LIPSTICK, A TIME-HONORED TECHNOLOGICAL FITNESS BOOSTER

THE MALE PEACOCKS EXUBERANT TAIL HAS NO FUNCTION BUT TO IMPRESSES POTENTIAL MATING PARTNERS

hair colors and a handful of races represented, this collection of women seems to have stepped out of an adolescent boy's fantasy. Yet their polished perfection also has a certain uncanny quality: The full lips and breasts, the flawless faces, the oh-so-perfect noses, the shiny skin and the sculpted bodies. Too perfect to be human, they cannot be trusted. Could the pop group perhaps be the latest models from Abyss Creations, manufacturer of the humanoid sex toy Real Doll? The name 'Pussycat Dolls' certainly adds to this suspicion, but then again, their hit songs with titles like 'Buttons' and 'Beep' point more in the direction of robotics. Of course these pop singers aren't replicants, silicone dolls, or robots of any kind, they're human. The problem is that they are so slick, so advanced, that us mediocre humans may start mistaking them for something else entirely.

Almost-But-Not-Quite Human

In 1970, Japanese robotics researcher Masahiro Mori introduced the uncanny valley hypothesis, which holds that when an artificial representation of a person looks and acts almost but not entirely human, actual people will react with discomfort. Mori's theory derives from German psychiatrist Ernst Jentsch's concept of *das unheimliche*, or 'the uncanny,' which describes the sensation that arises when something feels simultaneously familiar and foreign. Uncanny objects generate cognitive dissonance due to the paradoxical nature of feeling attracted and repulsed at the same time. There are various and overlapping reasons why an object might tumble into the uncanny valley, from how our brains perceive faces to the inborn aversion to physical markers of illness. A person's cultural background may also have an considerable

DETAIL FROM AN ADVERTISEMENT FOR THE BAY COSMETICS BRAND

SURVIVAL OF THE FINEST
CONTRARY TO THE PEACOCK, WHICH IS DEPENDENT ON GOOD GENES FOR ITS IMPRESSIVE TAIL, MANKIND DEVELOPED AN ADDITIONAL REGISTER THAT INTERVENES WITH OUR MUTUAL FITNESS ASSESSMENTS. THE MUNDANE LIPSTICK IS AMONG THE MOST ANCIENT TECHNOLOGIES AIMED AT INCREASING OUR FITNESS, AT LEAST IN THE MINDS OF OUR PEERS. AS THE COSMETIC-INDUSTRY ADVANCES AND BEAUTY IS TURNED INTO HYPER-BEAUTY (A SIMULATION OF A BEAUTY THAT NEVER REALLY EXISTED), OUR EVOLUTIONARY SENSIBILITIES FOR RECOGNIZING FITNESS ARE PUSHED TO THE LIMIT.

influence on how uncanny an object is perceived to be. For instance, a lifelike humanoid robot may call into question religious notions of the divinely-derived uniqueness of the human species.

Boomeranged Into The Uncanny Valley

The uncanny valley hypothesis is typically associated with humanoid robotic research and computer graphics that aim to create animated humans that could pass for real actors. However, a similar effect might occur when actual humans begin to augment their beauty, muscle strength, eyesight, or cognition so that they are no longer recognizable as normal people. As long as these enhancements remain within norms of acceptable human behavior, a negative reaction is unlikely. Once individuals supersede normal human variety, however, some disgust can be expected from the 'real' humans, who may feel they are falling behind the curve. The phenomenon of actual humans being mistaken for non-humans can be called the 'boomeranged uncanny valley effect.' The Pussycats Dolls, with their toylike hyper-beauty, surely fit the description. Some men might object that these pussycat lifeforms are simply splendidly desirable objects and that they would be happy to have sex with them anytime. Certainly from an evolutionary perspective, the sexual attraction of an upgraded woman is rather obvious. Recent genetic research may indicate that Neanderthals had the hots for *Homo sapiens* with their good looks and superior technology. After all, why be satisfied with a normal woman, if there is a more advanced hyper-woman available? | KVM

Who or what sets beauty standards?

It's Consumer Culture!

Consumer culture drives contemporary notions of beauty. Sex sells, sometimes too well. Advertisements play to insecurities by creating an idea of beauty that is plausible but perpetually out of reach. Yet when age and biology intrude into the marketer's vision, the effects can be unsettling for a consumer accustomed to a lifetime of flawless bodies. Reality becomes unreal when perfect beauty saturates visual culture. Aging is natural in humans, but in Barbie, it's astounding.

No, it's the Porn industry!

Mainstream culture feeds off its darker underbelly. Porn helped to popularize the ideal human body as hairless, barely legal, and surgically pumped and plumped. At least pornography makes no bones about its objectives. The bodies might be connected to personalities, but it's their sleek appearance and reliable performance that keep the consumers coming. Like a piece of hardware, though, planned obsolescence is also an issue. The lifespan of a standard performer is no more than that of the average iPod.

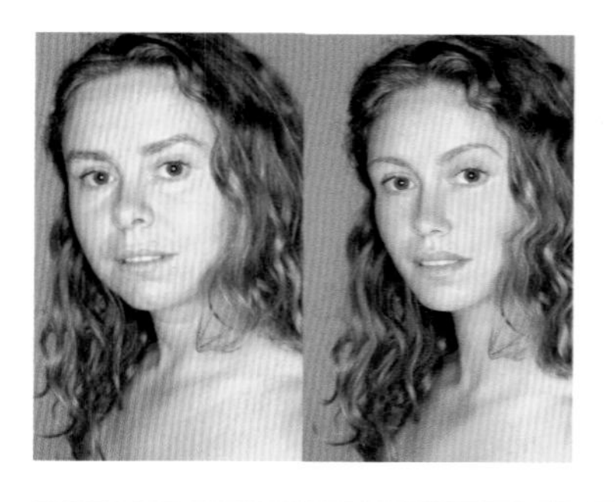

No, it's the Glossies!

Adobe's Photoshop may have replaced airbrushing as the tool of choice, but photographic manipulation is still pervasive in print media. Even ostensibly perfect stars and starlets are subjected to digital nips and tucks. Of course, blatant distortions of the truth inspire backlash. Some blogs now make a sport of spotting poorly Photoshopped images. Dove's 'Real Beauty' campaign positioned the brand as anti-glossy, using real people and (supposedly) eschewing retouching.

Software Maybe?

Researchers have developed software that purports to objectively rank the beauty of female faces. In contrast to the notion of physical beauty as a rare occurrence, the software rewards the most average looks. It is the Goldilocks point of physical beauty: the nose is not too big or too small, the eyes are not too close or too far apart. Variety may be the spice of life, but in digital life, it is unwelcome. | AG

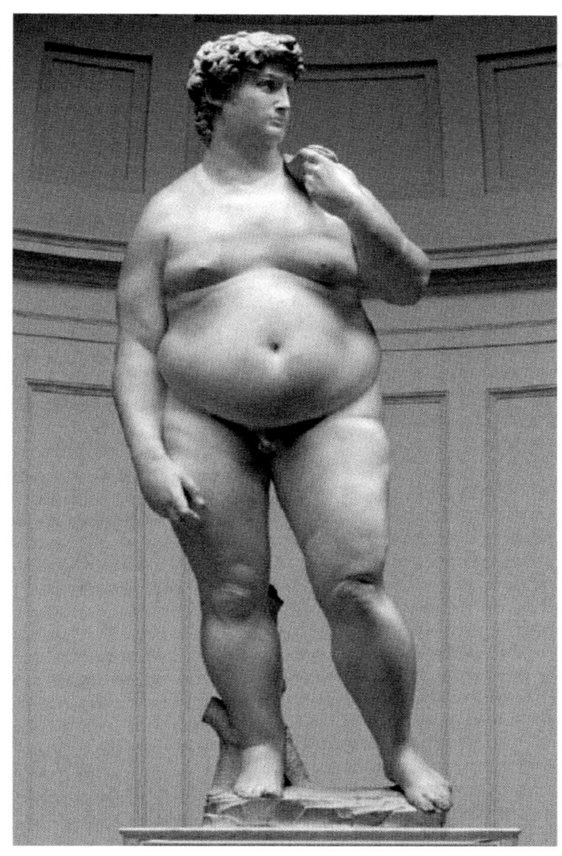

MICHELANGELO'S DAVID AFTER HIS STAY IN THE US. TAKEN FROM AN ADVERTISING CAMPAIGN CREATED FOR THE GERMAN OLYMPICS

What about the classics?

Has McDonald's caught up with Michelangelo? From the German advertising agency Scholz & Friends, the obese David is part of a campaign for the German Olympic Sports Federation that uses altered sculptures to warn: "If you don't move, you get fat."

Here, classical beauty contrasts with the sorry state of the modern physique. We may be able to dismiss the artificiality of the latest fashion magazine, but a masterpiece's patina of age brings with it a sense of respectability. The cache of the classics argues for the 'correct' human body as an athletic one. Ancient Greco-Roman depictions of the human body inspired an artistic renaissance in Italy and are still admired today for their exceptional realism. Yet Greek art is not so much real as it is hyperreal. The women may be a bit rounder than we like our models, but the men embody chiseled standards of masculine grace. They are the ancient-world equivalent of wishful digital manipulation. The desire to not only celebrate youth and beauty in art, but also to create impossibly perfect human forms, is not just a by-product of contemporary media culture. The same impulse guided the sculptor Praxiteles in the 4th century BC, and was appropriated by Michelangelo nearly two centuries later. Yet, with predictions estimating 2.3 billion overweight adults in the world by 2015, the chubby exception through much of human history is fast becoming the rule. Our bodies may be changing, but our standards are not. | AG

Golden Ratio People

Research learns that our physical attraction to another person increases if his or her body is symmetrical and in proportion. Designer Boudewien van den Berg manipulated the portraits of her friends to comply with golden ratio proportions. The result is a somewhat uncanny collection of beautiful people. Is this what we would look like if we all complied with utopian beauty standards?

MANIPULATED PHOTOGRAPHS OF PEOPLE BY BOUDEWIEN VAN DEN BERG

ANTROPO-MORPHISM & PRODUCT DESIGN

By Joran Damsteeg, Koert van Mensvoort and Hendrik-Jan Grievink

THE TORRE AGBAR IN BARCELONA (DESIGNED IN 2005 BY ARCHITECT JEAN NOUVEL) AND HIS ONE-YEAR OLDER BROTHER SWISS RE BUILDING (DESIGNED BY SIR NORMAN FOSTER) ARE BOTH ICONIC LANDMARKS IN THE SKYLINE OF BOTH CITES. THE LATTER IS COLLOQUIALLY REFERRED TO AS THE 'GHERKIN' ON ACCOUNT OF ITS RESEMBLANCE TO A CUCUMBER. HOWEVER, SOME PEOPLE SEE SOMETHING QUITE DIFFERENT IN ITS SUGGESTIVE SHAPE.

ELEVEN GOLDEN RULES

The focus of product design has shifted away from the simple desire to create functional mass-produced products, and moved toward intelligent products and emotional design. Increasingly, products aspire to maintain empathetic relationships with their users. In order to make a connection, products often try to mimic human behavior. However, we humans have such a complex society and so many unwritten social and behavioral rules that it's hard to copy them correctly. People are so expert at identifying and categorizing other humans that when we see something only slightly different from the norm, we're immediately put on guard. Anthropomorphic products have to 'be human' in exactly the right proportions, or consumers will be thrown off, confused, irritated, or downright offended. Anthropomorphism occurs in many shapes and sizes, and results in many different reactions. Matching the right form to the desired effect is a difficult job for a product designer. The following 11 Golden Rules give an overview of the benefits and risks involved with the creation of anthropomorphic products, and how designers should deal with the issues that arise.

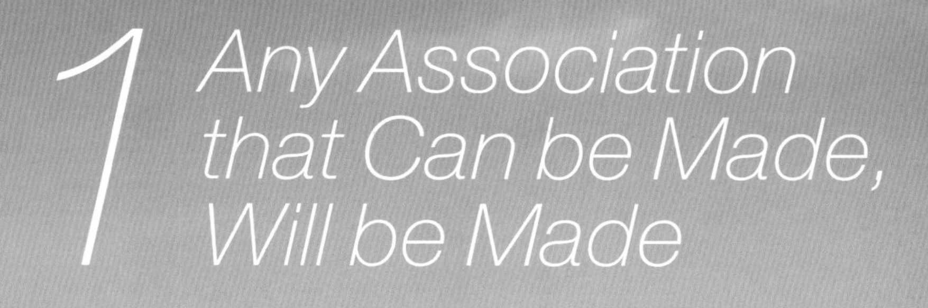

1 Any Association that Can be Made, Will be Made

People have evolutionarily built-in mechanisms that help us to recognize the human face and body, and what gender, race and mood those faces and bodies are projecting. We're so good at recognizing each other, we do it even when it's not applicable. When we look at animals, machines, and random objects we infer characteristics that aren't there. Especially when a product was intended to have certain human qualities, it's easy to imagine even more. Since designers can create anthropomorphic products without even realizing it, it's better to design these characteristics intentionally.

2 Different People Antropomorphize Differently

What people experience as anthropomorphic is highly personal. Tests have shown that when given a selection of products with anthropomorphic characteristics, people differ greatly in how human-like they perceive the objects to be. Even if an anthropomorphic product fits the cultural, social and ethical norms of a society, it's still possible some people just won't like it because they experience it differently.

THE BIG NETWORK HARD DRIVE BY LA CIE: JUST ANOTHER NETWORKED DRIVE WITH A BLINKING LIGHT, OR A SMALL VERSION OF HAL 9000 (THE KILLING COMPUTER FROM KUBRICK'S *2001: A SPACE ODYSSEY*) IN A PLASTIC WHITE BOX?

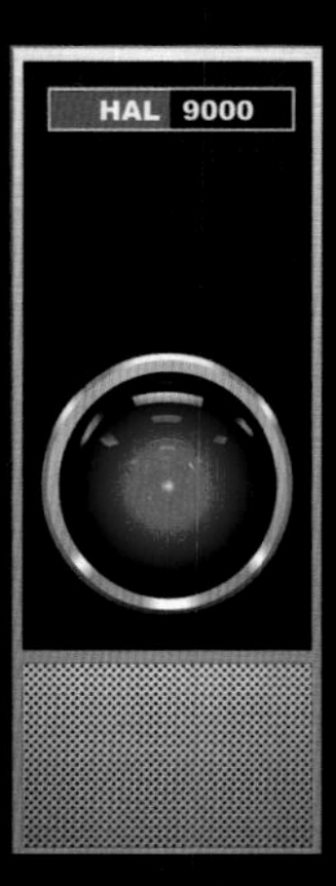

E MACHINE
ING
OU LIKE A
R?'

3 Keep it ASS: Abstract, Simple and Subtle

Making good use of anthropomorphism isn't easy. As you've probably already noticed people may dislike products purely because of their anthropomorphic elements One way to reduce this risk is to downplay the anthropomorphic qualities: keep it as simple, subtle and abstract as possible. When the implementation is so subtle that most people won't consciously notice it, they are less likely to be annoyed, while the product can still achieve the desired effect. Abstraction reduces the chance of directly evoking negative emotions, while preserving the positive associations

DON'T ANTHROPOMORPHIZE COMPUTERS, THEY HATE THAT. VIRAL INTERNET VIDEO ON OFFICE WORKER WHO TRASHES HIS COMPUTER.

4 Complex Products tend to Be Antropomorphized

hink about a spoon. Now think about a spoon with a face. What do you think it is? Most likely, you think t's a spoon with a face. Now think about a computer, vhich doesn't have a face. Are you more likely to swear at the spoon or the computer? Humans have a natural tendency to anthropomorphize things they can't explain. In the past, mysterious phenomena such as he weather, the sun or the moon were anthropomor-

products have advanced to such people don't understand them. T device by ascribing human emotior behavior. The more complex, capat a product is, the more likely it's gc morphized. Designers of techno products should anticipate how u morphize their product, and desig

5 Consider Zoomorphism as an Alternative

When a product imitates animal behavior, the strict social rules governing anthropomorphic products don't apply. People may be much more forgiving when a zoomorphic product makes an error, and fascinated rather than disturbed when it behaves other than expected. Similar to how we think a person walking in circles on the street is weird, but a dog chasing its tail is funny, Sony's robot dog Aibo is considered adorable, while Honda's humanoid robot Asimo seems clumsy and slow.

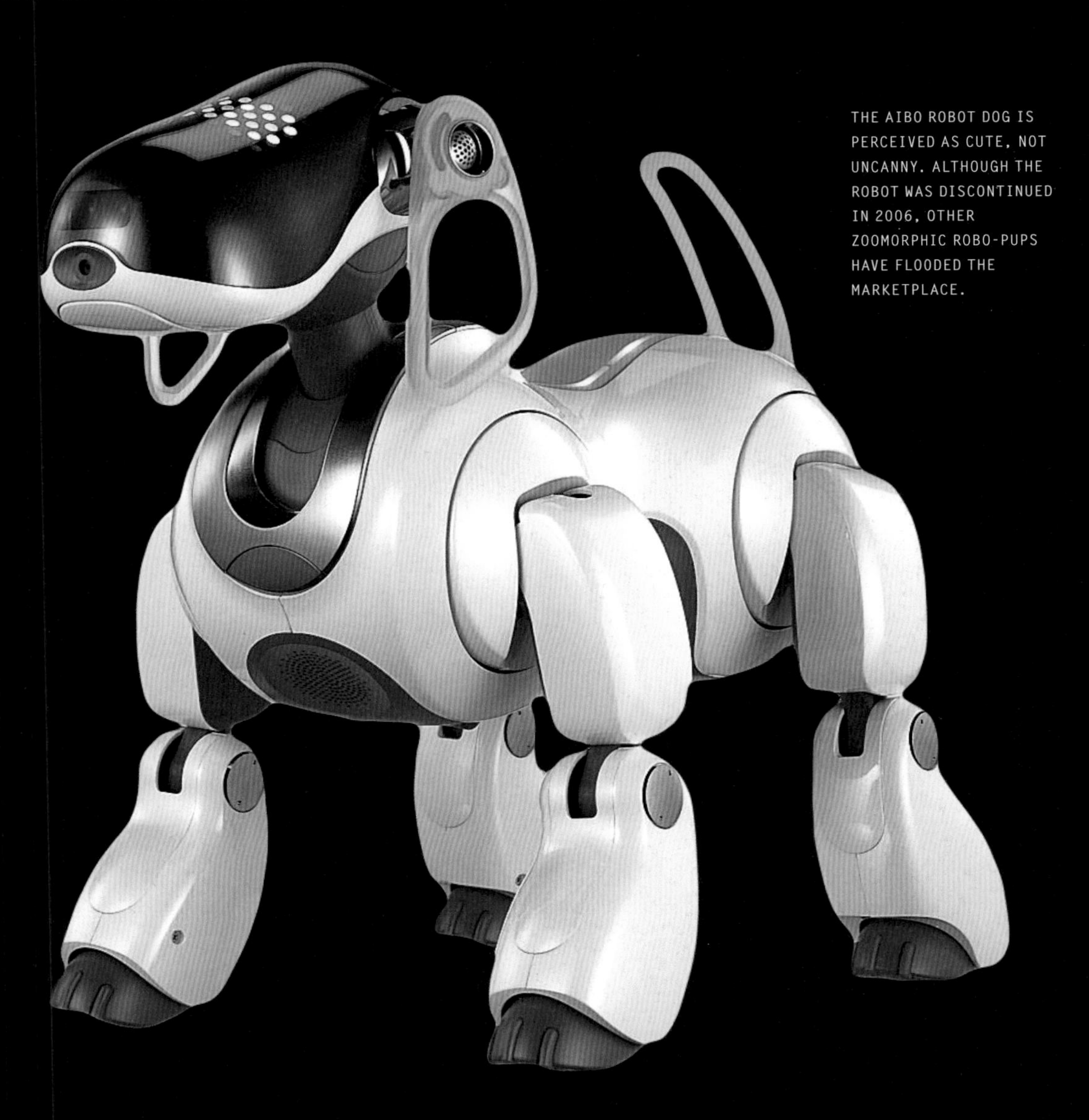

THE AIBO ROBOT DOG IS PERCEIVED AS CUTE, NOT UNCANNY. ALTHOUGH THE ROBOT WAS DISCONTINUED IN 2006, OTHER ZOOMORPHIC ROBO-PUPS HAVE FLOODED THE MARKETPLACE.

6 Meet People's Expectations

People expect many things from each other: Expect them to say hi in the morning; expect them to buy a ticket for the bus; expect them to watch out when driving a car; expect them to do their jobs well. People also expect certain behaviors from anthropomorphic products. When a product works differently than promised, this can cause confusion or anger. When a person gives commands to a product and the product ignores him, he becomes frustrated, because the product feels like a person who rudely turns his back. You wouldn't accept that behavior from a person, so why would you accept it from a product?

SAYA THE ROBOT MIGHT LOOK LIKE A HUMAN, BUT SHE DOESN'T EXACTLY ACT LIKE ONE.

7 Respect Social Standards

CLIPPIT WAS AN ANTHROPOMORPHIZED VERSION OF THE HELP FEATURE IN MICROSOFT OFFICE. THE PAPERCLIP WAS WIDELY REVILED FOR ITS DISTRACTING QUESTIONS AND IRRITATING NOISES.

Anthropomorphic products enter the human social space. Humans have the most complex social behavior of any organism on Earth. Anyone or anything trying to join in should be careful to do it right. Although an anthropomorphic product may function perfectly, if it crosses social boundaries it will still tick people off. This can cause the product to become a social reject, which won't do sales much good. Luckily, it's not hard to figure out why things go wrong. Imagine a scenario where a person and a product interact, then replace the product with a second person. If the actions of the second person and the product don't match up, then there's something off about the product's design.

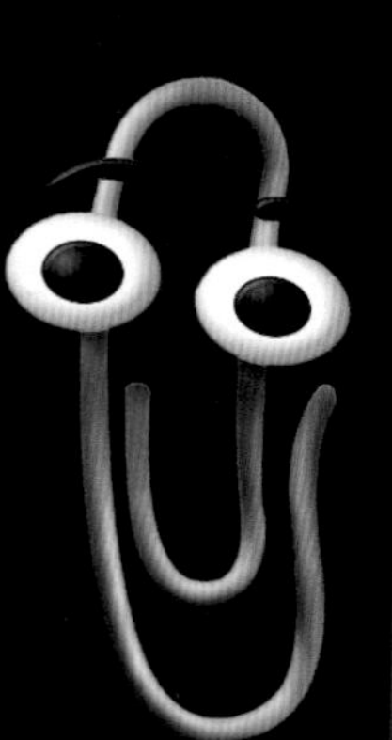

8 Use Human Ethics

IT'S NORMAL TO CHOP UP WOOD OR VEGETABLES, BUT NOT OTHER PEOPLE. HOPEFULLY NO CHEF WILL CONFUSE THIS KNIFE-BLOCK BY MAARTEN BAAS WITH AN ACTUAL HUMAN HEAD.

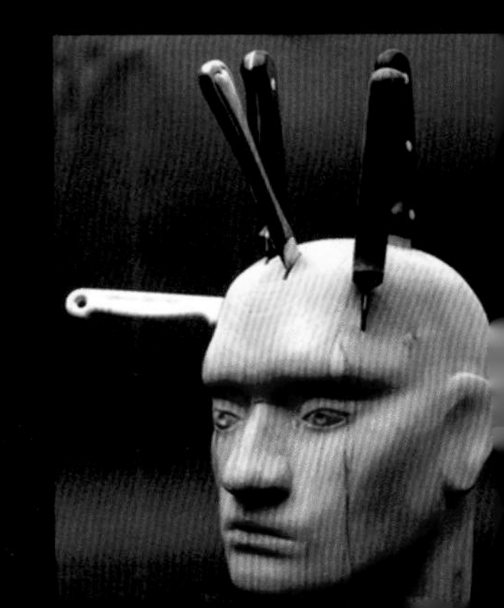

Anthropomorphic products blur the boundaries between products and people. Ethical norms for people don't usually apply to products and vice versa. For example, there's no need to apologize if you accidentally run into an object. But with an anthropomorphic product, you might instinctively say sorry, because it seems like the right thing to do. People can apply their attitude towards humans to products, which isn't necessarily a bad thing. But transferring attitudes from a product to a human might lead to problems, especially when the product induces abnormal social behavior. Don't make your product do what you wouldn't want a person to do.

9 Be aware of the Ecosystem you’re Invading

Vith most products, one wouldn’t normally worry about he environment that it enters. However, anthropomorphic oroducts inevitably elicit responses from others, even from non-human entities. This can have obvious advantages, or instance, when a human-shaped scarecrow frightens off the birds. But when daddy’s new toy frightens the children or the pets, there is a significant chance that it vill end up on the attic. Bringing home an anthropomorphic oroduct can be like introducing a new person into the household, which doesn’t always go as smoothly as the amily might hope.

PETS AND ROOMBA VACUUM CLEANERS HAVE A TUMULTUOUS RELATIONSHIP. WHAT SOME ANIMALS SEE AS A HOME INVADER, OTHERS SEE AS A MOBILE SLEEPING SPOT.

10 Enhance Human Experience, don’t Replace it

The hidden danger with interactive products is that they will become so good at fulfilling our needs that they start to replace actual humans. This is not a futuristic scenario: In an increasing number of locations, from supermarket self-scan checkouts to online bookstores, automatization has replaced human contact. Eventually this may lead to us becoming alienated from other people, which seems to contradict today’s rapidly increasing communication possibilities. Anthropomorphic products have the potential to support, stimulate and enhance human contact, but they may also eliminate it.

11 Don't use Anthropomorphism if it Does Not Serve any Purpose

Anthropomorphism can be a powerful tool in product design. But there are also risks involved that urge designers to be careful in their implementation. This final Golden Rule is also a warning: Don't use anthropomorphism simply to 'dress up' a product; it will make it distracting and confusing, and although it may increase the initial appeal of the product, people will soon lose interest for it, as the promise of human likeness is empty.

WHY WOULD A CHEAP DUSTPAN SMILE IF IT'S ONLY GOING TO GET COVERED IN DIRT? A JOKEY PRODUCT MIGHT BE AMUSING AT FIRST, BUT THE NOVELTY QUICKLY WEARS OFF IF THERE'S NO UNDERLYING PURPOSE FOR THE PRODUCT TO GRIN.

From Mainstreet to The Mansion

By Peter Lunenfeld

Nature demanded that we make a choice between immortality and sex, but the next nature of the 21st century may not. For help, we can look back at the 20th century, which had many storytellers playing with the parameters of the sex-equals-death equation.

None were more successful than two young men from the Midwest who ended up in Southern California, turning their dreams into reality, which Los Angeles always promises but rarely delivers. Walt Disney and Hugh Hefner, who seem miles apart, are in fact two sides of the same coin, flipping to decide what the next nature of sex and death will be. Life itself had a choice to make early on. Would it choose unchanging immortality, or infinite mutability punctuated by death and rebirth? Though single-celled organisms are still around, life in its wisdom abandoned self-replication and embraced sex, the intertwining of individuals to produce different offspring, which adapt to their environments, and grow into their own sexual maturity to repeat the process. In other words, life would rather fuck and evolve than endure the stasis of immortality. Life traded sex for death, and we are all the better for it.

Walt Disney achieved worldwide fame because he understood that children, and the stories they were told, needed a different model in the wake of the changes in the Western culture of death. The classic childrens' fairy tales the Brothers Grimm told were inundated with both sex and death. As their parents, siblings, aunts, uncles, cousins, grandparents, and others lived close at hand, often in the same room, the muffled sounds and musky smells of sex were inescapable. So too was the presence of death. The old died with their families, friends and relatives passed away young, many without any medical treatment, and most importantly, children lost their brothers and sisters. They were, therefore, in need of stories to explain and contain all that sex and death. This is why Red Riding Hood is eaten in the original story, why the Little Match Girl perishes in the cold, and, why the Little Mermaid suffers for her love of the prince and ultimately dissolves into foam. Walt Disney was born in 1901, close to those preceding centuries of misery but on the cusp of the 20th century in which death would be battled more ferociously than ever before in human history. Disney invented his most famous character, Mickey Mouse, in 1928; the year Alexander Fleming discovered penicillin.

Disney's fairy tales feature the off-screen deaths of many a mother, and just deserts to the villains, but they sanitize mortality. By the time of the corporate resurgence of Disney with the *Little Mermaid* in the 1980s, Ariel and the Prince marry and live happily ever after, an ending diametrically opposed to Hans Christian Andersen. Yet the Disney version befits a century's medical progress, the end of polio, tuberculosis, smallpox and even bubonic plague. But with the end of death, the Magic Kingdom saw no point to sex either, and banished it too, creating the squeaky clean image that conquered the world's children.

If Walt Disney came to understand the changing nature of death and used that to build a Magic Kingdom for children and the childlike, another Midwesterner, born a generation later, laid out a new set of illustrated stories for grownups. Hugh Hefner came back to Chicago from service in the last few months of WWII poised to make his contribution as a member of the Greatest Generation. After the requisite failures — in school, in marriage, in business — he borrowed $1,000 dollars from his mother, raised a few thousand more from friends, and founded a magazine he called *Stag Party*. After a trademark dispute Hefner changed the name to *Playboy*. He morphed his mascot from a stag to a rabbit, and the ubiquitous bunny was born.

The first issue featured Marilyn Monroe, "with nothing on but the radio." History rewards the prescient, and *Playboy* went to press virtually the same time as the first successful clinical trials of the oral contraceptive pill were held. Hefner's magazine became the public face of technological, mid-20th century hedonism, of sex severed from reproduction. Of course, for all its sophistication, *Playboy* isn't really about sex so much as it is about masturbation. And masturbation, as the story of Onan goes, is about wasting your seed. Five decades into the so-called *Playboy* revolution, we find ourselves awash in an online pornotopia, with the now octogenarian Hef popping Viagra the way that other people take vitamins, playing out his own fantasies of sex without reproduction, sex without death.

So what then, do these two multi-mediated, mid-century, midwestern megalomaniacs teach us about next nature? Walt Disney sees death receding and builds a Magic Kingdom out of Princesses who don't die, but don't get to fuck much either. Hugh Hefner, like Disney an avid cartoonist in his youth, fills a Mansion in the Hollywood Hills with the Girls Next Door who grew up wanting to be Princesses, but are instead stripped of their robes and have their breasts pumped up, creating a masturbatory simulation of sex without reproduction. Next nature deserves better.

THE MOST OVERTLY SEXUAL DISNEY CHARACTER, JESSICA RABBIT: "I'M NOT BAD, I'M JUST DRAWN THAT WAY."

PLAYBOY COVER OF HYPERREAL JESSICA RABBIT, 1988

HUGH HEFNER AND HIS PLAYBOY BUNNIES: WHO'S HYPERREAL?

Share
The Health.
• Real-time Body Monitoring System®
• Auto patient-to-doctor File Updating System®
• AA and Weight Watchers compatibility
• Medicare discount/penalty tracking®
THE TWITTER IMPLANT SUPPORTS YOUR HEALTHY LIFE-STYLE BY SIMPLY TWEETING EVERYTHING THAT HAPPENS IN YOUR BODY. THE TINY DEVICE COMES FREE OF CHARGE WITH SELECTED MEDICARE POLICIES, AUTOMATICALLY ALERTS YOUR DOCTOR IF SOMETHING IS WRONG AND CAN EVEN HELP YOU SAVE MONEY ON YOUR HEALTH-INSURANCE BILL...
The Social Medicare Twitter Implant®

THE THINGS WE DESIGN END UP DESIGNING US

DNA-World

By Niko Vegt

The DNA World map is an imaginative map of the emerging world of DNA-related technologies, application domains, opportunities, fears, risks and desires.

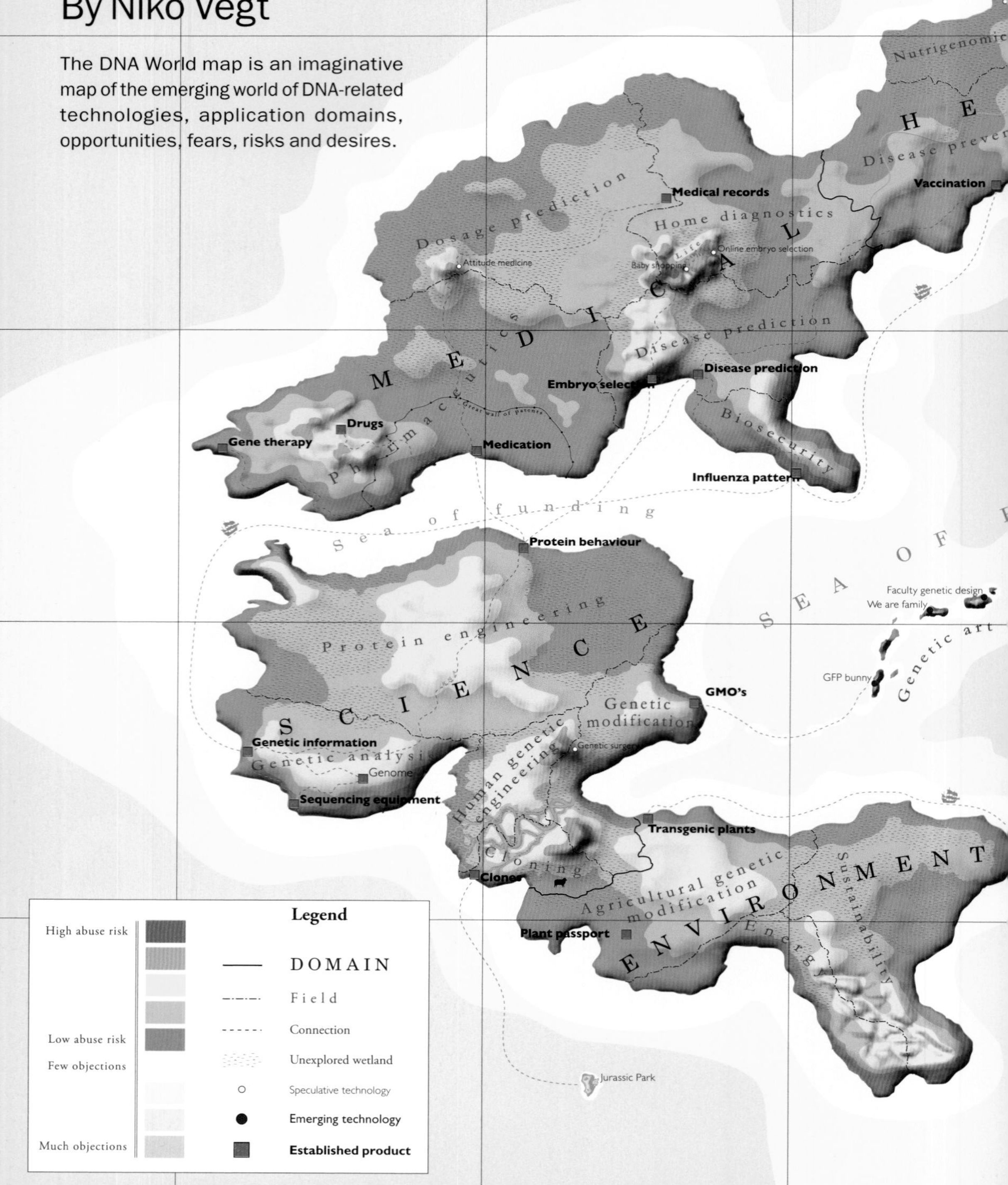

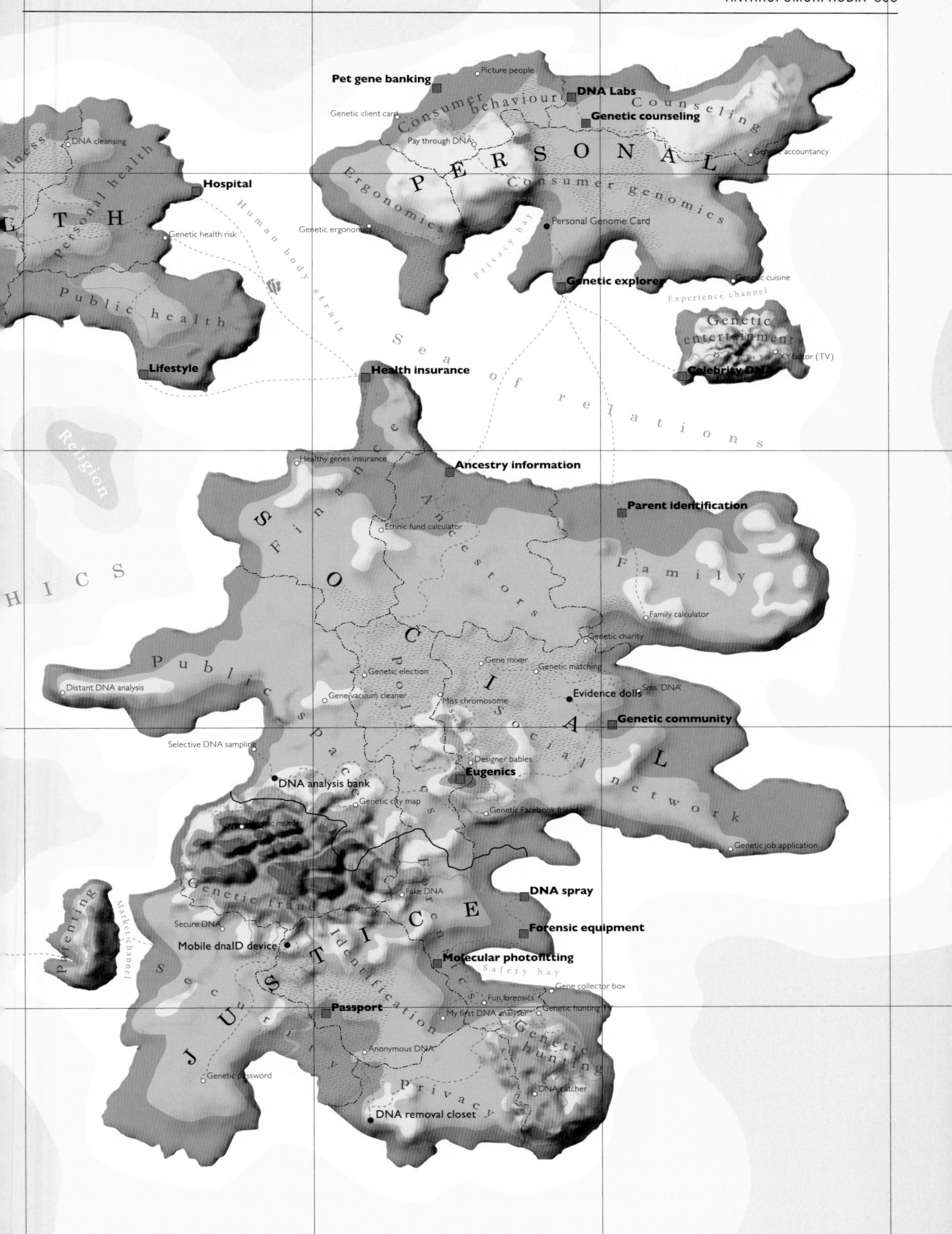

Pet gene banking
Picture people
DNA Labs
Genetic counseling
Genetic client card
Consumer behaviour
Counseling
Pay through DNA
PERSONAL
Genetic accountancy
Ergonomics
Consumer genomics
Genetic ergonomics
Personal Genome Card
Privacy bay
Genetic explorer
Genetic cuisine
Experience channel
Genetic entertainment
XY factor (TV)
Celebrity DNA
DNA cleansing
Personal health
Hospital
Human body strait
Genetic health risk
Public health
Lifestyle
Health insurance
Sea of relations
Religion
HICS
Healthy genes insurance
Ancestry information
Parent identification
Finance
Ancestors
Ethnic fund calculator
SOCIAL
Family
Family calculator
Genetic charity
Public space
Distant DNA analysis
Gene mixer
Genetic matching
Genetic election
Politics
Gene(vacuum) cleaner
Miss chromosome
Evidence dolls
Sms 'DNA'
Genetic community
Social network
Selective DNA sampling
Designer babies
Eugenics
DNA analysis bank
Genetic city map
Genetic Facebook friend
Genetic job application
Genetic fraud
Fake DNA
DNA spray
Forensic equipment
Patenting
Market channel
Secure DNA
Mobile dnaID device
JUSTICE
Identification
Forensics
Molecular photofitting
Safety bay
Gene collector box
Security
Fun forensics
Passport
My first DNA analyser
Genetic hunting TV
Genetic hunting
Anonymous DNA
Genetic password
Privacy
DNA catcher
DNA removal closet

SHOULD WE CLONE NEANDER-THALS?

By Zach Zorich

Revival of an evolutionary milestone or Frankenstein-style perversity? Cloning Neanderthals will bring us more ethical challenges than technological ones. Neanderthal cells could be the key to discovering treatments to diseases that are largely human-specific, such as HIV, polio, and smallpox. But, if we clone Neanderthals, will we also recreate their original habitat where they once lived?

If Neanderthals ever walk the earth again, the primordial ooze from which they will rise is an emulsion of oil, water, and DNA capture beads engineered in the laboratory of 454 Life Sciences in Branford, Connecticut. Over the past four years those beads have been gathering tiny fragments of DNA from samples of dissolved organic materials, including pieces of Neanderthal bone. Genetic sequences have given paleoanthropologists a new line of evidence for testing ideas about the biology of our closest extinct relative.

It would take about 10 million changes to make a modern human genome match the Neanderthal genome.

The first studies of Neanderthal DNA focused on the genetic sequences of mitochondria, the microscopic organelles that convert food to energy within cells. In 2005, however, 454 began a collaborative project with the Max Planck Institute in Leipzig, Germany, to sequence the full genetic code of a Neanderthal woman who died in Croatia's Vindija cave 30,000 years ago. As the Neanderthal genome is painstakingly sequenced, the archaeologists and biologists who study it will be faced with an opportunity that seemed like science fiction just 10 years ago. They will be able to look at the genetic blueprint of humankind's nearest relative and understand its biology as intimately as our own. In addition to giving scientists the ability to answer questions about Neanderthals' relationship to our own species – did we interbreed, are we separate species, who was smarter – the Neanderthal genome may be useful in researching medical treatments. Newly developed techniques could make cloning Neanderthal cells or body parts a reality within a few years. The ability to use the genes of extinct hominins is going to force the field of paleoanthropology into some unfamiliar ethical territory. There are still technical obstacles, but soon it could be possible to use that long-extinct genome to safely create a healthy, living Neanderthal clone. Should it be done?

At the 454 Life Sciences offices, Gerald Irzyk, Jason Affourtit, and Thomas Jarvie explain the process they use to read the chemicals that made up Neanderthal DNA and the genes that determined a large part of their biology. DNA has a shape, called a double helix, that makes it look like a twisted ladder. Each rung on the ladder is called a base-pair. The rungs are made up of a pair of chemicals called nucleotides–adenine, thymine, cytosine, and guanine, which are usually referred to by their first initials. The sequence of the nucleotides in the DNA determines what genes an organism has and how they function.

Although most of the Neanderthal genome sequencing is now being done by the San Diego-based company Illumina, the Max Planck Institute initially chose 454 because it had come up with a way to read hundreds of thousands of DNA sequences at a time. Genome-sequencing technology is advancing at a rate comparable to computer processing power. “Six years ago if you wanted to sequence E. coli [a species of bacteria], which is about four million base-pairs in length, it would have taken one or maybe two million dollars, and it would have taken a year and 150 people,” says Jarvie. “Nowadays, one person can do it in two days and it would cost a few hundred dollars.” Putting the fragments themselves in order can be a little tricky. “At first glance, it's just this completely random assemblage of As, Ts, Cs, and Gs,” says Irzyk. “But it turns out there are patterns and motifs, and sometimes these are very specific to a group of organisms.” For the Neanderthal sample, the human and chimpanzee genomes were used as references for checking the sequence.

Working with ancient DNA can be much more problematic than sequencing genetic material from living species. Within hours of death, cells begin to break down in a process called apoptosis. The dying cells release enzymes that chop up DNA into tiny pieces. In a human cell, this means that the entire three-billion-base-pair genome is reduced to fragments a few hundred base-pairs long or shorter. The DNA also goes through chemical changes that alter the nucleotides as it ages–C changes into T, and G turns into A–which can cause the gene sequence to be interpreted incorrectly. In the case of the Neanderthal sample, somewhere between 90 and 99 % of the DNA came from bacteria and other contaminants that had found their way into the bone as it sat in the ground and in storage. The contaminant DNA has to be identified and eliminated. Given the similarity between Neanderthal and modern human DNA, this can be especially difficult when the contamination comes from the people who excavated or analyzed the bone.

According to Stephan Schuster, a geneticist at the Pennsylvania State University, the first draft of the Neanderthal genome is likely to contain many errors. He estimates that getting a completely accurate DNA sequence will require taking five separate samples from the same individual, and sequencing that genome 30 times. Schuster sequenced the mammoth genome in 2007, and that approach might work for large animals, but taking five samples from a single Neanderthal would require the destruction of a large amount of valuable bone. Carles Lalueza-Fox, a paleogeneticist at Spain's University of Barcelona, believes the accuracy of the DNA could be checked by resequencing dozens or hundreds of times the areas of the Neanderthal genome that seem likely to have errors. Cloning a Neanderthal will take a lot more than just an accurately reconstructed genome. Artificially assembling an exact copy of the Neanderthal DNA sequence could be done easily and cheaply with current technology, but a free-floating strand of DNA isn't much good to a cell. "The bigger challenge is – how do you assemble a genome without a cell?" asks James Noonan, a geneticist at Yale University. "How do you package DNA into chromosomes, and get that into a nucleus? We don't know how to do that." The shape of the DNA within the chromosomes affects the way that genes interact with chemicals inside the cell. Those interactions control when, how much, and what types of proteins a cell's DNA produces. Those proteins are the building blocks of an organism, so the way a genome expresses itself is as important as the DNA. According to Schuster and Lalueza-Fox, the cellular damage that occurs after death makes it impossible to understand Neanderthal gene expression. This could mean that making a clone identical to someone who lived 30,000 years ago is impossible.

One way to get around the problems of working with an artificial genome would be to alter the DNA inside a living cell. This kind of genetic engineering can already be done, but very few changes can be made at one time. To clone a Neanderthal, thousands or possibly millions of changes would have to be made to a human cell's DNA. George Church, a professor of genetics at Harvard Medical School, is part of a research team that is developing a technique to make hundreds of alterations to a genome at the same time. The technique, multiplex automated genome engineering (MAGE), uses short strands of DNA called oligonucleotides to insert pieces of artificial genetic material into a cell's genome at specifically targeted sites. MAGE has been used successfully to make 24 alterations to the genomes of bacteria, mice, and, more recently, human cells. Church estimates that it would take about 10 million changes to make a modern human genome match the Neanderthal genome. Accomplishing this would be a matter of drastically scaling up the technique.

Neanderthal cells could be important for discovering treatments to diseases that are largely human-specific, such as HIV and polio.

Church believes the place to start with Neanderthal cloning is on the cellular level, creating liver, pancreas, or brain cells. "You can't really tell anything from just looking at the gene sequence," he says. "It's hard to predict physical traits; you have to test them in living cells." Neanderthal cells could be important for discovering treatments to diseases that are largely human-specific, such as HIV, polio, and smallpox, he says. If Neanderthals are sufficiently different from modern humans, they may have a genetic immunity to these diseases. There may also be differences in their biology that lead to new drugs or gene therapy treatments. So far, efforts to revive extinct species using cloning have a dismal track record. On January 6, 2000, a violent storm in northern Spain caused a tree branch to fall on Celia, the last Pyrenean ibex, crushing her skull. That would seem like a clear indication that the ibex's evolutionary luck had run out, but a tissue sample taken from Celia's ear provided DNA that a team of Spanish scientists used to reconstruct 439 eggs.

Only 57 developed into embryos, 52 did not survive the full term of the pregnancy, four were stillborn, and the one clone that survived birth died of lung failure within hours of delivery.

Even if nuclear transfer cloning could be perfected, it would likely require a horrifying period of trial and error.

The ibex clones were created using techniques pioneered by Advanced Cell Technology, a biotechnology company in Worcester, Massachusetts. The technique, called nuclear transfer, involves removing the nucleus, the part containing the cell's genetic material, of a donor egg cell and replacing it with a nucleus containing clone DNA. In the ibex's case, goat eggs were used because the species are closely related and goats have been successfully cloned many times, explains Robert Lanza, ACT's chief scientific officer. According to Lanza, species such as cows and goats are now routinely cloned with few problems.

Species that have not been repeatedly cloned still face risks. The nuclear transfer process disrupts the cell and often causes it to die. The number of sick and dead individuals produced by nuclear transfer cloning is the reason nearly all scientists are opposed to human reproductive cloning. But even if nuclear transfer cloning could be perfected in humans or Neanderthals, it would likely require a horrifying period of trial and error. There is, however, another option.

The best way to clone Neanderthals may be to create stem cells that have their DNA. In recent years, geneticists have learned how to take skin cells and return them to a state called pluripotency, where they can become almost any type of cell in the human body. Church proposes to use the MAGE technique to alter a stem cell's DNA to match the Neanderthal genome. That stem cell would be left to reproduce, creating a colony of cells that could be programmed to become any type of cell that existed in the Neanderthal's body. Colonies of heart, brain, and liver cells, or possibly entire organs, could be grown for research purposes. This technique could also be used to create a person. A stem cell with Neanderthal DNA could be implanted in a human blastocyst–a cluster of cells in the process of developing into an embryo. Then, all of the non-Neanderthal cells could be kept from growing. The individual who developed from that blastocyst would be entirely the result of Neanderthal genes. In effect, it would be a cloned Neanderthal. Church believes that after the earliest stages of development, the genes would express themselves as they did in the original individual, eliminating any influences from the modern human or chimpanzee cell.

The technique is new, and has only been tested in mice so far, but Church thinks it might work in humans. However, he points out that anyone cloned by this process would still be lacking the environmental and cultural factors that would have influenced how the original Neanderthals grew up. "They would be something new," Church says, "neo-Neanderthals."
In northern Spain 49,000 years ago, 11 Neanderthals were murdered. Their tooth enamel shows that each of them had gone through several periods of severe starvation, a condition their assailants probably shared. Cut marks on the bones indicate the people were butchered with stone tools. About 700 feet inside El Sidrön cave, a research team including Lalueza-Fox excavated 1,700 bones from that cannibalistic feast. Much of what is known about Neanderthal genetics comes from those 11 individuals.

Lalueza-Fox does not plan to sequence the entire genome of the El Sidrön Neanderthals. He is interested in specific genes. "I choose genes that are somehow related to individuality," he says. "I'd like to create a personal image of these guys." So far, his work has shown that Neanderthals had a unique variant of the gene for pale skin and red hair, which may mean their skin and hair color differed from modern humans. Lalueza-Fox tested the blood types of two Neanderthals and found they were both type O. He also discovered that modern humans and Neanderthals share a version of a gene called FOXP2, which is associated with language ability and means that Neanderthals probably spoke their own languages.

The Neanderthals broke away from the lineage of modern humans around 450,000 years ago. They evolved larger brains and became shorter than their likely ancestor, Homo heidelbergensis. They also developed a wider variety of stone tools and more efficient techniques for making them. On average, Neanderthals had brains that were 100 cubic centimeters (about three ounces) larger than those of people living today. But those differences are likely due to their larger overall body size. Those large brains were housed inside skulls that were broader and flatter, with lower foreheads than modern humans. Their faces protruded forward and lacked chins. Their arms and the lower part of their legs were shorter than modern humans', making them slower and less efficient runners, but they also had more muscle mass. Their bones were often thicker and stronger than ours, but they typically show a lot of healed breaks that are thought to result from hunting techniques requiring close contact with large game such as bison and mammoths. They had barrel-shaped chests and broad, projecting noses, traits some paleoanthropologists believe would have helped Neanderthals breathe more easily when chasing prey in cold environments. Recent studies comparing Neanderthal and modern human anatomy have created some surprising insights. "Neanderthals are not just sort of funny Eskimos who lived 60,000 years ago," says Jean-Jacques Hublin, a paleoanthropologist at Max Planck. "They have a different way to give birth to babies, differences in life history, shape of inner ear, genetics, the speed of development of individuals, weaning, age of puberty." A study comparing Neanderthal and modern children showed Neanderthals had shorter childhoods. Some paleoanthropologists believe they reached physical maturity at age 15.

As different as Neanderthals were, they may not have been different enough to be considered a separate species. "There are humans today who are more different from each other in phenotype [physical characteristics]," says John Hawks, a paleoanthropologist at the University of Wisconsin. He has studied differences in the DNA of modern human populations to understand the rate of evolutionary change in *Homo sapiens*. Many of the differences between a Neanderthal clone and a modern human would be due to genetic changes our species has undergone since Neanderthals became extinct. "In the last 30,000 years we count about 2,500 to 3,000 events that resulted in positive functional changes [in the human genome]," says Hawks. Modern humans, he says, are as different from *Homo sapiens* who lived in the Neolithic period 10,000 years ago, as Neolithic people would have been from Neanderthals. Clones created from a genome that is more than 30,000 years old will not have immunity to a wide variety of diseases, some of which would likely be fatal. They will be lactose intolerant, have difficulty metabolizing alcohol, be prone to developing Alzheimer's disease, and maybe most importantly, will have brains different from modern people's.

Clones will be lactose intolerant, have difficulty metabolizing alcohol, be prone to developing Alzheimer's disease, and will have brains different from modern people's.

Bruce Lahn at the University of Chicago studies the evolutionary history of the genes that control human brain development. One gene that affects brain size particularly interests him, a variant of the microcephalin gene, which Lahn thinks may have entered the human gene pool through interbreeding with Neanderthals. If that turns out to be true, roughly 75 % of the world's population has a brain gene inherited from Neanderthals. Lahn is excited to see what the Neanderthal microcephalin gene sequence looks like. "Is the Neanderthal sequence more similar to the ancestral version or the newer, derived version of the gene?" Lahn asks. "Or is the Neanderthal yet a third version that is very different from either of the two human versions? No matter how you look at it, it makes that data very interesting."

The Neanderthals' brains made them capable of some impressive cultural innovations. They were burying their dead as early as 110,000 years ago, which means that they had a social system that required formal disposal of the deceased. Around 40,000 years ago, they adopted new stone-tool-making traditions, the Châttelperronian tradition in Western Europe and the Uluzzian in Italy, that included a greater variety of tools than they had used in hundreds of thousands of years. But even if they were as adaptable as *Homo sapiens*, the question remains – if they were so smart, why are they dead? Chris Stringer of London's Natural History Museum believes our species hunted and gathered food so intensively that there simply was not enough room for the Neanderthals to make a living. In other words, they had the same problem as many species facing extinction today – they were crowded out of their ecological niche by *Homo sapiens*. Finding a place in the world for a Neanderthal clone would be only one dilemma that would have to be solved.

How much does a human genome need to be changed before the individual created from it is no longer considered human?

Bernard Rollin, a bioethicist and professor of philosophy at Colorado State University, doesn't believe that creating a Neanderthal clone would be an ethical problem in and of itself. The problem lies in how that individual would be treated by others. "I don't think it is fair to put people...into a circumstance where they are going to be mocked and possibly feared," he says, "and this is equally important, it's not going to have a peer group. Given that humans are at some level social beings, it would be grossly unfair." The sentiment was echoed by Stringer, "You would be bringing this Neanderthal back into a world it did not belong to...It doesn't have its home environment anymore."

There were no cities when the Neanderthals went extinct, and at their population's peak there may have only been 10,000 of them spread across Europe. A cloned Neanderthal might be missing the genetic adaptations we have evolved to cope with the world's greater population density, whatever those adaptations might be. But, not everyone agrees that Neanderthals were so different from modern humans that they would automatically be shunned as outcasts.

"I'm convinced that if one were to raise a Neanderthal in a modern human family he would function just like everybody else," says Trenton Holliday, a paleoanthropologist at Tulane University. "I have no reason to doubt he could speak and do all the things that modern humans do."

"I think there would be no question that if you cloned a Neanderthal, that individual would be recognized as having human rights under the Constitution and international treaties," says Lori Andrews, a professor at Chicago-Kent College of Law. The law does not define what a human being is, but legal scholars are debating questions of human rights in cases involving genetic engineering. "This is a species-altering event," says Andrews, "it changes the way we are creating a new generation." How much does a human genome need to be changed before the individual created from it is no longer considered human?

Legal precedent in the United States seems to be on the side of Neanderthal human rights. In 1997, Stuart Newman, a biology professor at New York Medical School attempted to patent the genome of a chimpanzee-human hybrid as a means of preventing anyone from creating such a creature. The patent office, however, turned down his application on the basis that it would violate the Constitution's 13th amendment prohibition against slavery. Andrews believes the patent office's ruling shows the law recognizes that an individual with a half-chimpanzee and half-human genome would deserve human rights. A Neanderthal would have a genome that is even more recognizably human than Newman's hybrid. "If we are going to give the Neanderthals humans rights... what's going to happen to that individual?" Andrews says. "Obviously, it won't have traditional freedoms. It's going to be studied and it's going to be experimented on. And yet, if it is accorded legal protections, it will have the right to not be the subject of research, so the very reasons for which you would create it would be an abridgment of rights."

Human rights laws vary widely around the world. "There is not a universal ban on cloning," says Anderson. "Even in the United States there are some states that ban it, others that don't." On August 8, 2005, the United Nations voted to ban human cloning. It sent a clear message that most governments believe that human cloning is unethical. The ban, however, is non-binding. The legal issues surrounding a cloned Neanderthal would not stop with its rights. Under current laws, genomes can be patented, meaning that someone or some company could potentially own the genetic code of a long-dead person. Svante Pääbo, who heads the Neanderthal genome sequencing project at Max Planck, refused to comment for this article, citing concerns about violating an embargo agreement with the journal that is going to publish the genome sequence. But he did send this statement: "We have no plans to patent any of the genes in the Neanderthal."

The ultimate goal of studying human evolution is to better understand the human race. The opportunity to meet a Neanderthal and see firsthand our common but separate humanity seems, on the surface, too good to pass up. But what if the thing we learned from cloning a Neanderthal is that our curiosity is greater than our compassion? Would there be enough scientific benefit to make it worth the risks?

"I'd rather not be on record saying there would," Holliday told me, laughing at the question. "I mean, come on, of course I'd like to see a cloned Neanderthal, but my desire to see a cloned Neanderthal and the little bit of information we would get out of it...I don't think it would be worth the obvious problems." Hublin takes a harder line. "We are not Frankenstein doctors who use human genes to create creatures just to see how they work." Noonan agrees, "If your experiment succeeds and you generate a Neanderthal who talks, you have violated every ethical rule we have," he says, "and if your experiment fails...well. It's a lose-lose." Other scientists think there may be circumstances that could justify Neanderthal cloning.

"If we could really do it and we know we are doing it right, I'm actually for it," says Lahn. "Not to understate the problem of that person living in an environment where they might not fit in. So, if we could also create their habitat and create a bunch of them, that would be a different story."

"We could learn a lot more from a living adult Neanderthal than we could from cell cultures," says Church. Special arrangements would have to be made to create a place for a cloned Neanderthal to live and pursue the life he or she would want, he says. The clone would also have to have a peer group, which would mean creating several clones, if not a whole colony. According to Church, studying those Neanderthals, with their consent, would have the potential to cure diseases and save lives. The Neanderthals' differently shaped brains might give them a different way of thinking that would be useful in problem-solving. They would also expand humanity's genetic diversity, helping protect our genus from future extinction. "Just saying 'no' is not necessarily the safest or most moral path," he says. "It is a very risky decision to do nothing."

But what if the thing we learned from cloning a Neanderthal is that our curiosity is greater than our compassion?

Hawks believes the barriers to Neanderthal cloning will come down. "We are going to bring back the mammoth... the impetus against doing Neanderthal because it is too weird is going to go away." He doesn't think creating a Neanderthal clone is ethical science, but points out that there are always people who are willing to overlook the ethics. "In the end," Hawks says, "we are going to have a cloned Neanderthal, I'm just sure of it."

THIS ARTICLE BY ZACH ZORICH ORIGINALLY APPEARED IN ARCHEOLOGY MAGAZINE

これらのレンズは衣裳プレーヤーおよび衣裳党のためのある普及した生気の特性の特徴の目を特に実現する。

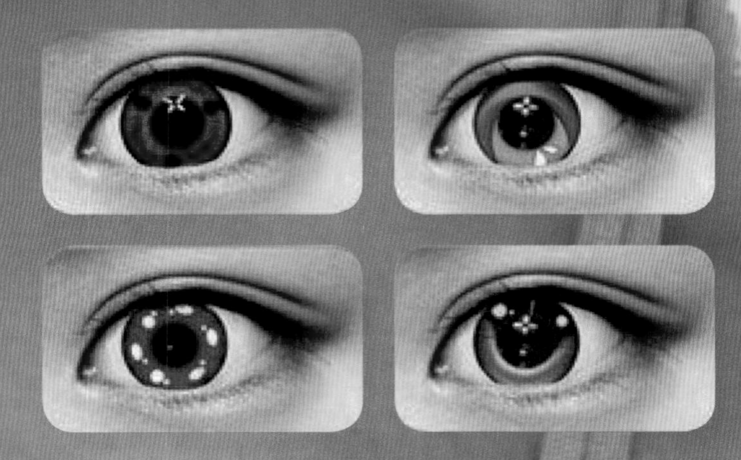

#6 ANTHROPOMORPHOBIA

wwww.nextnature.net/
themes/anthropomorphobia

NEXT NATURE

BACK TO THE TRIBE

ISBN 978-84-92861-53-8
9 788492 861538

FOR THOUSANDS OF YEARS, PEOPLE LIVED IN SMALL TRIBAL SETTINGS.

ACCORDING TO ANTHROPOLOGIST ROBIN DUNBAR, A TRIBE'S MAXIMUM SIZE IS SET BETWEEN 100 AND 250 PEOPLE. THIS NUMBER REPRESENTS THE UPPER LIMIT OF THE BRAIN'S ABILITY TO KEEP TRACK OF PERSONALITIES, SOCIAL RELATIONSHIPS, AND WHO PREFERS GIRAFFE TO WILDEBEEST FOR DINNER.

THE INVENTION OF MORE COMPLEX SOCIAL STRUCTURES IS RELATIVELY NEW.

AGRICULTURE HAS BEEN IN EXISTENCE FOR ABOUT 10,000 YEARS, AND CITIES WITH OVER ONE MILLION INHABITANTS HAVE ONLY BEEN AROUND FOR 2,000 YEARS. HUMANS HAVE RE-MADE THEIR SOCIETIES AND HABITATS IN THE BLINK OF AN EVOLUTIONARY EYE. HAVE OUR GENOMES HAD TIME TO CATCH UP?

GENETICALLY WE HAVEN'T CHANGED: WE ARE STILL ATTUNED TO A TRIBAL WAY OF LIVING.

WE'RE NOT AS WELL ADAPTED TO MODERNITY AS WE THINK WE ARE. WE WORK AT DESKS AND COMMUTE BY CAR, BUT OUR BODIES THRIVE ON HARD PHYSICAL ACTIVITY. ANCIENT APPETITES FOR ENERGY-RICH FOODS CAUSE US TO BINGE ON SUGAR, FAT, AND SALT. WE CAN BLAME HEART DISEASE, DIABETES AND ALLERGIES ON A GENOME THAT STILL THINKS IT'S LIVING IN A FAMINE-PRONE SAVANNAH.

**DIFFERENT TOTEMS.
SAME RITUALS.**

DIFFERENT TOTEMS, SAME RITUALS THE BOUNDARIES BETWEEN BRAND IDENTITY AND THE TRIBAL IDENTITY ARE MURKY, IF THEY WERE EVEN SOLID TO BEGIN WITH. TODAY'S CORPORATE BRANDS AND LOGOS MAY BE AS RICH IN CULTURAL ASSOCIATIONS AS THE RITUAL MASKS OF ANCIENT TRIBES. TECHNOLOGY CHANGES OUR CULTURAL TRAPPINGS, BUT IT DOES NOT CHANGE THE TRIBAL IMPULSES THAT GIVE RISE TO THEM.

SUCCESSFUL NEW TECHNOLOGIES OFTEN TRIGGER ANCIENT IMPULSES.

THE PRINTING PRESS DOMINATED HUMAN PSYCHOLOGICAL AND SOCIAL SPACE FOR ONLY A FEW HUNDRED YEARS. NOW TEXT MESSAGES, TWEETS AND SKYPE CALLS HAVE RETURNED US TO OUR TRIBAL DAYS. DIGITAL TECHNOLOGIES HAVE PLACED US IN THE IMMERSIVE, INSTANTANEOUS, AND HIGHLY SOCIAL WORLD OF THE TRIBE. THE FLICKER OF THE COMPUTER REPLACES THE FLICKER OF THE CAMP FIRE.

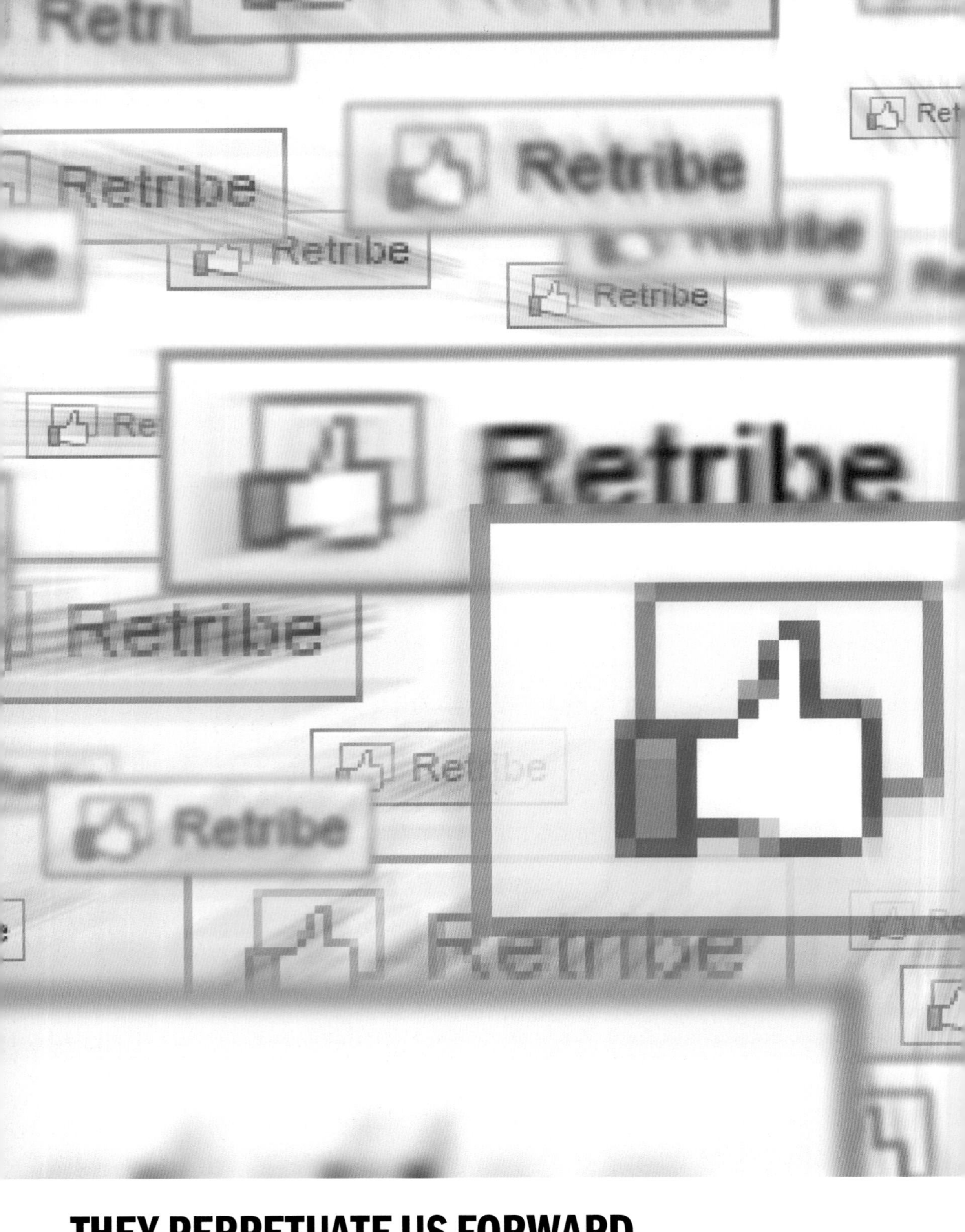

THEY PERPETUATE US FORWARD, RATHER THAN BACK TO NATURE.

DIGITAL COMMUNICATIONS ARE NOT REGRESSIVE; INDUSTRIALIZED SOCIETIES WILL NOT BE RETURNING TO THE HUNT ANYTIME SOON. INSTEAD, OUR RE-TRIBALIZED CULTURE IS A HYBRID OF ANCIENT AND ADVANCED INSTINCTS. FACEBOOK EXPANDS OUR TRIBES ACROSS DISTANCES UNIMAGINABLE TO OUR ANCESTORS. WE SHARE OUR BEDS WITH LOVERS AND LAPTOPS ALIKE. HUMAN NATURE HAS BEEN ABSORBED INTO NEXT NATURE.

BACK TO THE TRIBE: FORWARD TO NATURE

By Koert van Mensvoort

The internet is high-tech, but its social effects are decidedly retro. Are Twitter, Facebook and other social networking platforms triggering a second rise of the tribal lifestyle?

Traditionally, technology is seen as a force that diminishes our instincts and distances us from nature. Increasingly however, we realize technology can also energize and amplify our deepest human sensibilities – even some we had forgotten about. Propelling us not so much back to, but rather forward to nature.

Almost two decades ago, Brian Eno – artist, composer, inventor, thinker – gave an interview in which he stated the problem with computers was that there is not enough Africa in them.[1]

"Africa is everything that something like classical music isn't. Classical - perhaps I should say 'orchestral' - music is so digital, so cut up, rhythmically, pitch wise and in terms of the roles of the musicians. It's all in little boxes." [...] "Do you know what a nerd is? A nerd is a human being without enough Africa in him or her. I know this sounds sort of inversely racist to say, but I think the African connection is so important. I want so desperately for that sensibility to flood into these other areas, like computers." [...] "It uses so little of my body. You're just sitting there, and it's quite boring. You've got this stupid little mouse that requires one hand, and your eyes. That's it. What about the rest of you? No African would stand for a computer like that. It's imprisoning."

Twenty years ago, when Eno gave his interview, no one had a mobile phone. Today when you accidentally leave your house without your phone you feel amputated – as if you left a limb on the table – and you quickly run back into your house to get it. Social software networks like Facebook, MySpace, Qzone and Twitter reached the mainstream at an even faster pace. All these communication technologies have one thing in common: they restructure the social linkage between people. Arguably they bring a taste of Africa back to computing.

Re-tribalization

Study of the history of mankind shows that, for thousands of years, people lived in bands or tribal settings of no more than 150 people.[2] The invention of larger and more complex social structures – cities, corporations, nations, etc – is relatively new. Although we have proven able to live in more complex social settings, our tribal sensibilities were never entirely washed away. Examples? Think of what happens during a football championship, or the role of fashion brands in defining our identities. As the tribal setting is the structure in which mankind evolved, it's only logical that we still have a tendency towards it.

The incorporation of an older nature within a next nature, is a powerful evolutionary principle.

If we consider the parallels between newly emerging communication technologies and a tribal way of living, some striking similarities occur. In a tribal setting, your identity is entirely wrapped up in the question of how people know you. Looking at a social network like Facebook, we see the same pattern at work. People are shaping their identities by exhibiting their relationships to each other. By scrawling messages on each other's walls and exchanging totem-like visual symbols, you define yourself in terms of who your friends are. Media theorist Marshall McLuhan was the first to envision the re-tribalizing powers of electronic technology. According to McLuhan it was the phonetic alphabet and the printing press that caused a mechanical culture of industrial production and nation states, which consequentially resulted in the detribalization of Western man into a linear, specialized and detached professional. The introduction of the electronic media however, which saturates our sensory perception entirely, was elucidated by McLuhan as a 'break boundary' between the fragmented literate man and integral man, just as phonetic literacy was a break boundary between oral-tribal man and literate man.[3]

Secondary Orality

The growth of social networks and the Internet as a whole can be largely contributed to an outpouring of expression that feels more like 'talking' than writing: blog posts, comments, videos responses, tweets and status updates. We seem to be making up the rules as we go, but is this really the case? Researchers have been exploring the parallels between online social networks and tribal societies. In the collective pit-a-pat of profile-peeking, messaging and 'friending,' they see the resurgence of ancient patterns of oral communication.[4] An early student of electronic orality was Walter J. Ong, a professor at St. Louis University and former student of Marshall McLuhan. Ong coined the term 'secondary orality' in 1982 to describe the tendency of electronic media to echo the cadences of earlier oral cultures.[5] Oral cultures were characterized as being aggregative rather than analytic, additive rather than subordinate, close to the human lifeworld, redundant or 'copious,' conservative or traditionalist and more situational and participatory than the more detached and abstract literate cultures. Oral cultures operate on polychronic time, with many things happening at once. Socialization plays a great role.

New technologies, ancient impulses

Secondary orality is similar yet different from the original oral cultures, as it presumes and is dependent upon writing and digital technology. The revival of an older nature within a next nature – in order to eventually transform and supersede it – is a powerful evolutionary principle. Although the power of the newly emerging digital tribes lies in the revival of some deeply rooted human sensibilities, they are literally of a different nature than the ancient tribes they resemble. Their primary foundation is not the human social intuitions engraved in our DNA, but the digitalism of the database is the primary foundation they are built on. Hence, the next tribes are not so much about being tribal, as they are about being digital. The result is a marriage between old and new, between ancient and alien.

The power of the newly emerging digital tribes lies in the revival of some deeply rooted human sensibilities.

In tribal societies, people define their bond through direct, ongoing face-to-face contact. On the Internet, people connect for a variety of reasons ranging from family ties, to life-long friendships, to mutual interests to we-haven't-met-but-it-seems-cool-to-have-you-in-my-tribe. While traditionally one would belong to only one tribe, you are now linked into tightly knitted network of tribes that together constitute what McLuhan already called the 'global village'. And besides the dependence on digital technology, the next tribes are typically facilitated by corporations with commercial incentives that aren't necessarily geared at the wellbeing of the tribe members. Nonetheless, their brand territories are strong and may soon be competing with the geographical borders of countries.

It remains to be seen whether these next tribes are as durable as an organizational infrastructure. Will they replace nation states – a product of a diminishing print and writing culture – in due time? If so, will they be able to replace their functioning? Provide for our security? Democracy? Provide for public spaces and freedom of speech? Turn our living space into a shopping mall? Or bring totalitarian regimes? The question marks are numerous. Nonetheless, we rush to join. Why? Some of the most important people we know have joined. We intuitively don't want to miss out on the opportunity to reconnect with them in what could be the next social setting. Indeed, the success of a new technology often depends on its capacity to trigger an ancient impulse. Back to the tribe, forward to a next nature.

REFERENCES

1 KELLY, KEVIN (1995) GOSSIP IS PHILOSOPHY, INTERVIEW WITH BRIAN ENO, WIRED MAGAZINE, 03.05 MAY 1993.
2 DUNBAR, R.I.M. (1992). "NEOCORTEX SIZE AS A CONSTRAINT ON GROUP SIZE IN PRIMATES". JOURNAL OF HUMAN EVOLUTION 22 (6): 469-493.
3 NORDEN, ERIC (1969) PLAYBOY INTERVIEW: MARSHALL MCLUHAN, PLAYBOY: PP. 26-27, 45, 55-56, 61, 63, MARCH 1969.
4 WRIGHT, ALEX (2007) FRIENDING, ANCIENT OR OTHERWISE, NEW YORK TIMES, 2 DECEMBER 2007
5 ONG, WALTER J. (1982) ORALITY AND LITERACY: THE TECHNOLOGIZING OF THE WORD, (NEW. YORK: METHUEN, 1982)

NEW TECHNOLOGY TRIGGERS ANCIENT IMPULSES

ElectricMan Mar
After centuries
awareness is onc
42 years ago

ElectricMan Marshall McLuhan
Before the invention of the phonetic alphabet, man lived in a world where all the senses were balanced and simultaneous ...
ears ago

ElectricMan Mar
The family of m
the nation back
#tribalexistences.
42 years ago

ElectricMan Marshall McLuhan
For the past 3,500 years, the effects of media – whether it's speech, writing, printing, photography or television – have been overlooked by social observers ...
42 years ago

ElectricMan Marshall McLuhan
... an oral culture structured by a dominant auditory sense of life.
42 years ago

ElectricMa
Electronic
42 years ag

ElectricMan Marshall McLuhan
If the phonetic alphabet fell like a bombshell on #tribalman, the printing press hit him like a 100-megaton H-bomb.
42 years ago

ElectricMan Marshall McLuha
Type, the prototype of all m
ensured the primacy of the visua
finally sealed the doom of #trib
42 years ago

ElectricMan Marshall McLuhan
Most people cling to what I call the rearview-mirror view of their world.
42 years ago

ElectricMan Marshall McLuhan
Phonetic literacy propelled man from the #tribe. It gave him an eye for an ear.
42 years ago

ElectricMan Marshall McLuha
The typographic trance of t
endured until today, when t
media are at last demesmerizing us
42 years ago

ElectricMan Marshall McLuhan
Electronic media constitute a break boundary between fragmented man and integral man ...
42 years ago

ElectricMan Marshall McLuhan
... as phonetic literacy was a break boundary between #oral-tribalman and visual man.
42 years ago

ElectricMan Marshall McLuhan
An electrically imploded #tribal society discards the linear forward-motion of "progress."
42 years ago

ElectricMan Marshall McLuhan
Electric media bring man together in a #tribal village, where there is more room for diversity than within the homogenized society of Western man.
42 years ago

McLuhan's Tweets

As the new media theorist would say, "the medium (is more important than) the message." Twitter may be text, but it's really an oral and auditory system in disguise: instantaneous, encompassing, and social. As McLuhan would argue, micro-blogging is just another symptom of a society moving back to the tribe. Adapted from "The Playboy Interview: Marshall McLuhan," *Playboy Magazine*, March 1969.

Luhan
ciated sensibilities, human
ecoming integral and inclusive

ElectricMan Marshall McLuhan
Modern man's relationship to a computer is not by nature very different from prehistoric man's relationship to a boat or a wheel.
42 years ago

ElectricMan Marshall McLuhan
The Eskimo is a servomechanism of his kayak, the cowboy of his horse, the businessman of his clock, the cybernetician of his computer.
42 years ago

Luhan
ing from
titudinous

ElectricMan Marshall McLuhan
Our own Western time-space concepts derive from the environment created by phonetic writing.
42 years ago

ll McLuhan
n't enlarge social groups, they decentralize them.

ElectricMan Marshall McLuhan
Television is primarily responsible for ending the visual supremacy that characterized all mechanical technology...
42 years ago

d

ElectricMan Marshall McLuhan
I feel we're standing on the threshold of a liberating and exhilarating world in which the human #tribe can become truly one family
42 years ago

ElectricMan Marshall McLuhan
... it's creating a totally new type of national leader, a man who is much more of a #tribalchieftain than a politician.
42 years ago

as
onic

ElectricMan
Marshall McLuhan
The world is 'all-at-once.' Everything resonates with everything else in a total electrical field.
42 years ago

ElectricMan Marshall McLuhan
Now man is beginning to wear his brain outside his skull and his nerves outside his skin.
42 years ago

ElectricMan Marshall McLuhan
Automation will end the traditional concept of the job, replacing it with a role, and giving men the breath of leisure :)
42 years ago

ElectricMan Marshall McLuhan
Man becomes the sex organs of the machine world just as the bee is of the plant world, permitting it to reproduce and evolve to higher forms.
42 years ago

ElectricMan Marshall McLuhan
Literate man is alienated, impoverished man; #retribalized man can lead a far richer and more fulfilling life.
42 years ago

ElectricMan Marshall McLuhan
You could as easily ban drugs in a #retribalized society as outlaw clocks in a mechanical culture.
42 years ago

ElectricMan Marshall McLuhan
Electric technology is retrogressing Western man from the open plateaus of literate values into the heart of #tribal darkness, into what @JosephConrad termed "the Africa within."
42 years ago

ElectricMan
Marshall McLuhan
by NextNature.net
New technology breeds new man - and makes new man #tribal again.
just a minute ago

Brand New Worlds

DIFFERENT TECHNOLOGIES, DIFFERENT SHAPES OF NATION STATES: NATION STATE SHAPED BY THE GEOGRAPHICAL ENVIRONMENT; THE INFORMATION STATE, SHAPED BY MAPS; THE BRAND STATE, SHAPED BY SYMBOLISM AND MYTH?

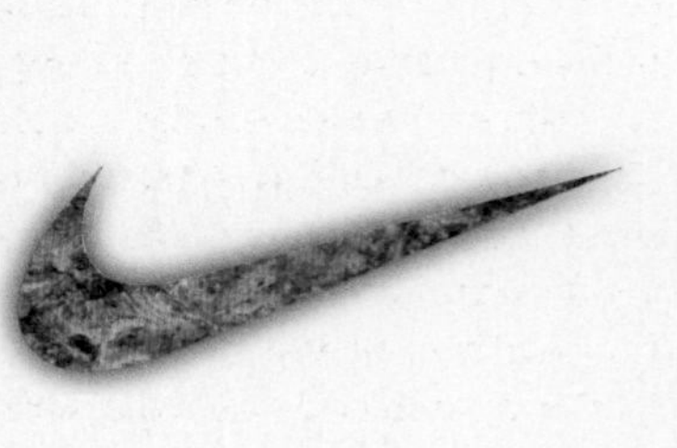

NOT ONLY ARE NATION STATES PRESENTING THEMSELVES MORE AND MORE AND MORE AS BRANDS, THE OPPOSITE IS ALSO TRUE. BRANDS PRESENT THEMSELVES AS COUNTRIES, SOMETIMES LITERALLY. AT THE SHANGHAI EXPO IN 2010, COCA-COLA HAD ITS OWN PAVILLION AND PLACED ITSELF AMONGST COUNTRY PAVILLIONS, CAREFULLY BLENDING TRADITIONAL CHINESE IMAGERY WITH BRAND VISUALS.

Brand Territories

Entering the store or restaurant of a multinational corporation is like entering a country. The locals all wear a national uniform, greet you in the same language, and serve you an ethnic meal. Any nods to local culture – the McArabia, the Croque McDo, the McRice – are insignificant compared to the consistency of the brand experience. Any McDonald's, be it in Manila or Moscow, is a sovereign embassy from the same corporate homeland. They exist in the 'brand space,' a place as much a state of mind as a physical location. Travelers often visit a corporate outpost to feel 'at home,' though they are homesick for the brand, not for their own country. One day, a Starbucks employee may come up to you and ask for your visa while you plug in your laptop and wait to buy a latte. | AG

Corporate Clans

With each purchase you buy your way into a corporation's clan. Our consumer choices are not just about our personal tastes, but also about our personalities. As tribal markers, corporate logos are just as recognizable to their customers as a warrior's face paint was to his enemies. Unsurprisingly we often inherit brand identities from our families or adopt them from our peer groups.

We consciously mark ourselves with branded clothing, electronics, and foods. But before the meaning expanded to include proprietary goods, 'brand' once meant a mark made by a hot iron. It only makes sense to return to the original meaning of the word by tattooing a permanent tribute to your favorite company on your skin-the ultimate symbol of tribal loyalty.

Brands and logos have become our global Esperanto. The dating website Branddating.nl already relies on the identification people have with brands. It replaces characteristics like sporty, spontaneous and funny with brands like Apple, Starbucks and Camel. No need to describe yourself as a chic nerd when your Apple computer already defines your tribe. Male - BMW, Armani, Durex - is looking for a Female – Dolce & Gabbana, New York Times, Victoria's Secret. However, if Brand Dating is successful, it could lead to some awkward outcomes for the clan members. Children of mixed-brand couples may struggle with their 'ethnic' identities. What soda will half Coca-Cola, half Pepsi children choose? With soft drinks as with human races, the packaging may look different, but the contents are nearly identical. | AG

Daring Logos Make Us Believe We Are Connected or Separate from Each Other

TRIBE COLORS
MEL GIBSON'S SCOTTISH TRIBES IN BRAVEHEART SHARE THE SAME COLOR SCHEME (AND SOCIAL PATTERNS) AS FACEBOOK. A FAN OF DR. PEPPER CAN MAKE A LASTING TRIBUTE TO THE BRAND, BUT THE MARRIAGE BETWEEN A COKE MAN AND A PEPSI LADY MIGHT PRESENT A PROBLEM. WILL THEIR CHILDREN DRINK *COKESI* OR *PEPOKE*?

COAT: 796.987,20

THIS COAT FROM SILKE WAWRO USES THE RECYCLED LABELS FROM OTHER CLOTHING TO PROCLAIM A COSMOPOLITAN AFFILIATION TO ALL BRANDS – OR TO NONE.

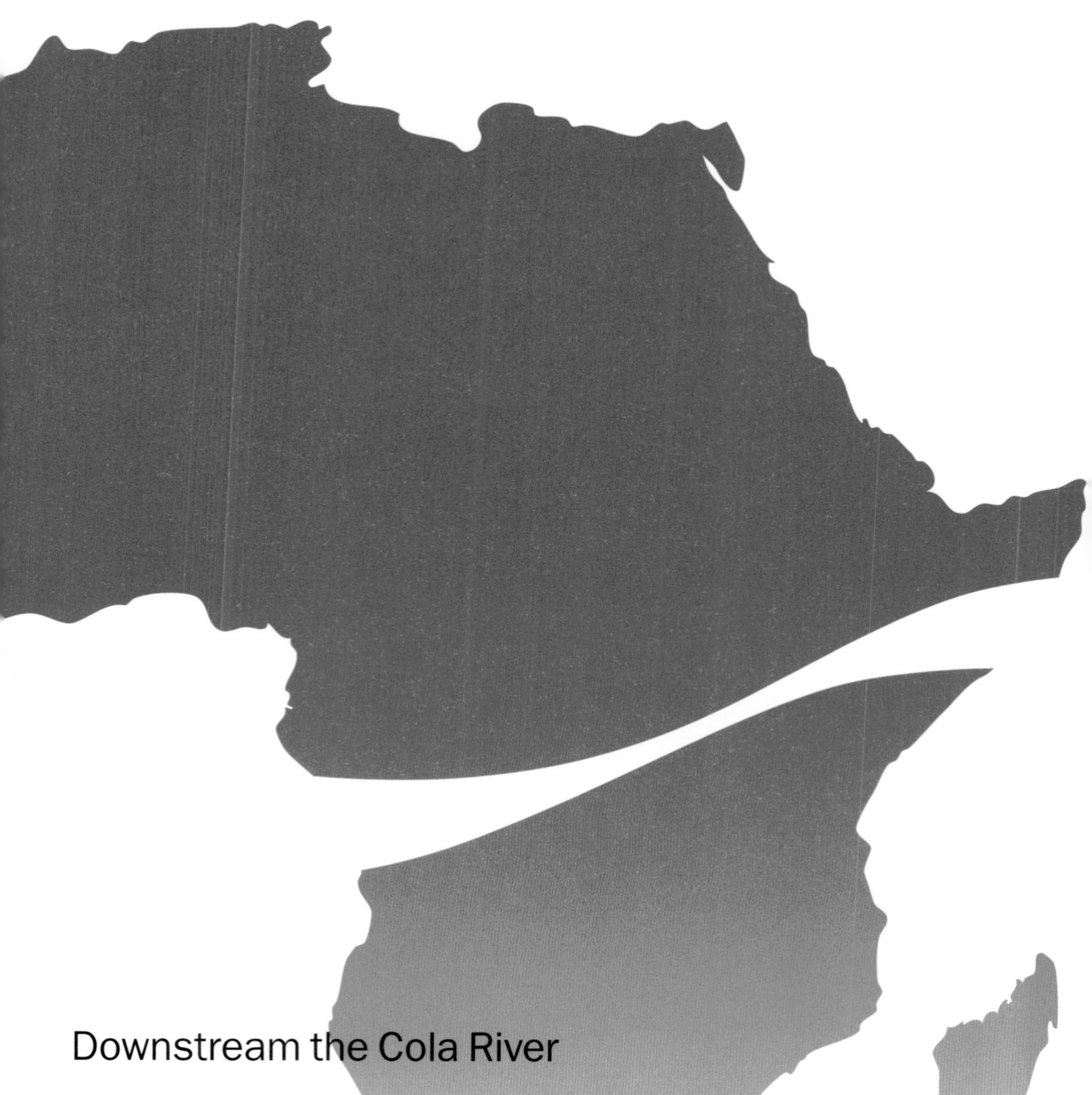

Downstream the Cola River

The Coca-Cola Company successfully distributes products to the most remote parts of Africa, an accomplishment that many NGOs struggle to replicate. For many people in these regions, a Coke is cheaper and easier to get than a bottle of drinking water. While we might cluck our tongues over the empty calories of sugary soda, it's more productive to take advantage of the system. Like following a river downstream or a pass through the mountains, ColaLife piggybacks on the Coca-Cola distribution chain to ship medicines to isolated villages. Parasitizing the crates of Coke, the ColaLife containers fit perfectly in the unused space between bottles. The contents of the aid pods are determined by health professionals to fit local needs. The crucial lesson from the project is that NGOs, rather than attempting to distribute their goods in Tanzania, Kenya, or Nigeria, could instead focus on how to spread their goods through brand territories like Coca-Cola, Nike or Google. | AG

UNEASY CART FELLOWS! NOTE HOW DIFFERENT THE COCA-COLA AND PEPSI CRATES ARE. THE COLALIFE PODS WOULD NOT WORK IN PEPSI CRATES. THE BOTTLES IN PEPSI CRATES MOVE AROUND A LOT MORE IN TRANSIT, SO THE POD WOULD NEED TO BE STRONGER.

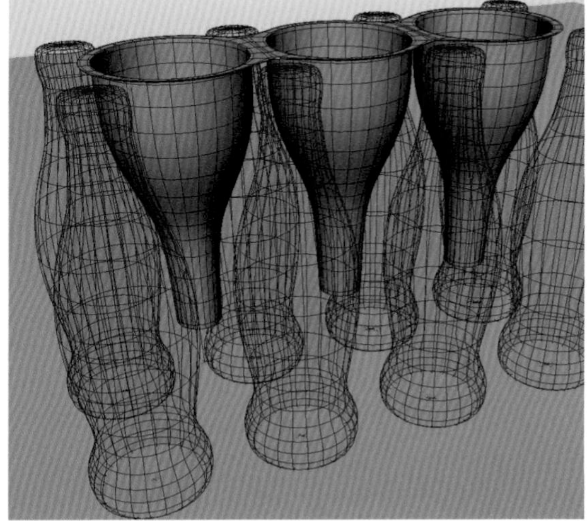

Real Virtuality

Almost every Internet user has at least one avatar running around somewhere on the web. We control then in numerous games and forums, but do they also control us? Researchers at the Virtual Human Interaction Lab at Stanford University have researched the effects of the appearance of avatars on the behavior of their controllers. People with more attractive avatars acted in more extroverted ways than players with more homely avatars. This effect, however, isn't as surprising as the fact that it doesn't disappear once the computer is turned off. In the test, the subjects were first asked to interact in a digital space and afterward, without knowing that it was still the same experiment, to split money with another person in real life. People who had been playing with taller avatars were more likely to make a proposal to split money in their favor, and less likely to accept a bad offer from the other side. This also works the other way around: shorter avatars resulted in less aggressive negotiations and increased acceptance of unfair offers. So, if you want to be a stronger negotiator, be prepared to take your digital booster shots before you go into a meeting. |TP

DIGITALLY ENHANCE YOUR EGO IN WORLD OF WARCRAFT

Boy Saves Sister from Moose Using Gaming Skills

Hans Jørgen Olsen, a 12-year-old Norwegian boy, saved himself and his sister from a moose attack using skills he picked up playing the online role playing game World of Warcraft. They got into trouble after trespassing the territory of the moose during a walk in the forest near their home. When the moose attacked them, Hans knew the first thing he had to do was 'taunt' and provoke the animal so that it would leave his sister alone so she could run to safety. 'Taunting' is a move players use in World of Warcraft to get monsters away from unarmed team members. Once Hans was a target, he remembered another skill he had picked up at level 30 in 'World of Warcraft' – feigning death. The moose lost interest in the inanimate boy and wandered off into the woods. When he was safely alone Hans ran home to share his tale of video game-inspired survival. | KVM

Virtual Characters Tortured for Science

In the classic Milgram Experiment conducted in the 1960s, volunteers were told by an authority figure to deliver electric shocks to another person as punishment for incorrect answers to a test. The other person wasn't really receiving the shocks, but the volunteers were tricked into thinking they were by shouts of pain and protest. Despite this feedback, some volunteers went on to deliver what would have been lethal shocks.

Professor Mel Slater of the Catalan Polytechnic University has recreated the Milgram experiment using a computer-generated woman, with some interesting results. "The main conclusion of our study is that humans tend to respond realistically at subjective, physiological, and behavioural levels in interaction with virtual characters notwithstanding their cognitive certainty that they are not real." Some part of the brain just doesn't distinguish real from virtual.

Avatar Funerals

The Internet is as much the site of second death as second life. *World of Warcraft* (WoW), a massively multi-player online role-playing game, is the most popular platform of its type. More than 12 million subscribers live and die as members of two warring factions. Players perish hundreds of times during the course of their virtual careers and spring back to life almost immediately. Sometimes, however, the logic of the computer game is reversed: the avatar remains 'alive' even after its real-life counterpart has died. This was the case for one avid WoW player who suffered a fatal stroke. A friend logged into her account and gathered with other in-game compatriots for a funeral in a combat zone. Members of the opposing team got word of the memorial and ambushed the unprepared mourners. The dead girl's avatar died again at her own funeral. This meta-event is either fascinating or silly, depending on the weight you give to fantasy warfare with elves. WoW players ascribe to opposing social rules. For the mourners, virtual life shares certain social mores with real life. A funeral is an acceptable way to commemorate a life, in the world and in the Warcraft world, and should be immune to interruption. For the attackers, WoW is foremost a game of war. They did nothing wrong by the game's laws, and for some commentators, their ambush was seen as a triumph of cunning over sentiment. Either way, life continues on Azeroth as it does on Earth. |KVM|AG

SCREENSHOT OF THE INFAMOUS FUNERAL AMBUSH IN WORLD OF WARCRAFT

SUICIDEMACHINE.ORG LOGO

A Webwill or Social Network Suicide?

My Webwill offers Internet users the assurance that their email and social network accounts will be managed according to their posthumous wishes. Subscribers can choose to send out emails to their friends and relatives from beyond the grave, or pass on important documents and passwords. Especially for sites like Facebook that usually keep 'dead' accounts active despite relatives' requests, My Webwill will ensure that you remain a digital ghost for as long or as little as you like. Those who want to get a real life might have to end the virtual one during their lifetime.

The Web2.0 Suicide Machine, developed by Moddr_ lab, allows users to kill off their entire social network presence. The 'machine' – which is less machine and more a collection of web scripts – automatically deletes every status update, tweet, and friendship. The suicide metaphor is a telling choice, making the online presence equivalent to the physical one. Here, it's not enough just to sign out of Twitter for the final time. The online self has a degree of autonomy, like a living organism, and has to be euthanized to keep it from mischief. The service has tapped into a rich vein of anti-social network sentiment that has inspired amongst other things an episode of South Park. Living a double life is fine for spies, but for normal people, it can be oppressive. What happens when you log off the mortal coil, but your online accounts remain stubbornly alive? | AG

SUICIDE MACHINE APPEARS BRIEFLY IN THE SOUTH PARK EPISODE 'YOU HAVE ZERO FRIENDS' (SEASON 14, EPISODE 4)

Progressive Nostalgia

TANG DYNASTY COKE
CHINESE ARTIST AI WEIWEI'S *COCA COLA VASE* (1997) IS DARKLY COMIC BUT UNCOMFORTABLE. THE ARTIST REAPPROPRIATES ANCIENT CHINESE URNS AND ROCKETS THEM FORWARD TO THE FUTURE, INSTANTLY MERGING MODERN AND THE ANCIENT SIGNS OF VALUE.

Designing the Future by Referring to the Past

As technology progresses, we constantly have to adapt ourselves to an ever-changing media landscape. While in our infant years, we almost automatically attune ourselves to whatever surrounds us; at an older age adapting tends to become more difficult. For many, the latest technological advancements merely feel alien, disruptive and inessential. 'Progressive Nostalgia' is a design strategy that aims to smoothen technological change, by linking newfangled technologies with familiar phenomena. The hypothesis behind this design methodology is, that people will feel more comfortable with technological changes when they are wrapped in a recognizable packaging by referring to accepted objects, habits, values, traditions or intuitions. The 'nostalgia' can refer to a phenomenon from your youth, your grandparents' life, or even the ancient history of humankind. The nostalgic element can be applied as a simple skeumorph, or more profoundly, as a reference to a preceding technology that provides the latest technology with a cultural and historical context and meaning. In the coming few pages, we explore five case studies of progressive nostalgic design. |KVM

WIFI DOWSING ROD

MIKE THOMPSON BASED THE DESIGN OF HIS WIRELESS INTERNET DETECTOR ON THE MAGICAL DOWSING ROD WHICH WAS ORIGINALLY EMPLOYED IN ATTEMPTS TO LOCATE GROUND WATER, BURIED METALS, GEMSTONES, OIL AND MANY OTHER OBJECTS AND MATERIALS, AS WELL AS SO-CALLED CURRENTS OF EARTH RADIATION. ALTHOUGH A BIT IMPRACTICAL, THE WIFI DOWSING ROD IS AN INTRIGUING ATTEMPT OF USING 'MAGIC' AS A CONSTRUCT TO COPE WITH THE TECHNOLOGICAL COMPLEXITY AROUND US. A GREAT GIFT FOR YOUR GRAND-GRAND-PARENTS.

Retrofit Lightbulb

Energy-saving light bulbs are typically as ugly as they are sustainable. Recently, however, light bulb designers seem to have discovered the benefits of a progressive nostalgic design approach by shaping the newest LED based lights in the shape of a classical light bulb. Unsure whether the banning of the classical light bulb in many countries – because they're such energy wasters – contributes to its status as a nostalgic object, which makes it a suitable candidate for mimicry by the less nostalgic LED lamps. Interestingly enough, at the time electric light was introduced, the now classic light bulb was still an alien technology. Light bulbs replaced candle lights and hence their design referenced candles. Manufacturers even made special versions that mimicked the flame flickering of a candle. Unsurprisingly, you can now also buy flickering flaming LED lights, which proves a progressive nostalgic design approach can easily turn into kitsch. | KVM

PHILIPS
PHILIPS

Cave House

One of the greatest qualities of architect Antonio Gaudí (1852-1926) was his profound understanding of the house as an artificial cave. We long ago left the Stone Age. Buildings today can be constructed out of steel, concrete or wood, in any imaginable shape or form, yet arguably the most natural housing is still a cave. Of all architects, Gaudí had the most intuitive sensibility for this. The aesthetics of his biomorphic cave tap deep into the human psyche and seem to tell us that this is what a house should look like, really. | KVM

Bone Chair

Joris Laarman's Bone Chair is inspired by how bones grow, forming thicker supports when strength is needed, and removing material where it is unnecessary. The chair was actually designed by a computer algorithm, originally developed by German scientist Claus Matthek, that mimics how bones and trees grow and respond to stress. According to Laarman, the result of his biomimetic technique is "far more efficient compared to modern geometric shapes."

Say goodbye to the harsh lines and minimal forms of modernism, and hello to design by next nature. Laarman's design is either a sneak preview into a shiny future of grown objects or merely the biomimic marketing of a clever stylist. It's not difficult to imagine living the futuristic primitive lifestyle in a grown-bone interior. After all, if chairs are already the extension of the backbone, why not extend it to the whole house?

| KVM

iPad Bookshelf

Flipping through the bookshelf on the iPad gives the user the familiar feeling of having an easily accessible library of physical books. The nostalgic reference to a bookshelf makes having a digital book collection feel tangible. Yet at the same time this model of storing books digitally is expected to have a huge impact on the publishing industry and the actual use of books. The first cars were designed as 'horseless carriages,' although they lead to the practical end of horse transport, while the envelope icon on email applications is part of a system that lead to a drastic reduction in paper mail. Similarly, the digital book cabinet is the first sign of extinction for the physical bookshelf it so elegantly simulates.

Retro iPhone dial

For those who want to dial the phone like grandma did, the iRetroPhone application replicates the long-defunct rotary phone. Both the iPad Bookshelf and the iRetroPhone app refer to a familiar technology, but while the bookshelf allows you carry an entire book collection in a tiny device, the iRetroPhone merely plays to nostalgia without adding anything useful. Having a rotary dial on your phone doesn't refer to an ancient human tradition or intuition; rather, it refers to an analog technology that was already pretty clumsy when it was still a norm. In the long run people may find the iPad Bookshelf metaphor equally corny, but before that happens, some of the essential benefits of traditional books – no eye strain, no glare, and no battery – will have to be adopted by its digital counterparts.

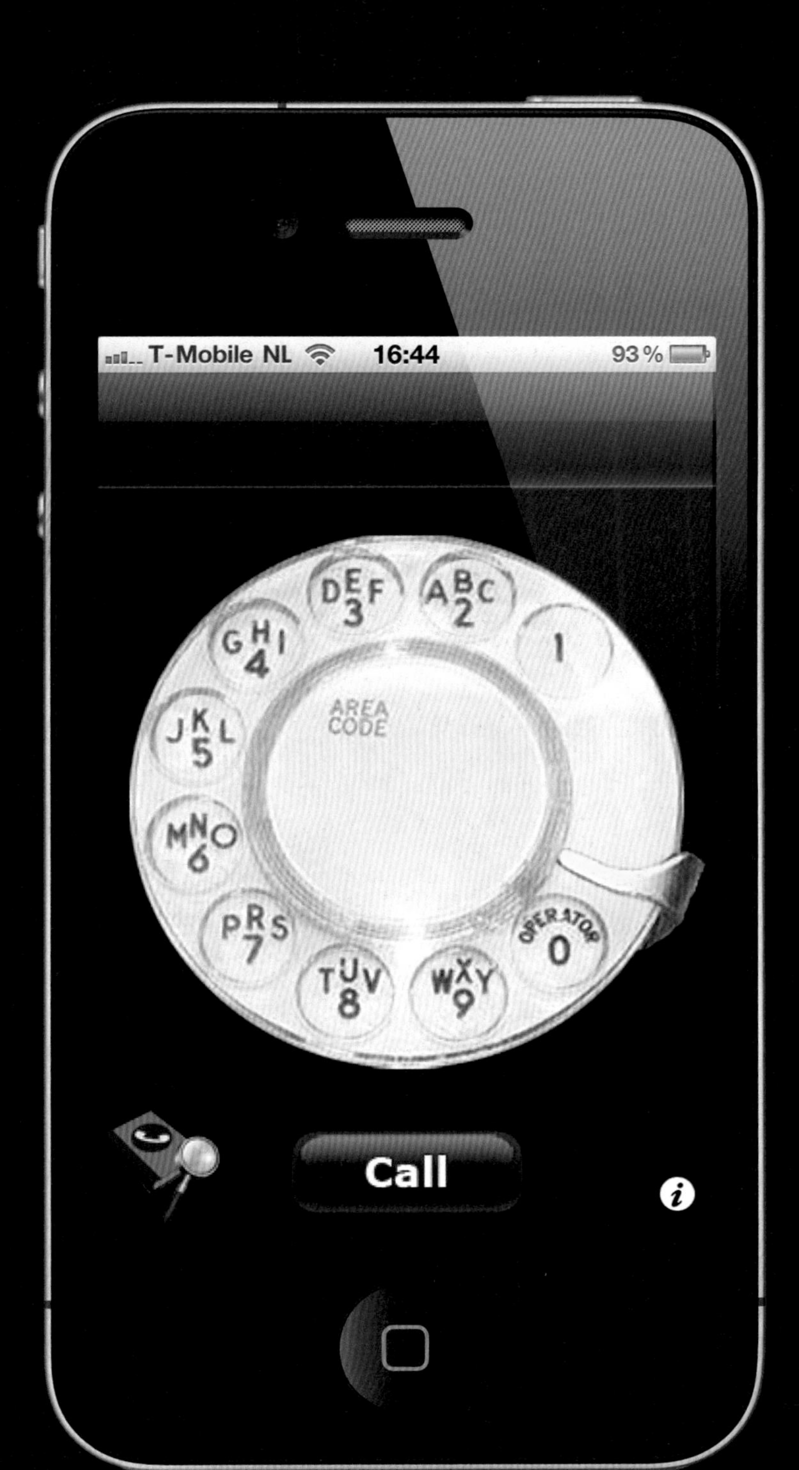
T-Mobile NL
16:44
93 %
DEF 3
ABC 2
1
GHI 4
AREA CODE
JKL 5
MNO 6
PRS 7
OPERATOR 0
TUV 8
WXY 9
Call

StoneAxe

The StoneAxe delivers a superb, user-controlled experience. Crafted from one piece of durable granite. Designed for social networking, it allows you to easily share meat with your friends and family. Cutting has never been easier. Say hello to the future.

Hunting Redefined.

MANKIND IS ARTIFICIAL BY NATURE

THE WORLD WITHOUT TECHNOLOGY

By Kevin Kelly

If a Technology Removal Beam swept across the earth and eliminated every invention from the world, would we survive? After all, the life of early *Homo sapiens* was typically “nasty, brutish and short.” Ultimately innovation allowed our forefathers to prosper. Tracing our relationship with technology back in time, we can understand how our genes have co-evolved with our inventions and especially how technology has domesticated us.

I remember the smoke the most. That pungent smell permeating the camps of tribal people. Everything they touch is infused with the lingering perfume of smoke – their food, shelter, tools, and art. Everything. Even the skin of the youngest tribal child emits smokiness when they pass by. I can hold a memento from my visits decades later and still get a whiff of that primeval scent. Anywhere in the world, no matter the tribe, steady wafts of smoke drift in from the central fire. If things are done properly, the flame never goes out. It smolders to roast bits of meat, and its embers warm bodies at night. The fire's ever-billowing clouds of smoke dry out sleeping mats overhead, preserve hanging strips of meat, and drive away bugs at night. Fire is a universal tool, good for so many things, and it leaves an indelible mark of smoke on a society with scant other technology.

Besides the smoke I remember the immediacy of experience that opens up when the mediation of technology is removed in a rough camp. Living close to the land as hunter-gatherers do, I got colder often, hotter more frequently, soaking wet a lot, bitten by insects faster, more synchronized to rhythm of the day and seasons. Time seemed abundant. I was shocked at how quickly I could dump the cloud of technology in my modern life for a cloud of smoke.

But I was only visiting. Living in a world without technology was a refreshing vacation, but the idea of spending my whole life there was, and is, unappealing. Like you, or almost anyone else with a job today, I could sell my car this morning and with the sale proceeds instantly buy a plane ticket to a remote point on earth in the afternoon. A string of very bumpy bus rides from the airport would take me to a drop-off where within a day or two of hiking I could settle in with a technologically simple tribe. I could choose a hundred sanctuaries of hunter-gatherer tribes that still quietly thrive all around the world. At first a visitor would be completely useless, but within three months even a novice could at least pull their own weight and survive. No electricity, no woven clothes, no money, no farm crops, no media of any type – only a handful of hand-made tools. Every adult living on earth today has the resources to relocate to such a world in less than 48 hours. But no one does. The gravity of technology holds us where we are. We accept our attachment. But to really appreciate the effects of technology – both its virtues and costs – we need to examine the world of humans before technology. What were our lives like without inventions? For that we need to peek back into the Paleolithic era when technology was scarce and humans lived primarily surrounded by things they did not make. We can also examine the remaining contemporary hunter-gatherer tribes still living close to nature to measure what, if anything, they gain from the small amount of technology they use.

Technology is anything designed by a mind

The problem with this line of questioning is that technology predated our humanness. Many other animals used tools millions of years before humans. Chimpanzees made (and of course still make) hunting tools from thin sticks to extract termites from mounds, or slam rocks to break nuts. Even termites themselves construct vast towering shells of mud for their homes. Ants herd aphids and farm fungi in gardens. Birds weave elaborate twiggy fabrics for their nests. The strategy of bending the environment to use as if it were part of your body is a billion-year-old trick at least.

Our humanoid ancestors first chipped stone scrapers 2.5 million years ago to give themselves claws. By about 250,000 years ago they devised crude techniques for cooking, or pre-digesting, with fire. Technology-assisted hunting, versus tool-free scavenging, is equally old. Archeologists found a stone point jammed into the vertebra of a horse and a wooden spear embedded in a 100,000 year old red deer skeleton. This pattern of tool use has only accelerated in the years since. To put it another way, no human tribe has been without at least a few knives of bone, sharpened sticks, or a stone hammer. There is no such thing as a total tool-free humanity. Long before we became the conscious beings we are now we were people of the tool. Hunters increased the power of a spear by launching it from a long swinging stick (the atlatl) which literally extended their arm. In fact all tools are extensions of our biological body, just as the artifact of a beehive is an extension of a bee. Neither honeycomb nor queen bee can exist alone. Same for us. Evolutionarily we've survived as a species because we've made tools, and we'd perish as a species without at least some of our inventions.

Although strictly speaking simple tools are a type of technology made by one person, we tend to think of technology as something much more complicated. But in fact technology is anything designed by a mind. Technology includes, not only nuclear reactors and genetically modified crops, but also bows and arrows, hide-tanning techniques, fire starters, and domesticated crops. Technology also includes intangible inventions such as calendars, mathematics, software, law, and writing, as these too derive from our heads. But technology also must include birds' nests and beaver dams since these too are the work of brains. All technology, both the chimp's termite fishing spear and the human's fishing spear, the beaver's dam and the human's dam, the warbler's hanging basket and the human's hanging basket, the leafcutter ant's garden and the human's garden, are all fundamentally natural. We tend to isolate human-made technology from nature, even to the point of thinking of it as anti-nature, only because it has grown to rival the impact and power of its home. But in its origins and fundamentals a tool is as natural as our life.

Tools and bigger brains mark the beginning of a distinctly human line in evolution two and a half million years ago. The first simple stone tools appeared in the same archeological moment that brains of the hominins who made them began to enlarge toward their current size. Thus hominins arrived on earth with rough chipped stone scrapers and cutters in hand. About a million years ago these large-brained, tool-wielding hominins drifted out of Africa and settle into southern Europe, where they evolved into the Neanderthal (with even bigger brains), and further into east Asia, where they evolved into *Homo erectus* (also bigger brained). Over the next several millions of years, all three hominin lines evolved, but the ones who remained in Africa evolved into the human form we see in ourselves. The exact time these proto-humans became fully modern humans is of course debated. Some say 200,000 years ago but the undisputed latest date is 100,000 years ago. By 100,000 years ago humans crossed the threshold where they were indistinguishable from us outwardly. We would not notice anything amiss if one of them were to stroll alongside one of us on the beach. However, their tools and most of their behavior were indistinguishable from their relatives the Neanderthals in Europe and Erectus in Asia.

For the next 50 millennia not much changed. The anatomy of African human skeletons remained constant over this time. Neither did their tools change much. Early humans employed rough-and-ready lumps of rock with sharpened edges to cut, poke, drill, or spear. But these hand-held tools were unspecialized, and did not vary by location or time. No matter where or when in this period (called the Mesolithic) a hominin picked up one of these tools it would resemble one made tens of thousands of miles away or tens of thousands of years apart, whether in the hands of Neanderthal, Erectus or *Homo sapiens*. Hominins simply lacked innovation. As scientist Jared Diamond put it, "Despite their large brains, something was missing."

Then about 50,000 years ago something amazing happened. While the bodies of early humans in Africa remained unchanged, their genes and minds shifted noticeably. For the first time hominins were full of ideas and innovation. These newly vitalized modern humans, which we now call Sapiens, charged into new regions beyond their ancestral homes in eastern Africa. They fanned out from the grasslands and in a relatively brief burst exploded from a few tens of thousands in Africa to an estimated eight million worldwide just before the dawn of agriculture 10,000 years ago.

We tend to isolate man-made technology from nature, even to the point of thinking of it as anti-nature

The speed at which Sapiens marched across the planet and settled every continent (except Antarctica) is astounding. In 5,000 years they overtook Europe. In another 15,000 they reached edges of Asia. Once tribes of Sapiens crossed the land bridge from Eurasia into what is now Alaska, it took them only a few thousand years to fill the whole of the New World. Sapiens increased so relentlessly that for the next 38,000 years they expanded their occupation at the average rate of one mile per year. Sapiens kept pushing until they reached the furthest they could go: land's end at the tip of South America. Less than 1,500 generations after their "great leap forward" in Africa, *Homo sapiens* had become the most widely distributed species in Earth's history, inhabiting every type of biome and every watershed on the planet. Sapiens was the most invasive alien species ever.

Language is a trick that allows the mind to question itself

Today the breadth of Sapiens occupation exceeds that of any other macro-species we know of; no other visible species occupies more niches, geographically and biological, than *Homo sapiens*. Sapiens overtake was always rapid. Jared Diamond notes that "after the ancestors of the Maori reached New Zealand" carrying only a few tools, "it apparently took them barely a century to discover all worthwhile stone sources; only a few more centuries to kill every last moa in some of the world's most rugged terrain." This sudden global expansion following millennia of steady sustainability is due to only one thing: technology and innovation. As Sapiens expanded in range they remade animal horns and tusks into thrusters and knives, cleverly turning the animals' own weapons against them. They sculpted figurines, the first art, and the first jewelry, beads cut from shells, at this threshold 50,000 years ago. While humans had long used fire, the first hearths and shelter structures were invented about this time. Trade of scarce shells, chert and flint rock began. At approximately the same time Sapiens invented fishing hooks and nets, and needles for sewing hides into clothes. They left behind the remains of tailored hides in graves. In fact, graves with deliberately interred burial goods were invented at this time. Sometimes recovered beads and ornaments in the burial site would trace the borders of the long-gone garments. A few bits of pottery from that time have the imprint of woven net and loose fabrics on them. In the same period Sapiens also invented animal traps. Their garbage reveals heaps of skeletons of small furred animals without their feet; Trappers today still skin small animals the same way by keeping the feet with the skin. On walls artists painted humans wearing parkas shooting animals with arrows or spears. Significantly, unlike Neanderthal and Erectus's crude creations, these tools varied in small stylistic and technological ways place by place. Sapiens had begun innovating.

The Sapiens mind's ability to make warm clothes opened up the Arctic regions, and the invention of fishing gear opened up the coasts and rivers of the world, particularly in the tropics, where large game was scarce. While Sapiens' innovation allowed them to prosper in new climates, the cold and its unique ecology especially drove innovation. More complex "technological units" are needed (or have been invented) by historical hunter-gatherer tribes the higher the latitude of their homes. Hunting oceanic sea mammals in Arctic climes took significantly more sophisticated gear that fishing salmon in a river. The ability of Sapiens to rapidly adapt tools allowed them to rapidly adapt to new ecological niches, at a much faster rate than genetic evolution could ever allow.

During their quick global takeover, Sapiens displaced (with or without interbreeding) the several other co-inhabiting hominin species on earth, including their cousins the Neanderthal. The Neanderthals were never abundant and may have only numbered 18,000 individuals at once. After dominating Europe for hundreds of thousands of years as the sole humanoid, the Neanderthals vanished in less than 100 generations after the tool-carrying Sapiens arrived. That is a blink in history. As anthropologist Richard Klein says, "this displacement occurred almost instantaneously from a geologic perspective. There were no intermediates in the archeological record. The Neanderthals were there one day, and the Cro-Magnons [Sapiens] were there the next." The Sapiens' layer was always on top, and never the reverse. It was not even necessary that the Sapiens slaughter the Neanderthals. Demographers have calculated that as little as a four percent difference in reproductive effectiveness (a reasonable expectation given Sapiens' ability to bring home more kinds of meat), could eclipse the lesser breeding species in a few thousands years. The speed of this several-thousand-year extinction was without precedent in natural evolution. Sadly it was only the first rapid species extinction to be caused by humans.

It should have been clear to Neanderthal, as it is now clear to us in the 21st century, that something new and big had appeared – a new biological and geological force. A number of scientists (Richard Klein, Ian Tattersall, William Calvin, among many others) think that the "something" that happened 50,000 years ago was the invention of language. Up until this point, humanoids were smart. They could make crude tools in a hit or miss way and handle fire – perhaps like an exceedingly smart chimp. The African hominin's growing brain size and physical stature had leveled off its increase, but evolution continued inside the brain. "What happened 50,000 years ago," says Klein, "was a change in the operating system of humans. Perhaps a point mutation effected the way the brain is wired that allowed languages, as we understand language today: rapidly produced, articulate speech." Instead of acquiring a larger brain, as the Neanderthal and Erectus did, Sapiens gained a rewired brain. Language altered the Neanderthal-type mind, and allowed Sapiens minds for the first time to invent with purpose and deliberation. Philosopher Daniel Dennet crows in elegant language: "There is no step more uplifting, more momentous in the history of mind design, than the invention of language. When *Homo sapiens* became the beneficiary of this invention, the species stepped into a slingshot that has launched it far beyond all other earthly species." The creation of language was the first singularity for humans. It changed everything. Life after language was unimaginable to those on the far side before it.

Language accelerates learning and creation by permitting communication and coordination. A new idea can be spread quickly by having someone explain it and communicate it to others before they have to discover it themselves. But the chief advantage of language is not communication, but auto-generation. Language is a trick that allows the mind to question itself. It is a magic mirror which reveals to the mind what the mind thinks. Language is a handle that turns a mind into a tool. With a grip on the slippery aimless activity of self-reference, self-awareness, language can harness a mind into a fountain of new ideas. Without the cerebral structure of language, we can't access our own mental activity. We certainly can't think the way we do. Try it yourself. If our minds can't tell stories, we can't consciously create; we can only create by accident. Until we tame the mind with an organization tool capable of communicating to itself, we have stray thoughts without a narrative. We have a feral mind. We have smartness without a tool.

A few scientists believe that, in fact, it was technology that sparked language. To throw a tool – a rock or stick – at an animal and hit it with sufficient force to kill it requires a serious computation in the hominin brain. Each throw requires a long succession of precise neural instructions executed in a split second. But unlike calculating how to grasp a branch in mid-air, the brain must calculate several alternative options for a throw at the same time: the animal speeds up, or it slows down; aim high, aim low. The mind must then spin out the results to gauge the best possible throw before the actual throw – all in a few milliseconds. Scientists like neurobiologist William Calvin believe that once a brain evolved the power to run multiple rapid throw scenarios, it hijacked this throw procedure to run multiple rapid sequences of notions. The brain would throw words instead of sticks. This reuse or repurposing of technology then became a primitive but advantageous language.

Until we tame the mind with an organization tool capable of communicating to itself, we have stray thoughts without a narrative. We have a feral mind.

The slippery genius of language opened up many new niches for spreading tribes of Sapiens. They could quickly adapt their tools to hunt or trap an increasing diversity of game, and to gather and process an increasing diversity of plants. There is some evidence that Neanderthals were stuck on a few sources of food. Examination of Neanderthal bones show they lacked the fatty acids found in fish and the Neanderthal diet was mostly meat. But not just any meat. Over half of their diet was woolly mammoth and reindeer. The demise of the Neanderthal may be correlated with the demise of great herds of these megafauna.

Sapiens thrived as broadly omnivorous hunter-gatherers. The unbroken line of human offspring for hundreds of thousands of years proves that a few tools will capture enough nutrition to create the next generation. We are here now because hunting-gathering in the past worked. Several analysis of historical hunter-gather diets show that they were able to secure enough calories to meet the US FDA requirements for folks their size. For example, the Dobe gathered on average 2,140 calories. Fish Creek tribe: 2,130. Hemple Bay tribe: 2,160. They had a varied diet of tubers, vegetables, fruit and meat. Based on studies of bones and pollen in their trash, so did the early Sapiens.

Foragers lived in the ultimate disposable culture. The best tools, artifacts, technology were all disposable

Philosopher Thomas Hobbes claimed the life of the savage – and by this he meant hunter-gatherers – was "nasty, short and brutish". But while the life of an early hunter-gatherer was short, and interrupted by nasty warfare, it was not brutish. With only a slim set of a dozen primitive tools humans not only secured enough to survive in all kinds of environments, but these tools and techniques will also afford them some leisure doing so. Anthropological studies confirm that hunter-gathers do not spend all day hunting and gathering. One researcher, Marshall Sahlins, concluded that hunter-gatherers worked only three to four hours a day on necessary food chores, putting in what he called "banker hours". The evidence for his surprising results are controversial: much of the research (by others) was based on time studies of groups who were previously hunting and gathering and returned to this mode only for a few weeks to demonstrate their efficiency. And the measurements lasted only a few weeks. Surveys of other tribes' yields gave daily calorie intakes of only 1,500 or 1,800 per day for their few hours of work. Furthermore the definition of what activities should be included as the work of food getting is not clear. For instance if a modern human goes shopping at a supermarket – to get food of course – is that classified as "work" time? Why not? Do the elaborate preparations for a community feast where food is exchanged, common to most forager tribes, count as food getting? All these variables shift the measure of how much work it takes to live as a hunter-gatherer with a low dose of technology.

A more realistic and less contentious average for food-gathering time among contemporary hunter-gatherer tribes based on a wider range of data is about six hours per day. That six-hour-per-day average belies a great variation in day-to-day routine. One to two-hour naps or whole days spent sleeping were not uncommon. As one anthropologist noted, when foragers set out to work, "they certainly did not approach it as an unpleasant job to be got over as soon as possible, nor as necessary evil to be postponed as long as possible". Outside observers almost universally noted the punctuated aspect of work among foragers. Gatherers may work very hard for several days in a row and then do nothing in terms of food getting for the rest of the week. This cycle is known among anthropologists as the "Paleolithic rhythm" – a day or two on, day or two off. An observer familiar with the Yamana tribe – but it could be almost any hunter tribe – wrote: "Their work is a more a matter of fits and starts, and in these occasional efforts they can develop considerable energy for a certain time. After that, however they show a desire for an incalculably long rest period during which they lie about doing nothing, without showing great fatigue". The Paleolithic rhythm actually reflects the "predator rhythm" since great hunters of the animal world, the lion and other large cats, exhibit the same style: hunting to exhaustion in a short burst and then lounging around for days afterward. Hunters, almost by definition, seldom go out hunting, and they succeed in getting a meal even less often. The efficiency of primitive tribal hunting, measured in the yield of calories per hour invested, was only half that of gathering. Meat is thus a treat in almost every foraging culture.

Then there are seasonal variations. Every ecosystem produces a "hungry season" for foragers. In higher cooler latitudes, this late-winter or early spring hungry season is more severe, but even in tropical latitudes, there are seasonal oscillations in the availability of favorite foods, supplemental fruits, or essential wild game. In addition, there are climatic variations:

extended periods of droughts, floods, and storms that can disrupt yearly patterns. These great punctuations over days, seasons, and years mean that while there are many times when hunter-gatherers are well fed, they also can – and do – expect many periods when they are hungry, famished and undernourished. Time spent in this state along the edge of malnutrition is mortal for young children and dire for adults.

Technology has domesticated us. As fast as we remake our tools, we remake ourselves.

The result of all this variation in calories is the Paleolithic rhythm at all scales of time. Importantly, these bursts in "work" are not by choice. When you are primarily dependent of natural systems to provide you foodstuffs, working more does not tend to produce more. You can't get twice as much food by working twice as hard. The time it takes a fig to ripen can neither be hurried, nor predicted exactly. Nor can the arrival of game herds. If you do not store surplus, nor cultivate in place, then motion must produce your food. Hunter-gatherers must be in ceaseless movement away from depleted sources in order to maintain production. But once you are committed to perpetual movement, surplus and its tools slow you down. In many contemporary hunter-gatherer tribes, being unencumbered with things is considered a virtue, even a virtue of character. You carry nothing, but cleverly make or procure whatever you need when you need it. "The efficient hunter who would accumulate supplies succeeds at the cost of his own esteem," says Robert Kelley. Additionally the surplus producer must share the extra food or goods with everyone, which reduces incentive to produce extra. For foragers food storage is therefore socially self-defeating. Instead your hunger must adapt to the movements of the wild. If a dry spell diminishes the yield of the sago, no amount of extra work time will advance the delivery of food. Therefore, foragers take a very accepting pace to eating. When food is there, all work very hard. When it is not, no problem; they will sit around and talk while they are hungry. This very reasonable approach is often misread as tribal laziness, but it is in fact a logical strategy if you rely on the environment to store your food.

We civilized modern workers can look at this leisurely approach to work and feel jealous. Three to six hours a day is a lot less then most adults any developed country put in to their labors. Furthermore, when asked, most acculturated hunter-gatherers don't want any more than they have. A tribe will rarely have more than one artifact, such as an ax, because why do you need more than one? Either you use the object when you need to, or more likely, you make one when you need one. Once used, artifacts are often discarded rather than saved. That way nothing extra needs to be carried, or cared for. Westerners giving gifts to foragers such as a blanket or knife were often mortified to see them trashed after a day. In a very curious way, foragers lived in the ultimate disposable culture. The best tools, artifacts, technology were all disposable. Elaborate hand-crafted shelters were considered disposable. When a clan or family travelled they might erect a home for only a night (a bamboo hut or snow igloo) and then abandoned it the next morning. Larger multi-family lodges might be abandoned after a few years rather than maintained. Same for food patches, which are abandoned after harvesting.

This easy just-in-time self-sufficiency and contentment led anthropologist Marshall Sahlins to declare hunter-gatherers as "the original affluent society." But while foragers had sufficient calories most days, and did not create a culture that continually craved more, a better summary might be that hunter-gatherers had "affluence without abundance." Based on numerous historical encounters with aboriginal tribes, they often, if not regularly, complained about being hungry. Famed anthropologist Collin Turnball noted, "The Mbuti complain of food shortage, although they frequently sing to the goodness of the forest." Often the complaints of hunter-gathers were about the monotony of a carbohydrate stable, like mongongo nuts for every meal; what they meant about shortages, or even hunger, was a shortage of meat, and a hunger for fat, and a distaste for periods of hunger. Their small amounts of technology gave them sufficiency for most of the time, but not abundance.

The fine line between average sufficiency and abundance matters in terms of health. When anthropologists measure the total fertility rate (the mean number of live births over the reproductive years) of women in modern hunter-gather tribes they find it relatively low – about five to six children in total – compared to the six to eight of agricultural communities. There are several factors behind this depressed fertility. Perhaps because of uneven nutrition, puberty comes late to forager girls at 16 or 17 years old (modern females start at 13). This late menarche for women, combined with a shorter lifespan, delays and thus abbreviates the childbearing window. Breastfeeding usually lasts longer in foragers, which extends the interval between births. Most tribes nurse till children are two to three years old, while a few tribes keep suckling for as long as six years. Also, many women are extremely lean and active, and like lean active women athletes in the west, often have irregular or no menstruations. One theory suggests women need a "critical fatness" to produce fertile eggs, a fatness many forager women lack – at least part of the year – because of a fluctuating diet. And of course, people anywhere can practice deliberate abstinence to space children, and foragers have reasons to do so.

Infanticide also contributed to small families. The prevalence of infanticide varied significantly among foraging tribes. It was as high as 30% of children in traditional Arctic tribes (in the early 1900s), and zero in others, with an average infanticide rate of 21% among the cross-culture sample of tribes anthropologists measured. In some cultures infanticide was biased towards females, perhaps to balance gender ratios, particularly in tribes where men contributed more food to the family than women (which was not the norm). In other tribes infanticide was practiced to space births, as Robert Kelley says, "in order to maximize reproductive success, rather than population control." In nomadic cultures mothers needed to carry not only their tools and household items, but also their small children. On frequent long migrations to find food a family had to carry all their possessions. For a pregnant woman to carry more than one small child would have endangered the older sibling. Better for all to have children spaced apart.

Child mortality in foraging tribes was severe. A survey of 25 hunter-gatherer tribes in historical times from various continents revealed that on average 25% of children died before they were one, and 37% died before they were 15. In one traditional hunter-gather tribe child mortality was found to be 60%. Most historical tribes have a population growth rate of approximately zero. This depression is made evident, says Robert Kelley in his survey of hunter-gathering peoples, because "when formerly mobile people become sedentary, the rate of population growth increases." All things equal, the constancy of farmed food breeds more people. While many children died young, hunting-gather elders did not have it much better. There are no known remains of a Neanderthal who lived to be older than 40. Because extremely high child mortality rates depress average life expectancy, if the outlier oldest is only 40, the median age of a Neanderthal was less than 20. It was a tough life. Based on an analysis of bone stress and cuts, one archeologist said the distribution of injuries on the bodies of Neanderthal were similar to those found on rodeo professionals – lots of head, trunk and arm injuries like the ones you might get from close encounters with large angry animals. Paleoanthropologist Erik Trinkaus discovered that the pattern of age-related mortality for hunter-gatherers and Neanderthal were nearly parallel, except historical foragers lived a little longer.

Grandparents are the conduits of culture, and without them, culture stagnates.

A typical tribe of hunters-gatherers had few very young children and no old people. This demographic may explain a common impression visitors had upon meeting intact historical hunter-gatherer tribes. They would remark that, "everyone looked extremely healthy and robust." That's in part because almost everyone was in the prime of life between 15 and 35. We might have the same reaction visiting a city or trendy neighborhood with the same youthful demographic. We'd call them young adults. Tribal life was a lifestyle for and of young adults.

A major effect of this short forager lifespan was the crippling absence of grandparents. Given that women would only start bearing by 17 or so, and die by their thirties, it would be common for parents to leave their children in their tweens. We tend to think that a shorter lifespan is rotten – for the individual – and no doubt it is. But a short lifespan is extremely detrimental for a society as well. Because without grandparents, it becomes exceedingly difficult to transmit knowledge over time. Grandparents are the conduits of culture, and without them, culture stagnates.

Imagine a society that not only lacked grandparents but also lacked language – as the pre-Sapiens did. How would learning be transmitted over generations? Your own parents would die before you were an adult and in any case, they could not communicate to you anything beyond what they could show you while you were immature. You would certainly not learn anything from anyone outside your immediate circle of peers. Innovation and cultural learning cease to flow.

Language upended this tight constriction by enabling both an idea to form, and then to be communicated. An innovation could be hatched and then spread across generations via children. Sapiens gained better hunting tools (like thrown spears which permitted a lightweight human to kill a huge dangerous animal from a safe distance), better fishing tools (barbed hooks and traps), and using hot stones to cook not just meat but to extract more calories from wild plants. And they gained all these within only 100 generations of using language. Better tools meant better nutrition.

The primary long-term consequence of this slightly better nutrition was a steady increase in longevity. Anthropologist Rachel Capsari studied the dental fossils of 768 hominin individuals from five million years ago, till the great leap, in Europe, Asia and Africa. She determined that there was a "dramatic increase in longevity in the modern humans" about 50,000 years ago. Increasing longevity allowed grand parenting, or what is called the "grandmother effect." In a virtuous circle, via the communication of grandparents and culture, ever more powerful innovations were able to lengthen life spans further, which gave more time to invent new tools, which increased population. Not only that, increased longevity "provides a selective advantage promoting further population increase," because a higher density of humans increased the rate and influence of innovations, which contributed to increased populations. Capsari states that the most "fundamental biological factor that underlies the behavioral innovations of modernity is the increase in adult survivorship." Increased longevity is probably the most measurable consequence of the acquisition of technology, and it is also the most consequential. By 20,000 years ago, as the world was warming up and its global ice caps retreating, Sapiens' population and tool kit expanded hand-in-hand. Sapiens used 40 kinds of tools, including anvils, pottery, and composites – complicated spears or cutters made from multiple pieces, such as many tiny flint shards and a handle. While still primarily a hunter-gatherer Sapiens also dappled in sedentism, returning to care for favorite food areas, and developed specialized tools for different types of ecosystems. We know from burial sites in the northern latitudes at this same time, that clothing also evolved from the general (a rough tunic) to specialized items such as a cap, a shirt, a jacket, trousers and moccasins. Henceforth the variety of human tools would become ever more specialized.

But most noticeably, without technology, your leisure was wasted.

The variety of Sapiens tribes exploded as they adapted into diverse watersheds and biomes. Their new tools reflected the specifics of their homes; river inhabitants had many nets; steppe hunters many kinds of points; forest dwellers many types of traps. Their language and looks were diverging. Yet they shared many qualities. Most hunter-gatherers clustered into family clans that averaged about 25 related people. Clans would gather in larger tribes of several hundred at seasonal feasts or camping grounds. One function of the tribes was to keep genes moving through intermarriage. Population was spread thinly. The average density of a tribe was less than .01 person per square kilometer in cooler climes. The 200-300 folk in your greater tribe would be the total number of people you'd meet in your lifetime. You might be aware of others outside of them because items for trade or barter could travel 300 kilometers. Some of the traded items would be body ornaments and beads, such as ocean shells for inlanders, forest feathers for the coast dwellers. Occasionally pigments were swapped for face painting, but these could also be applied on walls, or applied to carved wood figurines. The dozen tools you carried would have been bone drills, awls, needles, bone knives, a bone hook for fish on a spear, some stone scrapers, maybe some stone sharpeners. A number of your blades would be held by bone or wood handles, hafted with cane or hide cord. When you crouched around the fire, someone might play a drum or bone flute. The handful of your possessions might be buried with you when you die.

But don't take this for harmony. Within 20,000 years of the great march out of Africa, Sapiens helped exterminate 270 out of the 279 then existing species of megafauna. Sapiens used new innovations such as the bow and arrow, spear, and cliff-stampedes to kill off the last of the mastodons, mammoths, moas, woolly rhinos, giant camels – basically every large package of protein that walked on four legs. More than 80% of all large mammal genera on the planet were completely extinct by 10,000 years ago. Only a handful of species escaped this fate in North America including the bison, moose, elk and musk ox.

A world without technology had enough to continue life but not enough to transcend it.

Violence between tribes was endemic as well. The rules of harmony and cooperation that work so well among members of the same tribe, and are often the subject of envy of modern observers, do not apply to those outside of the tribe. Tribes would go to war over waterholes in Australia, or hunting grounds and wild-rice fields in the plains of the U.S, or river and ocean frontage along the coast in the Pacific Northwest. Commonly, without systems of arbitration, or even leaders, small feuds over stolen goods, or women, or signs of wealth such as pigs in New Guinea, could grow into multi-generational warfare. The death rate due to warfare was 5 times higher among hunter-gatherer tribes than in later agricultural-based societies (0.1% of the population killed per year in "civilized" wars versus 0.5% in war between tribes). Actual rates of warfare varied among tribes and regions, because as in the modern world, one belligerent tribe could disrupt the peace for many. But in general the more nomadic a tribe was, the more peaceful, since it would simply flee from conflict. But when fighting broke out it was fierce and deadly. When the numbers of warriors on both sides were about equal, primitive tribes usually beat the armies of civilization. The Celtic tribes defeated the Romans, the Tuareg smashed the French, the Zulus trumped the British, and it took the U.S Army three centuries to defeat the Apache tribes and only then because the Army hired traitor Apaches to quell their brothers. As Lawrence Keeley says in his survey of early warfare in War Before Civilization, "The facts recovered by ethnographers and archaeologists indicated unequivocally that primitive and prehistoric warfare was just as terrible and effective as the historic and civilized version. In fact, primitive warfare was much more deadly than that conducted between civilized states because of the greater frequency of combat and the more merciless way it was conducted...It is civilized warfare that is stylized, ritualized, and relatively less dangerous".

Before the singularity of language 50,000 years ago, the world lacked significant technology. For the next 40,000 years (four times as long as civilization has been around) every human who lived was a hunter-gatherer. During this time an estimated one billion people explored how far you could go with a handful of tools. This world without much technology provided "enough." There was leisure and satisfying work for humans. Happiness, too. Without technology, the rhythms and patterns of nature were immediate. Nature ruled your hunger and set your course. Nature was so vast, so bountiful and so close, few humans could separate from it. The attunement with the natural world felt divine. Yet, without technology, the recurring tragedy of child death was ever present. Accidents, warfare and disease meant your life, on average, was far less than half what it could have been. Maybe only a quarter of the natural lifespan you genes afforded. Hunger was always near.

But most noticeably, without technology, your leisure was wasted. You had much time for repetitions, but none for anything new. Within narrow limits you had no bosses. But the direction and interests of your life were laid out in well-worn paths. The cycles of your environment determined your life.

Turns out, the bounty of nature, though vast, does not hold all possibilities. The mind does, but it had not been fully unleashed yet. A world without technology had enough to continue life but not enough to transcend it. The mind, liberated by language, and enabled by the technium, transcended the constraints of nature, and opened up greater realms of possibility. There was a price to pay for this transcendence, but what we gained was civilization and progress. Those would disappear instantly if technology were to disappear. If a Technology Removal Beam swept across the earth and eliminated every scrap of invention from the world, and sent us into a world without technology – the bus ride no one wants to take – it would shake our foundations. Once the Beam pulverized all hunks of metal, metallic pieces, and slivers of iron and steel into rock dust, and vaporized all plastic, and of course zapped all electronics and modern medicines, and eroded highways and bridges till they disappeared into the landscape without a trace, then the first thing we'd want to do is re-manufacture some hand tools. But if the Technology Removal Beam prevented any constructed tools at all, humans, and humanity, would rapidly be endangered.

After the vast asphalt parking lots of suburban malls melted away, and the greasy blocks of industrial factories subsided beneath the dirt, and the endless sprawl containing hundreds of millions of homes vanished in dust, there would still not be enough bounty in the regrowth of the wilds to sustain six billion foragers. For one, the hugely productive herds of megafauna that supported millions of earlier hunters are gone for good. We devoured their easy pickings 20 millennia ago, and removing technology won't return them. For another the current global population of humans spread equally across all land mass on the planet, including the least hospitable areas of icy mountains and empty desert, would be 10 persons per square kilometer, which is several times more than the average density of sustainable foraging. Thus even a pristine environment could not support six billion Sapiens at the meager level of a hunter-gather.

Not that we'd last long anyway. Deprived of gun, spear, and knife, humans would no longer be the key predator. We in fact become prey. Any human lucky enough to eat well would become a desirable meal for newly revived packs of wolves and other alpha predators. The stress and inadequate nutrition of subsistence foraging would revert women's fertility to its earlier low rate. The growth rate of Homo sapiens would head quickly towards zero. The entirety of the species would retreat to a few remote havens, much as gorillas and chimpanzees have done.

Any human lucky enough to eat well would become a desirable meal for newly revived packs of wolves and other alpha predators.

We are not the same folks who marched out of Africa. Our genes have co-evolved with our inventions. In the past 10,000 years alone, in fact, our genes have evolved 100 times faster than the average rate for the previous six million years. This should not be a surprise. In the same period we domesticated the dog (all those breeds) from wolves, and cows and corn and more from their unrecognizable ancestors. We, too, have been domesticated. We have domesticated ourselves. Our teeth continue to shrink, our muscles thin out, our hair disappears, and our molecular digestion adjusts to new foods. Technology has domesticated us. As fast as we remake our tools, we remake ourselves. We are co-evolving with our technology, so that we have become deeply co-dependent on it. Sapiens can no longer survive biologically without some kind of tools. Nor can our humanity continue without the technium. In a world without technology, we would not be living, and we would not be human.

ADAPTED FROM *WHAT TECHNOLOGY WANTS* BY KEVIN KELLY, 2010.

Technology wants to be…

From stone-axe to mobile phone – throughout human history we've given birth to a wide range of technologies that extend our given capabilities. It is almost impossible to imagine a world without technology. Yet, despite our symbiotic relation with technology and the fact that we are irrevocably surrounded by it, we are still relatively unaware of how technology seeps into our lives. Inspired by the classical Maslow hierarchy of human needs that describes requirements like nutrition, security and love in subsequent stages, the pyramid of technology describes the six levels at which technology may operate. |KVM|HJG

MASLOV STYLE PYRAMID OF TECHNOLOGY

SIMILAR TO MASLOW'S PYRAMID, LOWER STAGES NEED TO BE FULFILLED BEFORE THE NEXT STAGE CAN BE REACHED. THE BASIC LEVELS SHOW THAT TECHNOLOGY NEEDS TO BE FUNCTIONAL BEFORE IT CAN BE APPLIED AND BECOME ACCEPTED. AT THE HIGHER LEVELS, A TECHNOLOGY MAY BECOME VITAL, OR EVEN INVISIBLE, DENOTING THAT IT IS NO LONGER EXPERIENCED AS TECHNOLOGY.

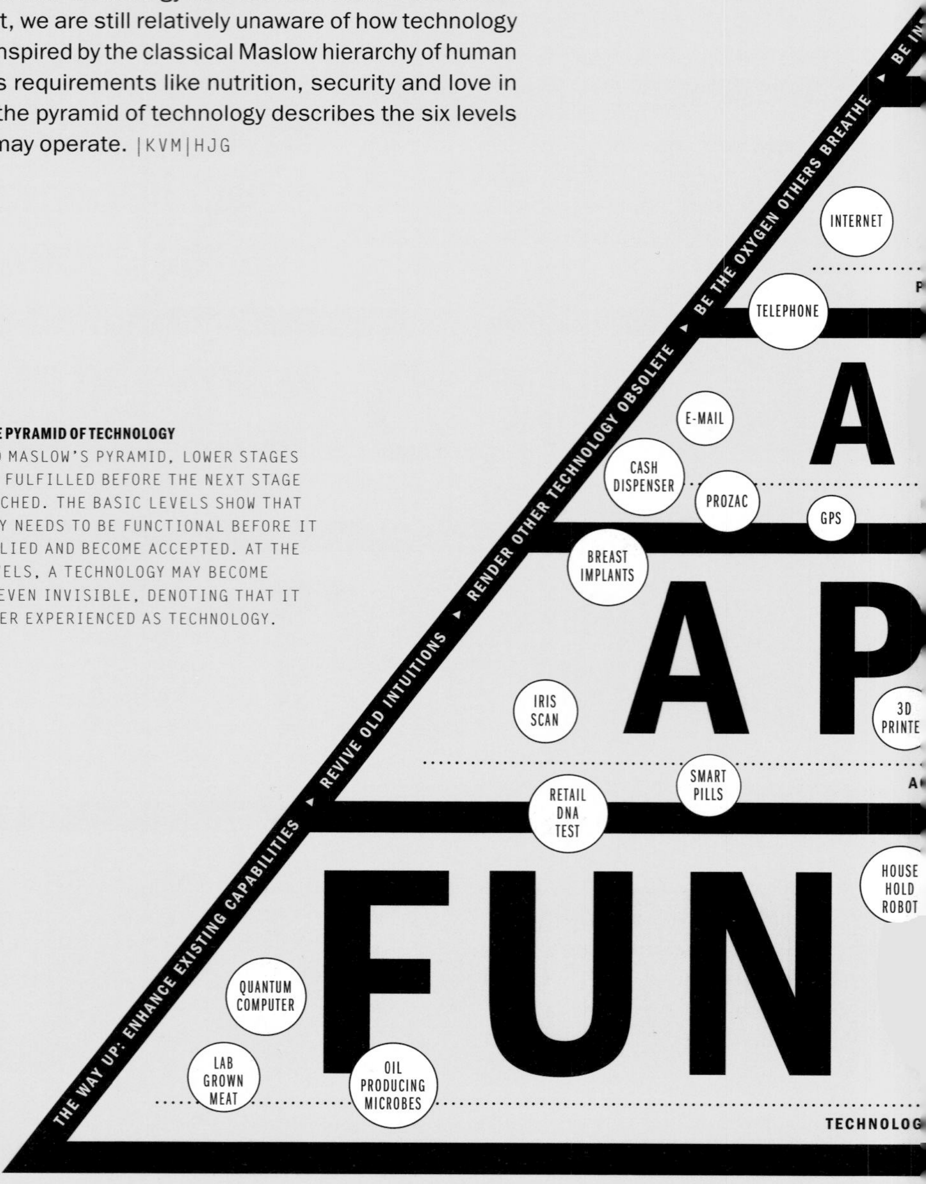

THE WAY UP: ENHANCE EXISTING CAPABILITIES ▸ REVIVE OLD INTUITIONS ▸ RENDER OTHER TECHNOLOGY OBSOLETE ▸ BE THE OXYGEN OTHERS BREATHE ▸ BE IN

INTERNET
TELEPHONE
A
E-MAIL
CASH DISPENSER
PROZAC
GPS
BREAST IMPLANTS
AP
IRIS SCAN
3D PRINTE
SMART PILLS
RETAIL DNA TEST
HOUSE HOLD ROBOT
FUN
QUANTUM COMPUTER
LAB GROWN MEAT
OIL PRODUCING MICROBES
TECHNOLOG

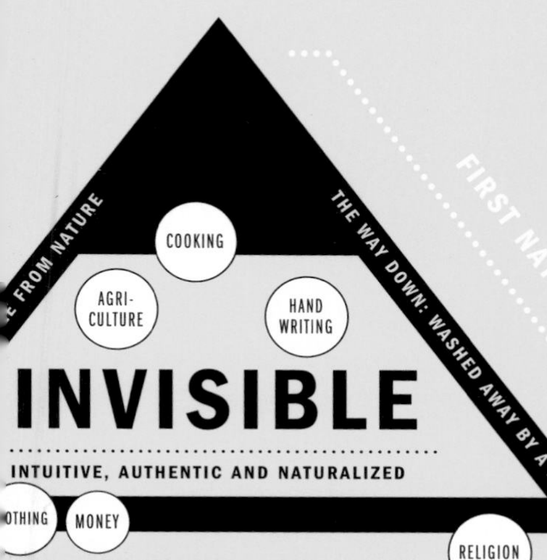

TECHNO NIRVANA?

AT THE TOP OF THE PYRAMID, TECHNOLOGY BECOMES INDISTINGUISHABLE FROM NATURE. LIKE IN MASLOW'S ORIGINAL PYRAMID, HOWEVER, THIS SUMMIT IS RARELY ATTAINED. MOST TECHNOLOGIES CLIMB NO HIGHER THAN HALFWAY THE PYRAMID BEFORE THEY STABILIZE OR ARE PUSHED BACK TO LOWER LEVELS BY NEWER, EMERGING TECHNOLOGIES.

IS INDIS

ANY
SUFFICIENTLY
ADVANCED
TECHNOLOGY
TINGUISHABLE
FROM NATURE

ANTHROPOCENE
THE GEOLOGICAL ERA THAT MARKS THE SIGNIFICANT GLOBAL IMPACT OF HUMAN ACTIVITIES ON THE EARTH'S ECOSYSTEMS. THE ANTHROPOCENE FOLLOWS ON THE HOLOCENE AND PLEISTOCENE.

ANTHROPOMORPHOBIA
THE FEAR OF ACKNOWLEDGING HUMAN QUALITIES IN NON-HUMAN THINGS. THIS PHOBIA CAN BE TRIGGERED BY PRODUCTS THAT LOOK AND BEHAVE LIKE PEOPLE, OR BY PEOPLE THAT LOOK TOO MUCH LIKE PRODUCTS.

BIOMIMICMARKETING
THE USE OF NATURAL IMAGES AND CONCEPTS TO MARKET AN IDEA, PRODUCT OR SERVICE.

BIONICS
THE APPLICATION OF BIOLOGICAL METHODS AND SYSTEMS FOUND IN OLD NATURE TO THE STUDY AND DESIGN OF ENGINEERING SYSTEMS AND MODERN TECHNOLOGIES. BIONICS IS ALSO KNOWN AS BIOMIMETICS OR BIOMIMICRY.

BIOPOLITICS
THE APPLICATION AND IMPACT OF POLITICAL POWER ON ALL FORMS OF LIFE.

BOOMERANGED METAPHORS
TRANSFERRING CONCEPTS AND IMAGES FROM THE DIGITAL REALM INTO THE PHYSICAL ENVIRONMENT, SUCH AS CREATING PHYSICAL OBJECTS THAT RESEMBLE PIXELATED ICONS, OR PLACING ENORMOUS GOOGLE MAPS MARKERS IN ACTUAL PLACES. BOOMERANGED METAPHORS CAN BE UNDERSTOOD AS A REVERSAL OF THE PROCESS OF PROGRESSIVE NOSTALGIA.

BRAND TERRITORY
THE VIRTUAL BRAND SPACE OF A CORPORATION THAT FUNCTIONS LIKE A TRIBE OR NATION, BUT DOES NOT EXIST IN ANY GEOGRAPHICAL OR LEGAL SENSE.

COEVOLUTION
A DRIVING FORCE IN EVOLUTION, HUMAN OR ANIMAL. POLLINATORS AND FLOWERS HAVE COEVOLVED TO WORK TOGETHER, WHILE PREY SPECIES HAVE CO-EVOLVED WITH THEIR PREDATORS TO AVOID BEING EATEN. HUMANS MAY BE COEVOLVING WITH THEIR TECHNOLOGY.

DESIGN BY EVOLUTION
THE APPROPRIATION OF EVOLUTIONARY PROCESSES TO OBTAIN A CERTAIN DESIGN RESULT.

DIGITAL NATIVE
ANYONE WHO HAS GROWN UP IMMERSED IN DIGITAL MEDIA TECHNOLOGIES, SPECIFICALLY THE INTERNET. DIGITAL NATIVES HAVE TROUBLE UNDERSTANDING LIFE BEFORE 24/7 ONLINE CONNECTIVITY.

DYNAMIC ARCHITECTURE
BUILDINGS THAT ADAPT TO THEIR ENVIRONMENT, OR ARE AT LEAST SHAPED BY IT. DYNAMIC ARCHITECTURE CAN BE A FORM OF GUIDED GROWTH, AS IS THE CASE IN TREE-ROOT BRIDGES OR SUPERMAN'S CRYSTAL FORTRESS.

ECONOMOLOGY
NEOLOGISM, COMPILED FROM THE WORDS 'ECONOMY' AND 'ECOLOGY', USED TO EMPHASIZE THE INTERCONNECTEDNESS BETWEEN THESE TWO TERMS. ECONOMY = ECOLOGY.

EVOLUTIONARY CATALYST
IN CHEMISTRY, A CATALYST IS A SUBSTANCE THAT INFLUENCES THE RATE OF A CHEMICAL REACTION. IN PARALLEL, 'CATALYSTS OF EVOLUTION' DESCRIBES THE ROLE OF HUMANS IN EITHER SPEEDING OR RESTRAINING NATURAL AND ARTIFICIAL SELECTION.

FAKE NATURE
A REPLICA OF REAL NATURE. FAKE NATURE PROJECTS THE ILLUSION OF LIFE, BE IT PLANT OR ANIMAL, ON HUMAN-MADE OR INANIMATE OBJECTS. PLASTIC FLOWERS AND ROBOTIC DOGS ARE AN EXAMPLE OF FAKE NATURE.

FIFTH LIMB
A PIECE OF TECHNOLOGY THAT IS SO NECESSARY OR FAMILIAR THAT IT BEGINS TO FEEL LIKE A NATURAL PART OF THE BODY. SEE ALSO: INTIMATE TECHNOLOGY.

FITNESS BOOSTERS
ANY MEANS, SUCH AS MAKEUP OR PLASTIC SURGERY, TO INCREASE APPARENT 'FITNESS.' THESE BOOSTERS ARE USED TO INCREASE THE PERCEPTION OF AN INDIVIDUAL AS A WORTHY MATE.

FOOD TECHNOLOGY
WAYS TO ENGINEER FOOD TO MAKE IT BETTER THAN WHAT WE CAN FIND IN OLD NATURE.

GENETIC SURPRISES
UNEXPECTED OUTCOMES AND CONSEQUENCES OF GENETIC ENGINEERING.

GEO-ENGINEERING
THE PURPOSEFUL ALTERATION OF THE PLANET'S CLIMATE TO IN ORDER TO COUNTER THE EFFECTS OF GLOBAL WARMING. PROPOSALS RANGE FROM INCREASING THE SOLAR REFLECTIVITY OF THE EARTH TO LARGE-SCALE CARBON SEQUESTRATION.

GUIDED GROWTH
DESIGN METHODOLOGY THAT FOCUSES ON STEERING OF GROWTH PROCESSES, WHILE RESPECTING THEIR AUTONOMY. IN CONTRADICTION TO THE 20TH CENTURY TOP-DOWN APPROACH OF MODULAR CONSTRUCTION AND BLUEPRINT-BASED ENGINEERING.

HUMANE TECHNOLOGY
TECHNOLOGY THAT ADAPTS TO HUMANS, RATHER THAN FORCING HUMANS TO ADAPT TO IT. IT WORKS WITH OUR BODIES, SENSES, AND INSTINCTS, AND TAKES HUMAN VALUES AS A CORNERSTONE OF ITS DEVELOPMENT.

HYPERNATURE
'SO-CALLED NATURE' THAT HAS BEEN EXAGGERATED AND ENHANCED BY HUMAN DESIGN, SUCH AS SEEDLESS GRAPES OR GENETICALLY MODIFIED SALMON THAT GROWS TWICE AS FAST A WILD SALMON. IT IS ARTIFICIAL, YET AUTHENTIC. ARGUABLY, HYPERNATURE IS A CULTURAL RATHER THAN NATURAL CATEGORY.

IMAGE CONSUMPTION
THE CONSUMPTION OF REPRESENTATIONS, RATHER THAN FROM THE THING ITSELF. IMAGE CONSUMPTION IS THE FIGURATIVE 'EATING' OF ATTRACTIVE IMAGES, SUCH AS PORNOGRAPHY OR FOOD PHOTOGRAPHY.

INFORMATION DECORATION
THE INCORPORATION OF DATA INTO UNOBTRUSIVE, DECORATIVE DISPLAYS, SUCH AS TRAINS THAT BLUSH BEFORE THEY LEAVE, OR WATER FOUNTAINS THAT RISE AND FALL ACCORDING TO STOCK PRICES. INFORMATION DECORATION IS A REACTION TO THE UBIQUITY OF SCREENS AS AN UNINSPIRED MEANS OF CONVEYING INFORMATION.

INTIMATE TECHNOLOGY
TECHNOLOGY ATTACHED TO THE SKIN OR SURGICALLY IMPLANTED INSIDE THE BODY. TECHNOLOGY CAN ALSO BE 'INTIMATE' IF IT IS CRUCIAL TO OUR SENSE OF IDENTITY.

MEDIA SCHEMA
KNOWLEDGE WE POSSESS ABOUT WHAT MEDIA ARE CAPABLE OF AND WHAT WE SHOULD EXPECT FROM THEM IN TERMS OF THEIR DEPICTIONS: REPRESENTATIONS, TRANSLATIONS, DISTORTIONS, ETC. MEDIA SCHEMAS ENABLE US TO REACT TO MEDIA IN A CONTROLLED WAY ("DON'T BE SCARED, IT'S ONLY A MOVIE."). THE TERM MEDIA SCHEMAS STEMS FROM THE CONCEPT OF SCHEMAS, WHICH IN PSYCHOLOGY AND COGNITIVE SCIENCES IS DESCRIBED AS A MENTAL STRUCTURE THAT REPRESENTS SOME ASPECT OF THE WORLD.

MEME
AN IDEA, BEHAVIOR OR STYLE THAT SPREADS BETWEEN INDIVIDUALS. WHILE GENES TRANSMIT BIOLOGICAL INFORMATION, MEMES ARE SAID TO TRANSMIT IDEAS AND BELIEF INFORMATION.

NEXT NATURE
THE NATURE CAUSED BY PEOPLE.

NONGENETIC EVOLUTION
EVOLUTIONARY PROCESSES THAT ARE NOT BASED ON CHANGES IN GENES. FOR INSTANCE, WHEN PRODUCTS OR IDEAS EVOLVE LIKE ORGANISMS, SUCH AS THE DEVELOPMENT OF THE RAZOR RAZORIUS GILLETTUS OR THE SHELL CORPORATE LOGO.

NOOSPHERE
THE SPHERE OF HUMAN THOUGHT ON EARTH, IN LEXICAL ANALOGY TO ATMOSPHERE AND BIOSPHERE. DERIVED FROM THE GREEK ⊠⊠⊠⊠ (NOUS 'MIND'), THE NOOSPHERE OR MINDSPHERE IS AN INTANGIBLE EXTENSION OF THE BIOSPHERE. JUST AS THE EMERGENCE OF LIFE FUNDAMENTALLY TRANSFORMED THE GEOSPHERE, THE EMERGENCE OF HUMAN COGNITION FUNDAMENTALLY TRANSFORMS THE BIOSPHERE.

OLD NATURE
WHEREVER HUMANS AREN'T: THE DEEP OCEAN, THE FAR SIDE OF THE MOON, PERHAPS SOME PARTS OF ANTARCTICA. INCLUDED EVERY ECOSYSTEM THAT PREDATES THE INTRODUCTION OF HUMAN-MADE NEXT NATURE, SUCH AS THE JURASSIC OR CAMBRIAN PERIODS.

POSTHUMAN
IN PHILOSOPHY, THE POSTHUMAN IS AN ENTITY THAT EXISTS AFTER OUR CURRENT DEFINITIONS OF 'HUMAN' HAVE BECOME OBSOLETE. IT IS SOMETIMES USED INTERCHANGEABLY WITH THE TERM 'TRANSHUMAN' TO DESCRIBE A PERSON THAT HAS BEEN MADE UNRECOGNIZABLE AS A HUMAN, USUALLY THROUGH SURGICAL OR TECHNOLOGICAL AUGMENTATIONS.

PROGRESSIVE NOSTALGIA
A DESIGN STRATEGY THAT SMOOTHS THE BUMPS OF TECHNOLOGICAL PROGRESS BY LINKING NEWFANGLED TECHNOLOGIES WITH FAMILIAR OR OLD-FASHIONED ONES. ACCORDING TO PROGRESSIVE NOSTALGIA, CONSUMERS FEEL MORE COMFORTABLE WITH NEW INVENTIONS THAT ARE WRAPPED IN RECOGNIZABLE PACKAGING. 'NOSTALGIA' CAN REFER TO A PHENOMENON FROM YOUR YOUTH, YOUR GRANDPARENTS' LIFE, OR EVEN ANCIENT HISTORY.

REAL VIRTUALITY
WHEN VIRTUAL WORLDS, SUCH AS MASSIVELY-MULTIPLAYER ONLINE ROLE-PLAYING GAMES (MMORPGS), BEGIN TO FEEL AS RICH, COMPLEX, AND WORTHWHILE AS NON-DIGITAL REALITY. IF YOU SPEND MORE TIME LIVING IN A GAME THAN IN REAL LIFE, THE VIRTUAL HAS BECOME YOUR NEW REALITY.

RETRIBALIZATION
WHEN AN INDUSTRIALIZED SOCIETY RETURNS TO A MORE TRIBAL MODE OF LIFE. COINED BY MEDIA THEORIST MARSHALL MCLUHAN IN 1969, RETRIBALIZATION IS TRIGGERED BY NEW MEDIA TECHNOLOGIES SUCH AS TELEVISION, CELL PHONES, AND ONLINE SOCIAL NETWORKS. THESE TECHNOLOGIES RETURN US TO A PRE-LITERATE MODE WHERE COMMUNICATION AND SOCIALIZATION IS IMMERSIVE AND INSTANTANEOUS.

SECONDARY ORALITY
THE RETURN TO A LARGELY ORAL, RATHER THAN WRITTEN, FORM OF EXISTENCE. SECONDARY ORALITY IS AN IMPORTANT ASPECT OF RETRIBALIZATION.

SENTIENT SPACES
SPACES THAT HAVE THE ABILITY TO FEEL OR PERCEIVE SUBJECTIVELY.

SINGULARITY
AN EVENT SO MOMENTOUS THAT PEOPLE WHO LIVE AFTER THE EVENT CANNOT UNDERSTAND LIFE BEFORE IT. THE INVENTION OF LANGUAGE WAS A SINGULARITY; HUMANS STRUGGLE TO UNDERSTAND WHAT IS IT LIKE TO HAVE CONSCIOUS THOUGHT WITHOUT THE USE OF WORDS. FUTURISTS LIKE RAY KURZWEIL POSIT THAT THE FUTURE MERGING OF INTELLIGENT COMPUTERS WITH HUMAN MINDS WILL GENERATE A SINGULARITY.

SKEUOMORPH
A REPLACEMENT OBJECT THAT MIMICS THE FORM OF THE ORIGINAL, EVEN WHEN THE ORIGINAL FORM IS NOT NECESSARY TO ENSURE FUNCTIONALITY. IN NEXT NATURE, DUTIFULLY PRODUCING SKEUOMORPHS OF NATURAL FORMS TENDS TO HOLD BACK INNOVATION, FOR INSTANCE, BY PRODUCING A FOOT-SHAPED PROSTHETIC WHEN A BLADE-SHAPED ONE MAY FUNCTION BETTER.

SOCIETY OF SIMULATIONS
A SOCIETY ONE WHERE REPRESENTATIONS OF THINGS ARE OFTEN MORE INFLUENTIAL, MEANINGFUL AND SATISFYING THAN THE THING ITSELF. FOR INSTANCE, A VIDEO GAME ENVIRONMENT MAY BE MORE SATISFYING THAN THE REAL ENVIRONMENT, OR A GENERIC SHIRT MAY BE INSTANTLY WORTH MORE BECAUSE OF A SEWN-ON LOGO. THERE IS NO DISTINCTION BETWEEN THE VIRTUAL AND THE REAL.

SUBURBAN UTOPIA
A COMPLETELY PLANNED ENVIRONMENT, WITH AGREEABLE HOUSES, MANICURED GARDENS AND TIDY PEOPLE, DESIGNED TO PROVIDE FOR EVERY NEED, YET ITS INHABITANTS MAY FEEL AS THOUGH A CERTAIN *JE NE SAIS QUOI* IS MISSING.

SYMBIOSIS
A LONG-TERM, MUTUALLY BENEFICIAL INTERACTION BETWEEN ANY TWO ENTITIES, SUCH AS BETWEEN A BEE AND A FLOWER. IN NEXT NATURE, SYMBIOSIS CAN ALSO OCCUR WITH TECHNOLOGIES, SUCH AS BETWEEN A HUMAN AND A CELLPHONE.

SYSTEM ANIMALS
ENTIRELY CULTURAL, TECHNOLOGICAL BEINGS, ALSO KNOWN AS HOMO SAPIENS. HUMANS ARE NOT EQUIPPED TO LIVE IN ANY ENVIRONMENT WITHOUT THE AID OF TOOLS, AND OUR INSTINCTS ALONE ARE NOT SUFFICIENT FOR US TO DEPEND ON THEM FOR SURVIVAL. HOWEVER, BECAUSE OUR TOOLS AND CULTURAL SYSTEMS ARE SO FLEXIBLE, WE HAVE BEEN ABLE TO COLONIZE JUST ABOUT EVERY TERRESTRIAL HABITAT ON EARTH.

TECHNO-RHETORIC
THE OVERLY OPTIMISTIC ASSUMPTION THAT ALL PROBLEMS CAN BE SOLVED WITH TECHNOLOGY.

WILD SYSTEMS
ANTHROPOGENIC PROCESSES THAT GO FERAL. A COMPUTER VIRUS, FOR EXAMPLE, CONTINUES RUNNING LONG AFTER ITS PROGRAMMER HAS HAD ANY DIRECT ROLE IN HOW IT FUNCTIONS OR WHAT COMPUTERS IT INFECTS.

NEXT NATURE

COVER IMAGE BY KOERT VAN MENSVOORT. ORIGINAL BABY PHOTO BY GETTY IMAGES

RECREATION

COVER IMAGE BY KOERT VAN MENSVOORT, FREE AFTER BOB ROSS

ROONGUTHAI, VIA WIKIMEDIACOMMONS.ORG
120 IMAGE OF MICROBE VIA NEWSWISE.COM
120 PHOTO OF SALMON: SOURCE UNKNOWN
120 FEATHERLESS CHICKEN VIA INHABITAT.COM
120 BANANA VIA WALLPAPERS-GRATUIT.COM
121 STREETLIGHT TREES BY MR_HAYATA, VIA FLICKR
121 RUSTY THE LABRADOODLE, VIA LABRADOODLES.AU.COM
121 PINEBERRY, VIA KABOCHAFASHION.WORDPRESS.COM
121 RAINBOW TULIP, VIA DEVIANTART.NET
122 DEAD COW VIA ARKNATURE.EU
123 TOURIST IN KAYAK MEETS HIGHLAND COW, PHOTO BY PETER VILLERIUS
123 TOURIST ON BICYCLE MEETS HIGHLAND COW, PHOTO BY CORRIE GERRITSMA
124 ILLUSTRATION OF AUROCHS, PUBLISHED IN 1556 IN A BOOK BY SIGISMUND VON HERBERSTEIN
124 DEPICTIONS OF AUROCHS IN BREHM'S LIFE OF ANIMALS / 1927 / ORIGINATOR UNKNOWN
132 FAKE FOR REAL CARD GAME. EDITORIAL TEAM: KOERT VAN MENSVOORT, HENDRIK-JAN GRIEVINK, ROLF COPPENS, ARNOUD VAN DEN HEUVEL, MIEKE GERRITZEN. DESIGN BY HENDRIK-JAN GRIEVINK. (BIS PUBLISHERS 2007, ISBN 978-90-6369-177-6)

WILD SYSTEMS

COVER IMAGE: *STORMY*, TAKEN FROM *MALWAREZ*, VISUALIZATIONS OF WORMS, VIRUSES, TROJANS AND SPYWARE CODE BY ALEX DRAGULESCU

136 *PLANET ROCK*, RENDERING BY RUBEN DAAS
138 *THE BLUE MARBLE*, PHOTOGRAPH OF THE EARTH TAKEN ON DECEMBER 7, 1972 BY THE CREW OF THE APOLLO 17 SPACECRAFT EN ROUTE TO THE MOON. IMAGE BY NASA, VIA WIKIMEDIA COMMONS
140 HIERARCHICAL STRUCTURE OF THE INTERNET (2007). RESEARCH BY SHAI CARMI ET AL, BAR ILAN UNIVERSITY, ISRAEL. VISUALISATION BY: LANET-VI PROGRAM OF I. ALVAREZ-HAMELIN ET AL.
142 VISA RENDERING BY RUBEN DAAS
150 *DOGGERLAND* (BACKGROUND), PHOTO MANIPULATION BY RUBEN DAAS
150 *DOGGERLAND* (INSET), A SPECULATIVE SATELLITE RENDERING OF WESTERN EUROPE AS IT (PRESUMABLY) WAS 10.000 YEARS AGO
152 MAPS FROM THE *ATLAS OF THE REAL WORLD*, BY DANIEL DORLING, MARK NEWMAN AND ANNA BARFORD, VIA WORLDMAPPER.ORG
154 HURRICANE CONTROL, PHOTO MANIPULATION BY RUBEN DAAS
154 *OBAMA SPEECH* BY HENDRIK-JAN GRIEVINK, IMAGE VIA CNN.COM
156 ERUPTION OF MOUNT PINATUBO IN 1991, IMAGE VIA SIMPLECLIMATE.WORDPRESS.ORG
158 *VACCINATORS WITHOUT BORDERS*, INFOTIZEMENT BY HENDRIK-JAN GRIEVINK
(ART) AND KOERT VAN MENSVOORT (COPY)
160 *MAPLE OF RATIBOR*, ORIGINALLY PUBLISHED IN THE PICTURE MAGAZINE (1893)
160 ROOT BRIDGE IN CHERRAPUNJI, INDIA.
161 PHOTOGRAPHER UNKNOWN
162 SPECULATIVE DESIGN FOR A HERBICIDE SPRAYER, PART OF THE *GROWTH ASSEMBLY* PROJECT BY ALEXANDRA DAISY GINSBERG AND SASCHA POHFLEPP
164 THE FORTRESS OF SOLITUDE, STILL TAKEN FROM FROM *SUPERMAN* (R. DONNER, 1978)
164 CRYSTAL CAVE DISCOVERED BY INDUSTRIAS PEÑOLES MINERS. IMAGE BY NATGEO.TV
165 *VENUS*, NATURAL CRYSTAL CHAIR BY TOKUJIN YOSKIOKA
165 ARCHITECTURE MODELS BY ARANDA & LASCH
166 VENICE RENDERING BY ARCHITECTS GMJ
167 ARTIFICIAL REEF UNDER VENICE FOUNDATIONS, COMPUTER RENDERING BY CHRISTIAN KERRIGAN
167 PROTOCELL OIL DROPLETS, IMAGES VIA RACHEL ARMSTRONG
169 COMPUTER VISUALIZATION OF THE VENICE REEF BY CHRISTIAN KERRIGAN
170 ADVERTISEMENT BY RAYFISHFOOTWEAR.COM
172 *PLASTIC GLOBE* BY HENDRIK-JAN GRIEVINK
174 PLASTIC NURDLES ON THE BEACH, PHOTO BY PAUL L. NETTLES, VIA FLICKR
175 STYROFOAM GRANULARS ON A BEACH ROCK, PHOTO BY RALPH HOCKENS, VIA FLICK
176 *PLASTIC FOOD CHAIN*, ILLUSTRATION BY HENDRIK-JAN GRIEVINK
177 PLASTIC FOUND IN THE STOMACH OF ONE DEAD LAYSAN ALBATROSS CHICK, PHOTO BY SUSAN MIDDLETON
177 BIO-DEGRADABLE PLASTIC BOTTLE, VIA TREEHUGGER.COM
179 *PLASTIC STYLE, LEATHER BAG*, INFOTIZEMENT BY HENDRIK-JAN GRIEVINK. IMAGE FROM *INDEX, A STATEMENT ABOUT CONTENT / A BAG COLLECTION* BY FEMKE DE VRIES
186 COW, SOURCE UNKNOWN
187 ANCIENT CHINESE SPADE CURRENCY (LEFT), VIA LEPARADOUX.FREE.FR
187 ANCIENT CHINESE SPADE CURRENCY (RIGHT), VIA T.K. MALLON-MCCORGRAY
188 COINS, CLOCKWISE STARTING LOWER LEFT:
1. VIA PUBLICDOMAINPICTURES.COM
2. VIA WWW.PRINCETON.EDU
3. SOURCE UNKNOWN
4. SOURCE UNKNOWN

189 CHINESE BANKNOTE VIA NUMISMONDO.COM
190 OBVERSE OF U.S. ONE DOLLAR BILL (SERIES 2003). VIA WIKIMEDIA COMMONS
190 CREDIT CARD, VIA AMERICAN EXPRESS
191 SIM CARD, SOURCE UNKNOWN
191 M-PESA TRANSACTION, VIA INTEGRO.CO.KE
191 M-PESA KENYA MOBILE CASH TRANSFER AT SAFARICOM SHACK, VIA BLOGS.DFID.GOV.UK
191 *MOBILE PHONE PROVIDERS BECOME BANKS*, IMAGES BY KOERT VAN MENSVOORT
194 RAINFOREST, PHOTO BY TROPIC - JOURNEYS IN NATURE, AN AWARD-WINNING ECOTOURISM COMPANY WHICH HAS BEEN OPERATING IN ECUADOR AND THE GALAPAGOS ISLANDS SINCE 1994. IMAGE VIA FLICKR
195 DETAIL FROM *CREDIT ON COLOR* BY FABIENNE VAN BEEK (GRAPHIC DESIGN MUSEUM, BREDA, THE NETHERLANDS), WITH CONTRIBUTIONS OF MAREN KOEHLER, ADAM EEUWENS & ARJAN MAASLAND
198 *ECO COIN*, DESIGN BY NEXT NATURE STUDIO
199 ECO CURRENCY INFOGRAPHICS BY MARCEL VAN HEIST, BILLY SCHONEBERG AND JOP JAPENGA
200 *THE WORLD CARD WOOD*, INFOTIZEMENT BY ESHAN BAHA AND HAKKI ALTUN

OFFICE GARDEN

COVER IMAGE: *AUGMENTED REALITY WINDSCREEN*, PHOTO BY ERIK PAWASSER, COURTESY OF CONDÉ NAST

204 *FARMERS COUPLE* BY HENDRIK-JAN GRIEVINK
206 *TIES* BY HENDRIK-JAN GRIEVINK
208 *OFFICE FACTORY* BY RUBEN DAAS
210 *C.B.D.* BY HENDRIK-JAN GRIEVINK
212 *TOUCH* BY HENDRIK-JAN GRIEVINK
220 *CLOCKS*, SOFTWARE BY MACSLOW, MIRCO MÜLLER. SCREENSHOT BY ATTILIO DREI
222 *ROLEX* INFOTIZEMENT BY GRRR (SWANNY MOUTON)
224 *GOOGLE HOME SEARCH* BY GRRR (AFONSO GONSALVES AND SWANNY MOUTON)
226 -- RENDERED IMAGERY OF HUMAN BRAINS
227 -- INFORMATION VISUALIZATION OF THE INNER WORKS OF THE HUMAN BRAIN
228 PLACEBO BUTTONS, SOURCE UNKNOWN
228 -- GIANT REMOTE, SOURCE UNKNOWN
229 MECHANICAL AND ELECTRONIC QWERTY KEYBOARD, SOURCES UNKNOWN
229 *TWITTER IMPLANT* BY HENDRIK-JAN GRIEVINK
229 STILLS FROM YOUTUBE VIDEO *HP COMPUTERS ARE RACIST*
231 *CITY SOUNDS* INFOTIZEMENT BY INGO VALENTE
232 *BOOMERANGED METAPHORS*, CLOCKWISE, STARTING TOP LEFT:
1. METAL TRASHCAN, SOURCE UNKNOWN
2. TRASHCAN ICON, APPLE OS7
3. *CARDBOARD PIXEL TRASHCAN* BY CODECO
4. *BIN* BY FRONT

234 PLAYSTATION AD © PLAYSTATION
234 *AFTER MICROSOFT*, PHOTOGRAPH OF HILL IN SONOMA VALLEY (CALIFORNIA, USA) BY CHARLES O'REAR
234 SCREENSHOT FROM COMPUTER GAME *SIM CITY*
234 *CITY SKYLINE* OF TORONTO
235 *WORLD OF WORLD OF WARCRAFT* BY PARODY NEWS SITE THEONION.COM
235 VIDEO STILLS FROM *COLLATERAL MURDER*, VIA WIKILEAKS.ORG
236 *FARMVILLE*, VIA FARMVILLEFREAK.COM
238 IMAGES OF VIRTUAL REAL ESTATE ON SECOND LIFE VIA ANSHECHUNG.COM
240 GOLD FARMERS IN NANJING, CHINA. IMAGE BY JULIAN, VIA FLICKR

SUPERMARKET

ANTROPOMORPHOBIA

364 *MAN ON A LAWN MOWER* BY DUANE HANSON
365 DUMPING SILICONE GIRLFRIEND, VIA WEIRDNEWSFILES
365 *GIMINOID HI-1* BY PROF. HIROSHI ISHIGURO
366 *BE NORMAL*, SOCIETY OF MEDICATION INFOTIZEMENT BY HENDRIK-JAN GRIEVINK
368 *OVERSTEPPING* BY JULIE RRAP (2001). IMAGE COURTESY THE ARTIST AND ROSLYN OXLEY9 GALLERY (SYDNEY, AUSTRALIA)
369 *THE FLUIDHAND*, CREATED BY RESEARCH CENTER KARLSRUHE
369 *TOOTH PHONE*, A SPECUALTIVE DESIGN BY JAMES AUGER AND JIMMY LUIZEAU
369 *CONDUCTIVE SKIN* BY BECKY PILDITCH, MATT JOHNSON, ISABEL LIZARDI AND BIBI NELSON
369 *USB PROSTHETIC* BY JARRY JALAVA
370 *PUSSYCAT DOLLS*, THE UNCANNY GIRL GROUP DESIGNED BY CHOREOGRAPHER ROBIN ANTIN
372 LIPSTICK, VIA FREAKINGNEWS.COM
372 PEACOCK, VIA FLICKR
373 DETAIL FROM AN ADVERTISEMENT FOR BAY COSMETICS
374 *AGING BARBIE*, CREATOR UNKNOWN
374 PAMELA ANDERSON, SOURCE UNKNOWN
374 PORTRAIT OF BEAUTY INDUSTRY ACTIVIST SUNNY BERGMAN BY MANON VAN DER ZWAAL
374 SOFTWARE THAT RANKS BEAUTY BY AMIT KAGIAN, M.SC. GRADUATE FROM THE SCHOOL OF COMPUTER SCIENCES AT TEL AVIV UNIVERSITY.
375 *DAVID*, BY MICHELANGELO
375 *IF YOU DON'T MOVE, YOU GET FAT* (2007), IMAGE FROM AD CAMPAIGN CREATED BY SCHOLZ & FRIENDS FOR THE GERMAN OLYMPIC COMMITTEE
376 *MY INTERPRETATION OF A GENERATION*, MODIFIED PHOTO PORTRAITS BY BOUDEWIEN VAN DEN BERG
379 *TORRE AGBAR* (2005), BY JEAN NOUVEL
380 BIG NETWORK HARD DRIVE BY LA CIE
380 *HAL 9000*, THE MURDEROUS, HEURISTICALLY PROGRAMMED ALGORITHMIC COMPUTER FROM *2001: A SPACE ODYSSEY* (STANLEY KUBRICK, 1968). IT IS OFTEN CONJECTURED THAT THE NAME HAL WAS BASED ON A ONE-LETTER SHIFT FROM THE NAME IBM.
381 *SENSEO*, A REGISTERED TRADEMARK FOR A COFFEE BREWING SYSTEM FROM DUTCH COMPANIES PHILIPS AND DOUWE EGBERTS, A SUBSIDIARY OF SARA LEE CORPORATION. THE SENSEO WAS DEVELOPED IN 2011 BY WELL DESIGN (THE NETHERLANDS)
382 STILLS FROM VIRAL VIDEO OF OFFICE WORKER SMASHING COMPUTER → WWW.YOUTUBE.COM/WATCH?V=ZZR6ARA8NTU
383 *AIBO ROBOT DOG* BY SONY, INITIAL DESIGN BY HAJIME SORAYAMA
384 *CLIPPIT* (OR NICKNAMED *CLIPPY*), THE ÜBER ANNNOYING PAPERCLIP WE KNOW OFFICIALLY AS THE 'MICROSOFT OFFICE ASSISTANT', DESIGNED BY KEVAN J. ATTEBERRY
385 *SAYA* THE ROBOT TEACHER, DEVELOPED BY PROFESSOR HIROSHI KOBAYASHI (TOKYO UNIVERSITY OF SCIENCE)
385 *HEAD KNIFE BLOCK* BY MAARTEN BAAS
386 *ROOMBA* BY IROBOT
386 *AUTIMACY*, BY SIMONA KICUROVSKA
387 *SMILEY DUST PAN* BY DULTON
388 *DISNEYLAND MEMORIAL ORGY* IMAGE © PAUL KRASSNER, ART BY WALLY WOOD. AVAILABLE ONLY AT PAULKRASSNER.COM
389 *MICKEY PFIZER* BY HENDRIK-JAN GRIEVINK
390 *JESSICA RABBIT* FROM *WHO FRAMED ROGER RABBIT* (ROBERT ZEMECKIS, 1988)
390 HYPERREAL JESSICA RABBIT ON THE COVER OF PLAYBOY, 1988
390 HUGH HEFNER AND HIS BUNNIES, PHOTOGRAPHER UNKNOWN
395 *DNA WORLD MAP* BY NICO VEGT
404 *ANIME EYE LENSES* INFOTIZEMENT BY HENDRIK-JAN GRIEVINK

BACK TO THE TRIBE

COVER IMAGE: SCREENSHOT OF *WORLD OF WARCRAFT*, BY BLIZZARD ENTERTAINMENT, INC.

408 SAN ARTISANS IN GRASHOEK (NAMIBIA), PHOTO BY BENJAMIN COBB, VIA FLICKR
410 RUSH HOUR IN TOKYO, PHOTO BY KEVIN JAAKO, VIA FLICKR
412 NAKED MAN, PHOTO BY CHARLES ROFFEY, VIA FLICKR
414 *CHINESE FACE* BY HENDRIK-JAN GRIEVINK
416 SCREENSHOT OF *WORLD OF WARCRAFT*, BY BLIZZARD ENTERTAINMENT, INC
418 *RETRIBE* BY NEXT NATURE STUDIO
424 *MCLUHAN'S TWEETS* BY HENDRIK-JAN GRIEVINK
426 COCA-COLA PAVILLION AT EXPO 2010 (SHANGHAI, PEOPLE'S REPUBLIC OF CHINA), PHOTO BY HENDRIK-JAN GRIEVINK
426 BRAND GEOGRAPHY INFOGRAPHIC BY HENDRIK-JAN GRIEVINK
426 *STARBUCKS PASSPORT STAMP* BY HENDRIK-JAN GRIEVINK
428 *FACEHEART* BY HENDRIK-JAN GRIEVINK
428 *COLA-PEPSI CAN* BY FIRMA
428 DR PEPPER TATTOO, CREATOR UNKNOWN
429 *THE MOST EXPENSIVE COAT IN THE WORLD* BY SILKE WAWRO (VOLKSWARE)
430 *COLA RIVER* ILLUSTRATION BY HENDRIK-JAN GRIEVINK
431 IMAGES BY THE *COLALIFE* PROJECT, INITIATED BY ENGINEER SIMON BERRY → WWW.COLALIFE.ORG
432 SCREENSHOT OF *WORLD OF WARCRAFT*, BY BLIZZARD ENTERTAINMENT, INC
433 *ANGRY MOOSE*, ILLUSTRATION BY HENDRIK-JAN GRIEVINK
433 VIRTUAL MILGRAM EXPERIMENT, CATALAN POLYTECHNIC UNIVERSITY
434 FUNERAL AMBUSH IN WORLD OF WARCRAFT, SOURCE UNKNOWN
435 *WEB 2.0 SUICIDE MACHINE* BY MODDR_LAB, PROJECT LOGO BY BURO DUPLEX
435 VIDEO STILL FROM 'YOU HAVE ZERO FRIENDS' (SOUTH PARK SEASON 14, EPISODE 4)
435 SOCIAL MEDIA ICONS, VARIOUS SOURCES
436 *COCA COLA VASE* (1997) BY AI WEIWEI
437 *WIFI DOWSING ROD* BY MIKE THOMPSON
438 COLLECTION OF LIGHTBULBS BY PHILIPS
440 *CASA MILA* (1905-1910), BY ANTONI GAUDI (ALSO KNOWN AS *LA PEDRERA*). PHOTO BY S3K VIA FLICKR
441 *BONE CHAIR* (2006) BY JORIS LAARMAN LAB
442 BOOKSHELF RUNNING ON AN APPLE IPAD, GRAPHIC USER INTERFACE BY APPLE
443 IDIAL APP → HTTP://REP.SOSIPHONE.COM

THE NEXT NATURE NETWORK EXPLORES THE CHANGING RELATION BETWEEN PEOPLE, NATURE AND TECHNOLOGY. WE FUNCTION AS A GLOBAL THINK- AND DESIGN TANK THAT AIMS TO VISUALIZE, RESEARCH AND UNDERSTAND THE IMPLICATIONS OF THE NATURE CAUSED BY PEOPLE. WE INITIATE PUBLICATIONS IN VARIOUS MEDIA AND ORGANIZE EVENTS. THIS BOOK COULD NOT HAVE BEEN REALIZED WITHOUT THE RELENTLESS EFFORT OF THESE GOOD PEOPLE.

RACHEL ARMSTRONG

IS AN INTERDISCIPLINARY PRACTITIONER WITH A BACKGROUND IN MEDICINE. HER WORK USES ALL MANNERS OF MEDIA TO ENGAGE AUDIENCES AND BRING THEM INTO CONTACT WITH THE LATEST ADVANCES IN SCIENCE AND THEIR REAL POTENTIAL THROUGH THE INVENTIVE APPLICATIONS OF TECHNOLOGY, TO ADDRESS SOME OF THE BIGGEST PROBLEMS FACING THE WORLD TODAY. RACHEL WROTE THE ESSAY 'SELF REPAIRING ARCHITECTURE'.

MATTIJS BLIEK

DIGITAL NATIVE, SKILLED IN DESIGN AS WELL AS PROGRAMMING. MATTIJS WORKS AT GRRR AND COLLABORATED ON THE REDESIGN OF NEXTNATURE.NET AND MUCH OF ITS UPDATES SINCE 2010.

ROLF COPPENS

DESIGNER, MULTIMEDIA CONSULTANT AND FOUNDING PARTNER OF GRRR, THE AMSTERDAM-BASED DESIGN AGENCY FOR INTERACTIVE MEDIA THAT DESIGNED THE NEXTNATURE.NET WEBSITE. ROLF AIMS TO CREATE MEANINGFUL INTERACTIONS BETWEEN PEOPLE AND SYSTEMS BY COMBINING TECHNOLOGY WITH STRONG VISUALS AND CONCEPTS. ROLF HAS BEEN A CONTRIBUTOR TO NEXTNATURE.NET FROM THE FIRST DAY. FOR THIS BOOK, ROLF CREATED SOME OF THE INFOTIZEMENTS. HE ALSO COLLABORATED ON THE FAKE FOR REAL MEMORY GAME.

RUBEN DAAS

FREELANCE GRAPHIC DESIGNER WHO, DURING HIS INTERNSHIP AT THE NEXT NATURE STUDIO IN AMSTERDAM, COLLABORATED ON THE DESIGN OF BOTH THIS BOOK AND THE NANO SUPERMARKET PROJECT.

NATALIE DIXON

FORMER MAGAZINE EDITOR AND DIGITAL PUBLISHER FOR BRANDS SUCH AS NATIONAL GEOGRAPHIC AND SEVENTEEN. SHE RECENTLY LEFT CAPE TOWN TO COMPLETE AN MA IN NEW MEDIA AT THE UNIVERSITY OF AMSTERDAM. HER AREA OF RESEARCH IS THE AFFECTIVE BANDWIDTH OF MOBILE-MEDIATED COMMUNICATION. NATALIE WORKED AS THE SUB-EDITOR ON THIS BOOK.

BERRY EGGEN

FULL PROFESSOR IN AMBIENT INTELLIGENCE AND VICE DEAN OF THE DEPARTMENT OF INDUSTRIAL DESIGN, TU EINDHOVEN. BERRY FOUNDED THE NEXT NATURE LAB AT EINDHOVEN UNIVERSITY OF TECHNOLOGY TOGETHER WITH KOERT VAN MENSVOORT. HE IS ALSO A BOARD MEMBER OF THE NEXT NATURE FOUNDATION. BERRY WROTE 'THE SOUND OF THE BLUE CANARY'.

MIEKE GERRITZEN

ONE OF THE FIRST DESIGNERS INVOLVED IN THE DEVELOPMENT OF DIGITAL MEDIA IN THE NETHERLANDS. INITIATOR OF THE PUBLICATIONS CATALOGUE OF STRATEGY, EVERYONE IS A DESIGNER, MOBILE MINDED AND STYLE FIRST. TOGETHER WITH KOERT VAN MENSVOORT SHE COMPILED THE FIRST NEXT NATURE POCKET BOOK AND ORGANIZED THE NEXT NATURE POWER SHOW EVENTS IN AMSTERDAM, ESSEN AND LOS ANGELES. MIEKE IS CURRENTLY DIRECTOR OF THE GRAPHIC DESIGN MUSEUM IN BREDA AS WELL AS CHAIR OF THE NEXT NATURE FOUNDATION.

HENDRIK-JAN GRIEVINK

DESIGNER, WRITER AND TUTOR IN EDITORIAL DESIGN. HIS WORK IS BEST DESCRIBED AS VISUAL CULTURE CRITICISM THROUGH POPULAR AND RECOGNIZABLE IMAGES. HENDRIK-JAN JOINED THE NEXT NATURE NETWORK IN 2006 AND DESIGNED THE FAKE FOR REAL MEMORY GAME IN 2007. SINCE THEN, HE HAS BEEN RESPONSIBLE FOR THE ART DIRECTION OF MOST NEXT NATURE RELATED PROJECTS, INCLUDING THE NANO SUPERMARKET AND THIS BOOK. EVERY PAGE OF THIS BOOK IS LOVINGLY CONCEIVED BY HENDRIK-JAN AND KOERT VAN MENSVOORT.

ALLISON GUY

WRITER, EDITOR, AND ILLUSTRATOR. IN 2009, SHE MOVED FROM NEW YORK TO THE NETHERLANDS TO STUDY NEW MEDIA AT THE UNIVERSITY OF AMSTERDAM. HER RESEARCH INTERESTS INCLUDE URBAN ORGANISMS, CYBORG ART, AND THE BLURRING BOUNDARIES BETWEEN THE HUMAN AND THE ANIMAL. SHE WORKED AS A SUB-EDITOR ON THIS BOOK, AND CONTRIBUTES TO THE NEXT NATURE WEBSITE.

KARL GRANDIN

ARTIST, DESIGNER AND CO-FOUNDER OF THE CLOTHING BRAND CHEAP MONDAY. TOGETHER WITH ROLF COPPENS, KARL DESIGNED THE FIRST VERSION OF NEXTNATURE.NET, WHICH ALREADY FEATURED VISUAL ELEMENTS LIKE THE PEAR LOGO AND THE CORPORATE ANIMAL LOGO PATTERN, KEY ELEMENTS OF THE VISUAL IDENTITY OF NEXT NATURE.

BAS HARING

PHILOSOPHER, TV PRESENTER AND WRITER OF POPULAR SCIENCE LITERATURE. AMONG HIS BOOKS ARE CHEESE AND THE THEORY OF EVOLUTION, THE IRON WILL, AND FOR A SUCCESSFUL LIFE. BAS IS PROFESSOR IN THE PUBLIC UNDERSTANDING OF SCIENCE AT UNIVERSITY OF LEIDEN. FOR THIS BOOK BAS WROTE THE ESSAY NEXT NATURE SERVICES.

ARNOUD VAN DEN HEUVEL

SELF-EMPLOYED MULTIMEDIA-DESIGNER, OUT-OF-THE-BOX THINKER, AND CO-EDITOR OF NEXTNATURE.NET. HE IS INTERESTED IN SYSTEMS, CONSUMERISM, THE FUNCTION AND DEFINITION OF NATURE, WITH A SPECIAL FASCINATION IN QUANTITY, APPEARANCES AND COLORS OF OBJECTS PRODUCED BY PEOPLE. ARNOUD CREATED ANIMATIONS FOR NEXT NATURE POWER SHOW EVENTS AND -DVD AND HAS BEEN A CONTRIBUTOR TO NEXTNATURE.NET FROM THE FIRST DAY. HE ALSO COLLABORATED ON THE FAKE FOR REAL MEMORY GAME.

MINKE KAMPMAN

DIGITAL NOMAD THAT WANDERS BETWEEN DIFFERENT AREAS OF EXPERTISE AND INTERESTS. MINKE HOLDS DEGREES IN GRAPHIC DESIGN (BDES) AND NEW MEDIA (MA) AND TEACHES NEW MEDIA THEORY. FOR THIS BOOK, MINKE ARRANGED THE IMAGE RIGHTS.

KEVIN KELLY

WRITER, PHILOSOPHER AND SENIOR MAVERICK AT WIRED MAGAZINE. HE CO-FOUNDED WIRED IN 1993, AND SERVED AS ITS EXECUTIVE EDITOR FROM ITS INCEPTION UNTIL 1999. HE AUTHORED THE BOOKS WHAT TECHNOLOGY WANTS, NEW RULES FOR THE NEW ECONOMY AND THE CLASSIC BOOK ON DECENTRALIZED EMERGENT SYSTEMS, OUT OF CONTROL. KEVIN KELLY PRESENTED AT THE NEXT NATURE POWER SHOW IN LOS ANGELES AND WROTE THE ESSAY THE WORLD WITHOUT TECHNOLOGY IN THIS BOOK.

MICHAEL KLUVER

FREELANCE DESIGNER WHO, DURING HIS INTERNSHIP AT THE NEXT NATURE STUDIO, COLLABORATED ON THE DESIGN OF BOTH THIS BOOK AND THE NANO SUPERMARKET PROJECT.

TIJN KOOIJMANS

INTERACTION DESIGNER AT STUDIO SOPHISTI, A DESIGN COMPANY BASED IN AMSTERDAM SPECIALIZED IN INTERACTIVE PRODUCT DESIGN AND RELATED CONSULTANCY. TIJN CREATED THE NEXT NATURE SPOTTER APP FOR IPHONE, WHICH ALLOWS PEOPLE TO SHARE 'NEXT NATURE' THEY ENCOUNTER IN THEIR ENVIRONMENT.

PETER LUNENFELD

IS A PROFESSOR IN THE DESIGN AND MEDIA ARTS DEPARTMENT AT UCLA. HIS BOOKS INCLUDE THE DIGITAL DIALECTIC, SNAP TO GRID, USER, AND THE SECRET WAR BETWEEN DOWNLOADING AND UPLOADING. HE IS THE CREATOR AND EDITORIAL DIRECTOR OF THE MIT PRESS MEDIAWORK PROJECT. PETER PRESENTED AT THE NEXT NATURE POWER SHOW IN LOS ANGELES AND WROTE THE ESSAY FROM MAINSTREET TO THE MANSION FOR THIS BOOK.

KOERT VAN MENSVOORT

ARTIST, DESIGNER, WRITER, ENGINEER. INVESTING THE STRANGE AND THE BEAUTIFUL IS HIS MAIN INTEREST. THE DISCOVERY OF NEXT NATURE HAS BEEN THE MOST PROFOUND EXPERIENCE OF HIS LIFE SO FAR. HE WROTE THE FIRST ESSAY EXPLORING NEXT NATURE IN 2004, ORGANIZED THE FIRST NEXT NATURE EVENTS TOGETHER WITH MIEKE GERRITZEN AND GAVE OVER 50 LECTURES ON FOUR DIFFERENT CONTINENTS ON THE TOPIC. HE CURRENTLY DIRECTS THE NEXT NATURE FOUNDATION IN AMSTERDAM AND NEXT NATURE LAB AT EINDHOVEN UNIVERSITY OF TECHNOLOGY. EVERY SINGLE PAGE OF THIS BOOK IS LOVINGLY CONCEIVED BY KOERT AND HENDRIK-JAN GRIEVINK.

TRACY METZ

IS A JOURNALIST WITH THE DUTCH DAILY NEWSPAPER NRC HANDELSBLAD, WHERE SHE WRITES ABOUT ARCHITECTURE, LANDSCAPE AND URBAN DESIGN. SHE IS ALSO AN INTERNATIONAL CORRESPONDENT FOR ARCHITECTURAL RECORD AND A CONTRIBUTOR TO METROPOLIS, DOMUS AND GRAPHIS. TRACY PRESENTED AT NEXT NATURE EVENTS IN AMSTERDAM AND GERMANY. EARLY 2011, TRACY AND HER HUSBAND JEAN-BAPTISTE HOSTED AN INSPIRING PEER PREVIEW OF THIS BOOK FOR A SELECT AUDIENCE AT HER HOME IN AMSTERDAM. FOR THE BOOK SHE WROTE THE ESSAY NATURE IS AN AGREEMENT.

JOS DE MUL

FULL PROFESSOR OF PHILOSOPHY OF MAN AND CULTURE AT THE FACULTY OF PHILOSOPHY, ERASMUS UNIVERSITY IN ROTTERDAM, AND SCIENTIFIC DIRECTOR OF THE RESEARCH INSTITUTE PHILOSOPHY OF INFORMATION AND COMMUNICATION TECHNOLOGY. AMONG HIS BOOK PUBLICATIONS ARE: ROMANTIC DESIRE IN (POST) MODERN ART AND PHILOSOPHY, THE TRAGEDY OF FINITUDE AND DILTHEY'S HERMENEUTICS OF LIFE AND CYBERSPACE ODYSSEE. JOS GAVE A LEGENDARY PRESENTATION CALLED 'POETRY OF GENETICS' DURING FIRST NEXT NATURE POWER SHOW IN 2005. FOR THE BOOK HE WROTE THE ESSAY THE 'TECHNOLOGICAL SUBLIME'.

CAROLINE NEVEJAN

IS A RESEARCHER AND DESIGNER FOCUSING ON THE IMPLICATIONS OF TECHNOLOGY ON SOCIETY. SHE HAS BEEN INVOLVED IN INTERDISCIPLINARY PROJECTS FOR MORE THAN 20 YEARS. CURRENTLY SHE WORKS INTERNATIONALLY WITH PROFESSIONALS, ACADEMICS AND ARTISTS ON THE RESEARCH PROJECT WITNESSED PRESENCE.

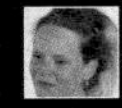

MAARTJE SOMERS

STUDIED CLASSICAL LANGUAGES IN AMSTERDAM, FOLLOWED BY INTERNATIONAL RELATIONS AND DEVELOPMENT STUDIES. FOR MANY YEARS SHE WORKED AS A CULTURE EDITOR FOR DUTCH NEWSPAPER HET PAROOL. SHE IS CURRENTLY EDITOR OF THE BOOKS SECTION OF DUTCH DAILY NRC HANDELSBLAD, REVIEWING NON-FICTION PUBLICATIONS ABOUT GLOBALIZATION, FOOD, INTERNATIONAL DEVELOPMENT AND ECONOMICS. MAARTJE CONTRIBUTED 'THE STORY OF OUR FOOD'.

BRUCE STERLING

AMERICAN SCIENCE FICTION AUTHOR, BEST KNOWN FOR HIS NOVELS AND HIS WORK ON THE MIRRORSHADES ANTHOLOGY, WHICH HELPED DEFINE THE CYBERPUNK GENRE. ACCORDING TO TIME MAGAZINE HE IS, 'ONE OF AMERICA'S BEST-KNOWN SCIENCE-FICTION WRITERS AND PERHAPS THE SHARPEST OBSERVER OF OUR MEDIA-CHOKED CULTURE WORKING TODAY IN ANY GENRE.' BRUCE WROTE THE PREFACE FOR THIS BOOK AND IS A BOARD MEMBER OF THE NEXT NATURE FOUNDATION.

CHRISTINE MITCHELL

RESEARCHER IN COMMUNICATIONS STUDIES AND LECTURER IN NEW MEDIA FROM MONTREAL. HER WORK DEALS WITH THE NATURALNESS OF LANGUAGE AND TRANSLATION BY MACHINES. CHRISTINE WAS A PROOFREADER FOR SOME OF THE ESSAYS IN THIS BOOK.

RONALD VAN TIENHOVEN

IS AN ARTIST AND DESIGNER. THE WIDE FOCUS AND CONTEXT OF HIS WORK ALLOWS HIM TO COOPERATE EXTENSIVELY WITH ARCHITECTS, URBAN DESIGNERS, LANDSCAPE DESIGNERS, ENGINEERS AND FILMMAKERS. VAN TIENHOVEN IS A BOARD MEMBER OF THE NEXT NATURE FOUNDATION AND COACHES AT THE NEXT NATURE LAB AT THE EINDHOVEN UNIVERSITY OF TECHNOLOGY.

TIES VAN DE WERFF

RESEARCHER, WRITER, CURATOR AND TEACHER. AS A CULTURAL SCIENTIST, TIES IS CONSTANTLY HAUNTING THE ZEITGEIST BY USING PHILOSOPHY, HISTORY, AND THE EPOCHAL TRIAD OF ART, SCIENCE AND TECHNOLOGY AS HIS FAVOURITE INSTRUMENTS. BESIDES WRITING AND TEACHING AT MAASTRICHT UNIVERSITY, TIES WORKS IN HIS OFFLINE BLOGGALLERY TANTE NETTY, BASED IN EINDHOVEN. TIES WORKS AS A PROGRAMMER FOR THE NEXT NATURE POWER SHOW AND WROTE THE ESSAY CLOUDING THE BRAIN IN THIS BOOK.

ZACH ZORICH

SENIOR EDITOR AT ARCHAEOLOGY MAGAZINE, FORMER REPORTER AND RESEARCHER AT DISCOVER MAGAZINE. ZACH WROTE THE ESSAY 'SHOULD WE CLONE NEANDERTHALS?'

SPECIAL THANKS TO

AAF VAN ESSEN, AARNOUT BROMBACKER, AD VAN MENSVOORT, ALEXANDRA DAISY GINSBERG, AMIR ADMONI, ANNA TETAS, ANNE BURDICK, ARNE HENDRIKS, ARNOUD TRAA, BAPTIST BRAYÉ, BAS GOUDSMIT, BILLY SCHONENBERG, BRUNO SETOLA, CAROLINE HUMMELS, CAROLINE F. STRAUSS, CHRISTIAN BRAMSIEPE, CORALIE VOGELAAR, DANIEL VAN DER VELDEN, DAVID MENTING, DAVID SPREEKMEESTER, DEBBIE MOLLENHAGEN, DENNIS LODEWIJKS, DICK RIJKEN, EDO PAULUS, EHSAN BAHA, ELLY GERRITZEN, ERIK DAVIS, FIONA RABY, FLIP ZIEDSES DES PLANTES, FLORIS KAAYK, GEERT LOVINK, GERBRAND OUDENAARDEN, GORDON TIEMSTRA, HAKKI ALTUN, HELENA MUSKENS, HENK OOSTERLING, JACK VAN WIJK, JAMES KING, JAN GRIEVINK, JAN VAN DER ASDONK, JEFFREY BRAUN, JOP JAPENGA, JORAN DAMSTEEG, JORIS VAN GELDER, JUDITH DE LEEUW, JULIAN BLEECKER, JURRIAN TJEENK WILLINK, LEV MANOVICH, KEVIN WINGATE, LUKE JOHNSON, KOEN VAN DER HAM, KRISTA TE BRAKE, LUNA MAURER, MARCEL HEIST, MARCO ROZENDAAL, MARCO VAN BEERS, MARGIT LUKACS, MARTIJN VAN MENSVOORT, MARTIJN VERKUIJL, MAZE DE BOER, MENNO STOFFELSEN, MICHIEL SCHWARZ, MIKE THOMPSON, MONIQUE VAN DUSSELDORP, NICOLAS NELSON, NIKO VEGT, OLGA BUSH, OLIVIER OTTEN, ORESTIS TSINALIS, PERSIJN BROERSEN, PETER SIJMONS, QUIRINE RACKÉ, RICARDO PORTHILIO, ROB SCHRÖDER, ROEL WOUTERS, RUDOLF VAN WEZEL, SACHA POHPLEP, SEBASTIAAN PIJNAPPEL, SABINE VAN GENT, SARA KOLSTER, SELBY GILDEMACHER, STEPHAN HOES, SUNNY BERGMAN, SUSANA SOARES, TACO STOLK, TEUN CASTELEIN, TIMO BLEEKER, TINKEBELL, TOBIE KERRIDGE, TOMAS PIETERS, TON MEIJDAM, VERA WINTHAGEN, VINCA KRUK AND WERNER LIPPERT.

CONCEPT, EDITING AND DESIGN
KOERT VAN MENSVOORT
HENDRIK-JAN GRIEVINK

GUEST ESSAYS BY
BRUCE STERLING
TRACY METZ
JOS DE MUL
BAS HARING
CAROLINE NEVEJAN
TIES VAN DE WERFF
BERRY EGGEN
MAARTJE SOMERS
PETER LUNENFELD
ZACH ZORICH
KEVIN KELLY

SHORT CONTRIBUTIONS BY
KOERT VAN MENSVOORT (KVM)
ALLISON GUY (AG)
HENDRIK-JAN GRIEVINK (HJG)
NATALIE DIXON (ND)
ROLF COPPENS (RC)
ARNOUD VAN DEN HEUVEL (AVDH)
MATTIJS BLIEK (MB)
MARCO VAN BEERS (MVB)
TOMAS PIETERS (TP)
JORAN DAMSTEEG (JD)
NIKO VEGT (NV)

SUBEDITING
NATALIE DIXON
ALLISON GUY

NEXT NATURE STUDIO
KOERT VAN MENSVOORT
HENDRIK-JAN GRIEVINK
RUBEN DAAS
MICHAEL KLUVER

PROOFREADING
NATALIE DIXON
ALLISON GUY
CHRISTINE MITCHELL

IMAGE RIGHTS
MINKE KAMPMAN

TRANSLATIONS
LAURA MARTZ
LIESBETH NIEUWENWEG

CONTACT
WWW.NEXTNATURE.NET
WWW.FACEBOOK.COM/NEXTNATURE
INFO@NEXTNATURE.NET

PUBLISHED BY
ACTAR (BARCELONA/NEW YORK)
PART OF ACTARBIRKHÄUSER

ISBN 978-84-92861-53-8

DISTRIBUTION
ACTARBIRKHÄUSERD
BARCELONA – BASEL – NEW YORK
WWW.ACTARBIRKHAUSER-D.COM

ROCA I BATLLE 2
E-08023 BARCELONA
T +34 93 417 49 93
F +34 93 418 67 07
SALESBARCELONA@ACTARBIRKHAUSER.COM

VIADUKTSTRASSE 42
CH-4051 BASEL
T +41 61 5689 800
F +41 61 5689 899
SALESBASEL@ACTARBIRKHAUSER.COM

151 GRAND STREET, 5TH FLOOR
NEW YORK, NY 10013, USA
T +1 212 966 2207
F +1 212 966 2214
SALESNEWYORK@ACTARBIRKHAUSER.COM

PRINTING
SPINHEX & INDUSTRIE (AMSTERDAM)

BINDING
BINDERIJ HEXSPOOR BV (BOXTEL)

PAPER
130 GR ARCTIC VOLUME HIGHWHITE 1.12
300 GR HELLO SATINATED

REALIZED WITH THE SUPPORT OF
MONDRIAAN FOUNDATION
FONDS BKVB
STICHTING DOEN
OVERVOORDE-GORDON FOUNDATION
/ PAUWHOF FUND
EINDHOVEN UNIVERSITY OF TECHNOLOGY

Pauwhof Fund

TU/e
Technische Universiteit
Eindhoven
University of Technology

spinhex&
industrie
drukkerij